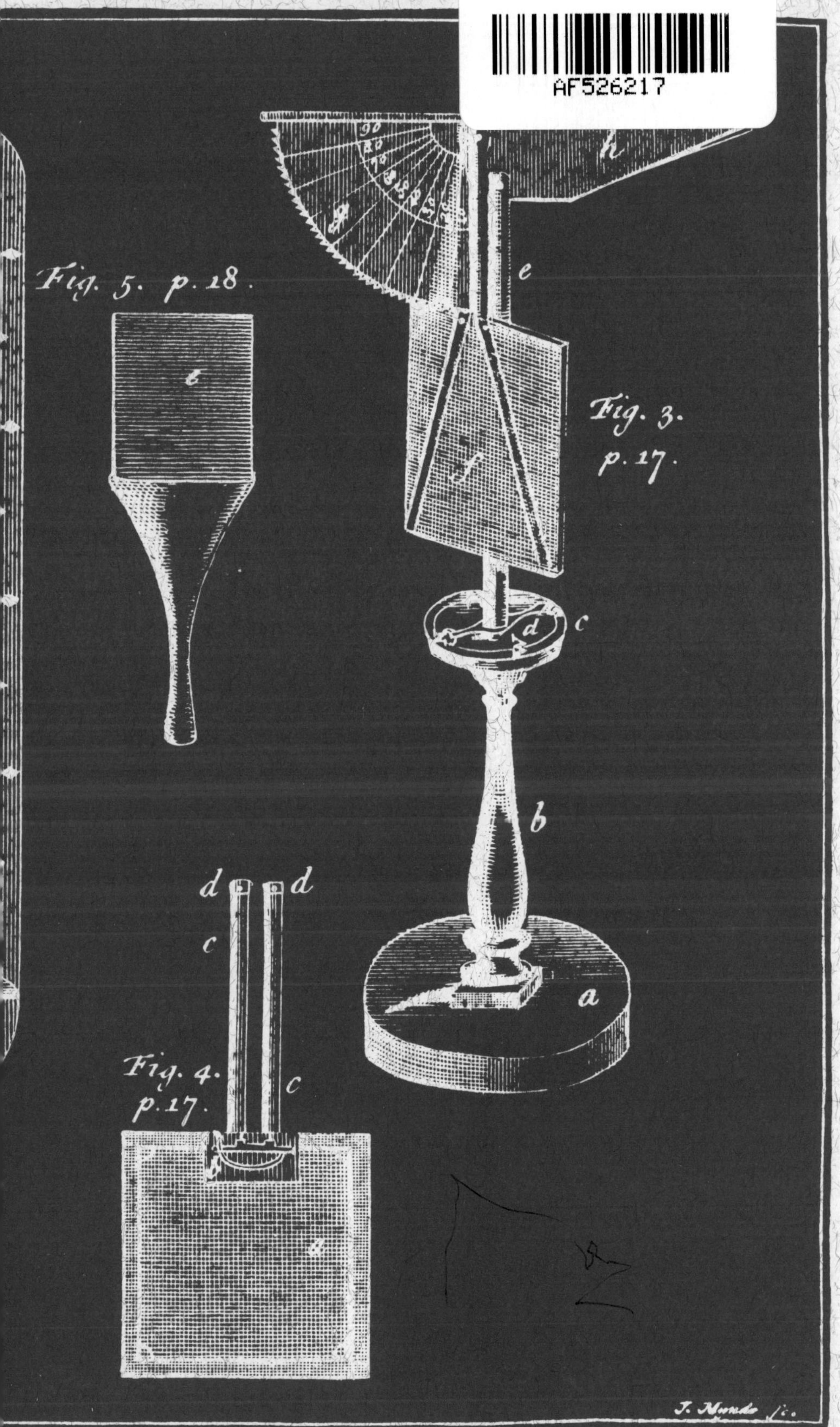
Fig. 5. p. 18.
Fig. 3.
p. 17.
Fig. 4.
p. 17.
h
e
f
d
c
b
a
d
d
c
c
J. Mynde sc.

Invention of the Meteorological Instruments

Invention of the

Meteorological Instruments

by W. E. Knowles Middleton

The Johns Hopkins Press, Baltimore

Baltimore, Maryland 21218

Manufactured in the United States of America

Library of Congress Catalog Card Number 68-31640

Preface

The seventeenth century saw the beginning of the modern period in science, marked by the first great generalizations associated with such names as Kepler, Galileo, Harvey, and Newton. It also witnessed the invention of a number of the most important scientific instruments, which led to a great advance in the exact knowledge of phenomena, and while the precise relationship between the theoretical progress and the technical advances may be disputed, it cannot be denied that the discipline in which most progress had been made by the year 1600, namely astronomy, was the one best provided with measuring instruments. It is notorious that Kepler's laws of planetary motion grew out of his work on observations made with the superior instruments of Tycho Brahe.

Meteorology as a science can hardly be said to have existed before the invention of the barometer and thermometer. As I have tried to show in *A History of the Theories of Rain* (London: Oldbourne, 1965; New York: Franklin Watts, 1966), the first impetus to meteorological theory was given by the puzzling behavior of the barometer in relation to the weather. The rain gauge, the windvane, and an elementary form of hygroscope are much older, but of these only the first could give numerical data, and it had not been widely used. In more recent times instruments have been devised to measure sunshine, the motion and height of clouds, upper winds, and in the last seventy-five years, the distribution of temperature and water vapor in the upper air. Perhaps I may be pardoned for believing that the history of meteorological instruments is an important part of the history of meteorology.

The title of this book will have informed the reader that no attempt has been made to treat the subject with the relative completeness that I attempted in *The History of the Barometer* (Baltimore: Johns Hopkins Press, 1964). To do so would require several large volumes. Now the history of any instrument or machine divides itself naturally into two parts: that of the actual invention, and that of the subsequent technical progress. In planning this book I have taken the view that the former is of greater interest to the nonspecialist; but on the other hand I have tried to recount the first technical advances following the actual invention of each instrument.

The concept of "early history" must, of course, have a different meaning for each class of instrument. The story of the windvane began in ancient times, for example, while the entire history of the radiosonde has occurred during the lifetime of many of my readers. Some instruments, notably the hygrometer, take several distinct forms, which had their origin at widely different dates. As a general rule I have endeavored to bring the history of each instrument up to the point where its technical development has reached a mature stage, with the proviso that I have not discussed improvements originating after the war of 1939–45. Many of these last are very complex and sophisticated indeed, and almost impossible to make clear to anyone who has not a very sound knowledge of electrical engineering, in fact, a sounder knowledge than I can pretend to. The history of radiosondes, radio direction-finding, and meteorological radar is still being added to, and I have not referred to the last-mentioned at all.

I have visited a number of museums and meteorological services in North America and in Western Europe, and the reader will find references to specific instruments. Museums, however, are notably poor in the sorts of instruments that are exposed out of doors, especially the parts that have to be on top of a mast, and the story is not told with anything like completeness in any museum that I have examined. I should like to express my appreciation of the great courtesy I have received from the curators of these institutions. At the same time I should like to record my opinion that many of their galleries might be enhanced by a freer use of models of early instruments, where specimens in good condition cannot be found.

Part of the research for this book was made possible by a grant from the Leverhulme Trustees, which is gratefully acknowledged. I also wish to thank the Keeper of the Department of Manuscripts in the British Museum, the Director of the Biblioteca Nazionale Centrale at Florence, and the Council of The Royal Society, for permission to quote from several manuscripts in their libraries; The Royal Meteorological Society for permission to reproduce certain illustrations from the *Quarterly Journal* of that Society, and the Harvard University Press for permission to quote from a translation of Vitruvius in Chapter 5. I wish to thank Dr. and Mrs. D. S. Poole of Esher and Dr. M. D. Poole of Newcastle-upon-Tyne for giving me the opportunity of examining many papers of W. H. Dines and L. H. G. Dines; and Mr. J. S. Dines of Hermitage for a fruitful discussion of his father's work. From Mr. C. Harmantas of the United States Weather Bureau I received much specialized information about radiosondes. Certain other obligations are acknowledged in the captions of some of the figures.

My wife cannot possibly be thanked enough for her part in the work, which included typing the entire manuscript twice, and much help with proofs and index, besides providing constant encouragement. Finally I wish to thank the officers and staff of The Johns Hopkins Press for their unfailing cooperation.

Vancouver, Canada W.E.K.M.

Table of Contents

List of Figures

Invention of the Meteorological Instruments

I

The Barometer

1. Prehistory. Though it is not the oldest meteorological instrument, the barometer is one of the most important. The science of meteorology began in the seventeenth century with attempts to find a relation between the weather and the puzzling fluctuations of the height of the mercury column,[1] and the possibility of forecasting was first seriously explored by drawing isobars on synoptic maps. But the invention of the barometer was of more than meteorological importance, for it finally disposed of the long-cherished dogma of the impossibility of a vacuum. A brief discussion of the ideas about vacuum at the beginning of the seventeenth century is a necessary preliminary to a consideration of the invention of the barometer.

One of the ideas held by Aristotle was that the existence of a vacuum anywhere in the universe was not only actually, but also logically, impossible. Somehow or other Aristotle was even able to believe that the universe has a finite size, without admitting the existence of an empty space outside it. He also believed that light cannot penetrate a vacuum, so that nothing can be seen through one, and hence there cannot be a vacuum between the earth and the stars.

Nineteen hundred years after Aristotle's death his great philosophical system was still accepted by nearly everyone. It is true that it had been amended in certain respects by many commentators, partly to bring it into better agreement with Christian dogma.[2] One respect in which Aristotle's ideas were altered was in regard to the weight of the air. He had believed that air has weight,[3] but this had come to be denied, and even the great Galileo stated in 1615 his belief that air weighs nothing at all,[4] with the corollary that it exerts no

[1] See W. E. K. Middleton, *A history of the theories of rain* (London, Oldbourne Press, 1965; New York, Franklin Watts, 1966), chap. 4.

[2] An idea of the extent of this literature as it applies to only one of Aristotle's works, the *Meteorologica*, can be obtained from G. Hellman's "Bibliographie der gedrückten Ausgaben, Ubersetzungen und Auslegungen der Meteorologie des Aristoteles," *Beiträge zur Geschichte der Meteorologie*, Vol. 2 (1917), pp. 3–45; note that this lists only printed works.

[3] Aristotle, *On the heavens*, translated by W. K. C. Guthrie (Loeb Classical Library, London and Cambridge, Mass., 1939), p. 355.

[4] Galileo, *Le opere*, ed. naz., Vol. IV (Florence, 1894), p. 167.

pressure. Indeed, at the beginning of the seventeenth century the very idea of the pressure of a fluid was scarcely understood by anyone, although in the thirteenth century one Jordanus had been quite clear about the vertical and lateral pressure exerted by liquids.[5] However wrong Galileo may have been about this, he was nevertheless quite sure, at least as early as 1612, that a vacuum is possible.[6] At about the same time, Isaac Beeckman of Middelburg in Holland showed in his private notes[7] that he understood the pressure of the air and also how it acts to prevent the occurrence of a vacuum when a body is moved. He compared air to a sponge, which can be compressed, but springs back to its previous state when allowed to do so.[8]

The philosopher René Descartes, who had an immense influence on seventeenth-century science even though most of his physical ideas have turned out to be wrong, had made up his mind that all space must be filled with something or other, making a vacuum impossible. Thus all bodies in the world are in contact, and can move only by each replacing the next. All bodies are also permeated by a subtle matter or ether, which can pass freely, even through solid substances.

In 1731 a pupil apparently wrote to Descartes asking him why mercury in a tube closed at its upper end does not fall out. From Descartes' reply, which has been preserved,[9] it is clear that the difficulty was with the subtle matter—why did it not come through the tube and push the mercury out? The explanation given by Descartes[10] is anything but convincing today, requiring the subtle matter that should do so to come all the way from "the sky above the air." Finally he gives the warning: "And so that you will make no mistake, you must not think that this quicksilver cannot be separated from the top of the tube by any force, but only that it needs as much force as is required to raise all the air that there is from that point to above the clouds."

On the strength of this statement, E. Gerland claimed that Descartes had had the idea of the barometer.[11] It may be objected, I think, that there is no indication that he realized that if the tube were long enough, part of the quicksilver would come out.

At this time, the behavior of suction pumps and siphons was causing

[5] See Cornelis de Waard, *L'expérience barométrique, ses antécédents et ses explications. Etude historique* (Thouars, 1936), pp. 68–69.

[6] Galileo, *Le opere*, ed. naz., Vol. IIIa (Florence, 1892), p. 350.

[7] Isaac Beeckman, *Journal tenu par lui de 1604 à 1634 publié avec une introduction et des notes par Cornelis de Waard* (4 vols., The Hague, 1939–1953), Vol. 1, p. 36.

[8] *Ibid.*, p. 46.

[9] Descartes, *Oeuvres*, eds. Chas. Adam and Paul Tannery (13 vols., Paris, 1897–1913), Vol. 1, pp. 205–8.

[10] I have translated it in *The history of the barometer* (Baltimore, 1964), pp. 7–8. This work will be referred to as *Hist. Bar.*

[11] E. Gerland in *U.S. Weather Bureau Bull.*, no. 11, pt. 3 (Washington, 1896), p. 690.

much curiosity. On July 27, 1630, Giovanni Batista Baliani of Genoa wrote very deferentially to Galileo to ask why a siphon, led over a hill about twenty-one meters high, failed to work. It was filled with water, and when one end was opened, the level in that limb dropped to about ten meters above the reservoir. When the other end was opened, the level in that limb also fell.[12]

Galileo's reply, dated August 6, 1630, said in effect that this was easy, and attributed the phenomenon to an energy or attractive force belonging to the vacuum.[13] Not at any time did phenomena of this sort convince him of the pressure exerted by the air. Discussing the suction pump at a later date, he thought that when the water column reached a certain height it simply broke off, like a cord to which too large a weight had been attached.[14] However, there was no doubt in his mind that a vacuum had been produced in the top part of Baliani's siphon, and by the end of the 1630's this opinion seems to have been widely held among those Italians who were following the new developments in physics. There was a notable group of these in Rome, including Rafaello Magiotti, Evangelista Torricelli, and Gasparo Berti. Galileo's *Discorsi* arrived in Rome in December, 1638, and De Waard,[15] after examining all the relevant documents, believed that the book stimulated Magiotti to suggest an experiment for studying the production of a vacuum, though I think that the idea might have come directly from Galileo. This experiment was carried out by Berti, who was said to be "especially clever at making experiments,"[16] and witnessed by several people, including Magiotti, as well as the Jesuits Athanasius Kircher and Niccolo Zucchi. Although all these three wrote accounts of the experiment later, the fullest description—and by far the most clear-sighted—was given by Berti's friend Emmanuel Maignan or Magnanus, who was told about it by Berti and shown drawings a few days after the event. His long account[17] was written after he had returned to Toulouse, his birthplace, in 1650.

Berti's experiment is shown in Figure 1.1, taken from the *Cursus philosophicus*. To the wall of his house he fastened the long lead tube *AB*, with the upper end *A* opposite a tower window. The lower end was provided with a stopcock *R*, which was under the surface of the water in the cask *EF*. To the upper end was cemented a large glass flask with two openings, the top one *C* having a lead plug into which brass screw *D* could be screwed. The remaining parts will be referred to later.

[12] Galileo, *Le opere*, ed. naz., Vol. XIV (Florence, 1904), pp. 124–25.
[13] *Ibid.*, pp. 127–30.
[14] Galileo, *Discorsi . . . intorno a due nuove scienze* [etc.] (Leyden, 1638), p. 17.
[15] *L'expérience barométrique*, pp. 101–10.
[16] Niccolo Zucchi, *Nova de machinis philosophia* [etc.] (Rome, 1649), p. 102.
[17] Maignan, *Cursus philosophicus concinnatus ex notissimis cuique principiis, ac praesertim quoad res physicas instauratas ex lege naturae sensatis experimentis passim comprobata* (4 vols. paged as one, Toulouse, 1653), pp. 1925–36. I have translated this passage in *Hist. Bar.*, pp. 10–15.

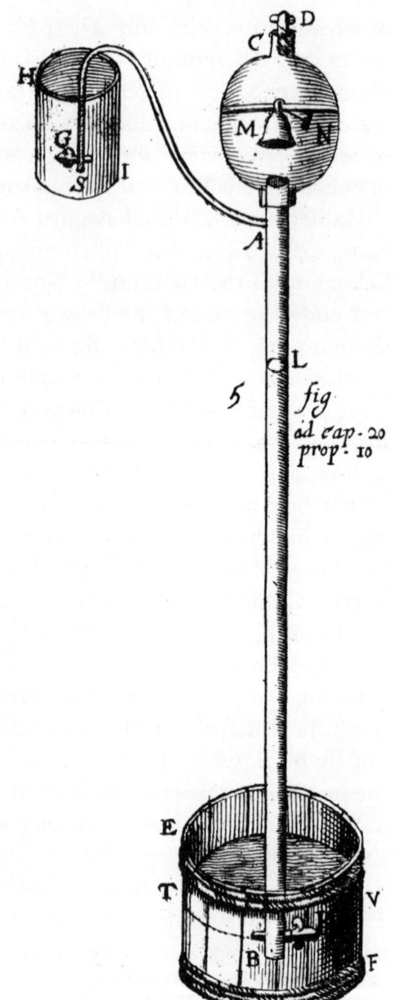

Fig. 1.1 Berti's experiment.

When the cask had been about half filled with water and the tap *R* had been closed, the tube and the flask were entirely filled through *C*, after which the screw *D* was put in. The first time this was done the flask broke, but finally everything was ready, and when the tap *R* was opened,

the water flowed (contrary to the hope of many)[18] out of the pipe into the cask *EF*, to an easily observable height; but not all of it flowed out, and it soon stood quite still. This was clear, because a mark was made in the cask at the surface of the water, and next day it was found that the water in it had remained exactly at the mark, although the tap *R* had been open all the time. Then, when this tap *R* had been carefully closed again, the screw *D* was taken out above. And as soon as it was removed, behold! the air rushed in with a loud noise, filling the space previously abandoned by the water. Then

[18] Presumably the Jesuits who were there. Nothing could show more clearly than this phrase what a tender plant experimental science was at the time.

by lowering a sounding line it was determined how much water was inside, or rather to what height the water stood in the tube, and it was found to stand about 18 cubits above the level of the water in the cask, at the mark *L*. I remember indeed that this was the number of cubits, or *braccia* as the Tuscans say, because it was this number and this distance to which at that time, according to an observation made by Galileo, it was said that water could rise in a pump . . . .

Then Berti devised another way of doing the experiment, soldering on a side tube *AS* that terminated in a tap *G* under some water in the vessel *HI*. After the water had been allowed to go down to *L* and the tap *R* had been closed, the flask was shown to be empty by opening the tap *G* under the water in *HI*, whereupon the water came in through the side tube and filled the flask completely, except for a small bubble under the mouth *C*, which was then shown to be air by raising the vessel *HI* still further.

The bell *M* and its hammer *N* were suggested by Athanasius Kircher as a further test for the vacuum, and the experiment seems to have been tried later, the iron hammer being raised by an external magnet. Maignan was acute enough to point out that with the bell solidly supported by the flask, its sound would certainly be heard, so that such an experiment would not be at all decisive.

A further important point was the appearance of the flask when it was supposed to be empty. Aristotle would have said that it should be completely opaque, because light, being what the philosophers called an "accident," would be without a "subject" in the flask, and so could not exist there. Maignan believed that light is a substance, and corpuscular in nature, so that it would readily be transmitted.

It is quite remarkable that we cannot establish the date of such an important experiment at all precisely. It can scarcely have been before 1639, and on the other hand Berti died before January 2, 1644, on which date his death was mentioned in a letter.[19] It is probable that it was performed before Galileo's death on January 8, 1642, for there exists a letter from Magiotti in Rome to Marin Mersenne in Paris, written in 1648, in which there is the phrase, "Berti believed that he could convince Galileo with this experiment."[20] As Magiotti also said in this letter that he had written to Torricelli about the experiment, there is some reason to think that it may have been done after the latter moved to Florence in 1641, but we cannot be sure; in fact, we do not know whether or not Torricelli witnessed it. What is beyond question is that it preceded the more famous experiment performed by Viviani at the suggestion of Torricelli. This was clearly stated by Zucchi in an anonymous pamphlet published in 1648,[21] and Zucchi saw the experiment made.

[19] De Waard, *L'expérience barométrique*, p. 109.

[20] Vienna, *Nationalbibliothek*, ms. 7049, letter CXXVII. "Il Sig[r] Berti credeva con questa esperienza convincere il Sig[r] Galileo."

[21] *Magno amico nonnemo ex Collegio Rom. S.J.S.D.—Experimenta vulgata non vacuum*

Even several years afterward, when Maignan, Zucchi,[22] and Kircher[23] published their accounts of it, there was no unanimity about its interpretation. The interest was in the presence or absence of a vacuum, and clerics like Zucchi and Kircher, though they had a genuine intellectual interest in the new science, could not help interpreting Berti's experiment in such a way as to avoid the admission that a vacuum could be produced. The supposed inability of light to travel through a vacuum was a stumbling block for many people for most of the seventeenth century; and while Maignan's remarks about the conduction of sound through the solid mounting of the bell were correct, the mere fact that a sound was heard must have greatly encouraged those who were determined not to believe in a vacuum. Thus the immediate results of the experiment were as favorable to the plenists as to the vacuists.

It will not do to say that "the barometer was invented" by Berti and his friends. A barometer is an instrument to measure the pressure of the air, and, quite apart from the fact that their apparatus had no scale, there is no indication that they were trying to construct a measuring instrument. Indeed, it is very unlikely that they suspected that the column of water was held up by the pressure of the air. They were bent on producing a vacuum, and—if we forget about water vapor—they succeeded.

2. The Torricellian experiment. In the letter from Magiotti to Mersenne already referred to, there is the following passage: "Now I wrote to Signor Torricelli about this experiment, stating my belief that if the water had been from the sea, and therefore heavier, it would not have stopped at *N*,[24] but lower. They made the experiment and finally arrived at quicksilver . . . ."[25] "They" were presumably Torricelli and Viviani.

Although neither Magiotti's letter to Torricelli nor any reply that the latter may have made has come down to us, it is most unlikely that Magiotti would have claimed the credit of this suggestion if he had not actually made it. But it is probable that it was not needed, for there exists in Florence a copy of Galileo's *Discorsi* with many manuscript additions and corrections in the hand of his young pupil Vincenzio Viviani.[26] In his edition of Galileo's works,[27] E. Alberi

probare, sed plenum et antiperistasim instabilire (Rome, 1648). There is a copy in the BN at Paris, R.25606. Its authorship was revealed when Zucchi reprinted it in his *Nova de machinis philosophia* [etc.], 2nd ed. (Rome, 1649), pp. 101–15.

[22] *Ibid.*

[23] Kircher, *Musurgia universalis sive ars magna consoni et dissoni in X libros digesta* [etc.] (Rome, 1650), pp. 12–13.

[24] This refers to Magiotti's drawing, not reproduced here.

[25] Vienna, *Nationalbibliothek*, ms. 7049, letter CXXVII, verso.

[26] Florence, *Bibl. Naz.*, ms. Gal. 79.

[27] *Le opere. Prima edizione completa condotta sugli autentici manoscritti Palatini* (16 vols., Florence, 1842–56).

DEL GALILEO. 17

per la sustanza, ò porosità del vetro, ò del legno, aria, ò altra più tenue, e spiritosa materia, si vedrà radunare (cedendogli l'acqua) nell' eminenza V, le quali cose, quando non si scorgano, verremo assicurati l'esperienza esser con le debite cautele stata tentata; e conosceremo l'acqua non esser distraibile, nè il vetro esser permeabile da veruna materia benche sottilissima,

Sagr. Et io mercè di questi discorsi ritrouo la causa di vn'effetto, che lungo tempo m'hà tenuto la mente ingombrata di marauiglia, e vota d'intelligenza. Osseruai già vna Citerna, nella quale per trarne l'acqua fù fatta fare vna Tromba, da chi forse credeua, mà vanamente, di poterne cauar con minor fatica l'istessa, ò maggior quantità, che con le secchie ordinarie; & hà questa tromba il suo stantuffo, e animella sù alta, si che l'acqua si fà salire per attrazzione, e non per impulso, come fanno le Trombe, che hanno l'ordigno da basso. Questa sin che nella Citerna vi è acqua sino ad vna determinata altezza, la tira abbondantemente, mà quando l'acqua abbassa oltre à vn determinato segno, la Tromba non lauora più. Io credetti, la prima volta che osseruai tale accidente, che l'ordigno fusse guasto, e trouato il Maestro, acciò lo raccomodasse, mi disse che non vi era altrimente difetto alcuno fuor che nell' acqua, la quale essendosi abbassata troppo, non patiua d'esser' alzata à tanta altezza; e mi soggiunse nè con Trombe, nè con altra machina, che solleui l'acqua per attrazzione, esser possibile farla montare vn capello più di diciotto braccia, e siano le Trombe larghe, ò strette, questa è la misura dell' altezza limitatissima. Et io sin hora sono stato così poco accorto, che intendendo, che vna corda, vna mazza di legno, e vna verga di ferro si può tanto, e tanto allungare, che finalmente il suo proprio peso la strappi, tenendola attaccata in alto, non mi è souuenuto, che l'istesso molto più ageuolmente accaderà di vna corda, ò verga di acqua. E che altro è quello, che si attrae nella Tromba, che vn Cilindro di acqua, il quale hauendo la sua attaccatura di sopra, allungato più, e più, finalmente arriua à quel' termine, oltre al quale tirato dal suo già fatto souerchio peso, non altrimente che se fusse vna corda, si strappa? Salu.

Fig. 1.2 Page 17 of the Discorsi, *with Viviani's additions.*

pointed out that these were made "with the approval of Galileo himself, as is drawn to our attention more than once in these marginal notes."[28] In Figure 1.2 I have reproduced one of the pages of this document,[29] the page which contains the passage about the breaking of the column of water like a cord too heavily loaded.[30] The addition at the bottom may be translated as follows: "And it is my belief that the same result will follow in other liquids, such as quicksilver, wine, oil, etc., in which the rupture will take place at a lesser or greater height than 18 *braccia*, according to the greater or lesser specific gravity of these liquids in relation to that of water, reciprocally; always measuring such heights perpendicularly, however."

One must suppose that the grand old man was thinking of a revised edition, and that the young pupil was giving joyful aid. In the first days of 1642 Galileo died, and Viviani, not yet twenty, became the pupil of Evangelista Torricelli. He was the "laboratory assistant" in one of the most important experiments in history.

Evangelista Torricelli was born somewhere near Faenza on October 15, 1608, the son of Gasparo Torricelli.[31] Orphaned at an early age, he was brought up by an uncle, Alessandro, a man of some learning who had become a monk. After studying mathematics with the local Jesuits, Evangelista went to Rome, probably in 1627, to study under Benedetto Castelli, a notable worker in hydrostatics and a friend of Galileo. Torricelli, a modest man of great personal charm, soon became a very accomplished mathematician indeed, and in October, 1641, he moved to Florence at the invitation of Galileo. Since Galileo died three months later, Torricelli cannot be called a pupil of Galileo except in a figurative sense.[32]

After Galileo's death, the Grand Duke Ferdinand II of Tuscany, a notable scientific man in his own right, made Torricelli his Philosopher and Mathematician, a post which he held until he died in 1647. In 1644 there appeared Torricelli's only published book, a collection of papers in mathematics and theoretical physics,[33] but he left much in manuscript, some of it later published by Tommaso Bonaventuri,[34] but much more having to wait until the twentieth century.[35] He published nothing whatever about the barometer, and our knowledge of his part in its history is derived entirely from three letters exchanged

[28] *Ibid.*, Vol. XIII, p. xii.
[29] Page 17.
[30] See p. 5 above.
[31] See Giovanni Ghinassi, *Lettere fin qui inedite di Ev. Torricelli, precedute della vita de lui* (Faenza, 1864).
[32] In their adulation of Galileo, Italian writers have tended to regard most of the younger contemporary scientists as his pupils.
[33] Torricelli, *Opera geometrica* [etc.] (Florence, 1644).
[34] Torricelli, *Lezioni accademiche* (Florence, 1715).
[35] *Opere*, eds. G. Loria and G. Vassura (5 vols., Faenza, 1919).

between Torricelli and Michelangelo Ricci of Rome in June, 1644.[36]

On June 11, 1644, Ricci had written to his friend, "I live in a great desire to know the success of those experiments you indicated to me."[37] On that same day, Torricelli was writing him the famous letter describing the experiment,[38] which must be presented in full, even in an account as condensed as the present one:

Most Illustrious and Reverend Sir:

Some weeks ago I asked Signor Antonio Nardi for some of my demonstrations on the area of the cycloid, requesting him to direct them to you, or rather to Signor Magiotti, after he had seen them. I have already hinted to you that some sort of philosophical experiment was being done concerning the vacuum, not simply to produce a vacuum, but to make an instrument which might show the changes of the air, now heavier and coarser, now lighter and more subtle. Many have said that there cannot be a vacuum; others that it occurs, but with the repugnance of nature, and with difficulty. I really do not remember that anyone has said that it may occur with no difficulty, and with no resistance from nature. I reasoned thus: if I found a very obvious cause of this resistance that is felt in trying to produce a vacuum, it would seem vain to try to attribute the resistance to the vacuum itself, as it would clearly derive from the other cause. On the contrary, making some very easy calculations, I find that the cause I adopted (i.e., the weight of the air) ought by itself to produce a greater resistance than it does when we attempt to make a vacuum. I say this because some philosopher, seeing that he could not escape confessing that the gravity of the air is the cause of the resistance that is felt in producing a vacuum, would not say that he conceded the operation of the weight of the air, but would persist in his assertion that Nature also helps by her repugnance to the vacuum. We live submerged at the bottom of an ocean of elementary air, which is known by incontestable experiments to have weight, and so much weight, that the heaviest part near the surface of the earth weighs about one four-hundredth as much as water. Then writers have observed regarding the twilight that the vaporous air is visible above us for about fifty or fifty-four miles. But I do not think it is as much as this, because I should then admit that the vacuum ought to produce a much greater resistance than it does, even if there is this escape for these writers, that this weight, given by Galileo, refers to the lowest air, frequented by man and animals, but above the peaks of high mountains the air begins to be very pure, and of much less weight than the four-hundredth part of the weight of water.[39]

We have made many glass vessels such as those shown at *A* and *B*, wide, and with necks two ells long [Fig. 1.3]. When these were filled with quicksilver, their mouths stopped with the finger, and then turned upside-down in a vase *C* which had some quicksilver in it, they were seen to empty themselves, and nothing took the place of the quicksilver in the vase which was being emptied. Nevertheless the neck *AD* always remained full to the height of an ell and a quarter and a finger more. To show that the vessel was

[36] *Opere*, vol. 3, pp. 186–88; 193–95; 198–201. The two by Torricelli, and part of the one from Ricci, were first printed by Carlo Dati in his *Lettere a Filaleti di Timauro Antiate, Della vera storia della cicloide, e della famosissima esperienza dell'argento vivo, 24 gennaio 1662* (Firenze, 1663).

[37] *Opere*, Vol. 3, p. 189.

[38] *Ibid.*, pp. 186–88. Dati, *Lettere a Filaleti*, pp. 20–21 (my translation).

[39] Galileo's value was too great by a factor of more than two.

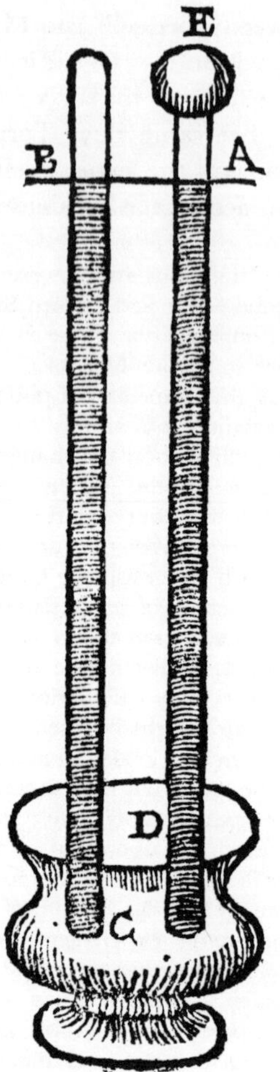

Fig. 1.3 Illustrating the Torricellian experiment.

perfectly empty, the basin was filled up to *D* with water; and on raising the vessel little by little, it was observed that when the mouth of the vessel reached the water, the quicksilver in the neck came down, and the water rushed in with horrible violence and filled the vessel completely up to *E*.

While the vessel *AE* was empty, and the quicksilver, though very heavy, was sustained in the neck *AC*, we discussed this force that held up the quicksilver against its natural tendency to fall down. It has been believed until now[40] that it was something inside the vessel *AE*, either from the vacuum, or from that extremely rarefied stuff; but I assert that it is external, and that the force comes from outside. On the surface of the liquid in the basin presses a height of fifty miles of air; yet what a marvel it is, if the quicksilver enters the glass *CE*, to being in which it has neither inclination nor repugnance, and rises there to the point at which it is in balance with the weight of the external air that is pushing it! Water, then, in a similar vessel but very much longer, will rise to about eighteen ells, that is to say, as much higher than the quick-

[40] Torricelli must be referring to Galileo's belief in the attraction exerted by the vacuum. Neither Beeckman (of whom he had probably never heard) nor Baliani (of whom he certainly had) would have subscribed to this belief.

silver as quicksilver is heavier than water, in order to come into equilibrium with the same cause, which pushes the one and the other.

This reasoning was confirmed by making the experiment at the same time with the vessel *A* and with the tube *B*, in which the quicksilver always stopped at the same level *AB*; an almost certain sign that the force was not within; because the vase *AE* would have had more force, there being more rarefied and attracting stuff, and this much more vigorous by virtue of its greater rarefaction than that in the very small space B.

With this principle I then tried to preserve all the kinds of repugnance which are felt to be in the various effects attributed to the vacuum, and up to this moment I have not met one of them that does not go well with it. I know that many objections will occur to you, but I hope also that thinking about the matter will appease them.

I have not been able to succeed in my chief intention, to find out with the instrument *EC* when the air is coarser and heavier and when more subtle and light; because the level *AB* changes from another cause (which I never thought of), that is, it is very sensitive to heat and cold, exactly as if the vase *AE* were full of air.

And humbly I pay my respects.

From your devoted and most deeply obliged servant.

V. Torricelli

Florence, 11 June, 1644.

At the beginning of this remarkable letter we find Torricelli far in advance of his predecessors in that he was trying not only to produce a vacuum, but also to make an instrument to "show the changes of the air." He has no doubt about the possibility of making a vacuum, but he has to show that it is the weight (*il peso* or *la gravità*) of the air, and this alone, that is responsible for the resistance we feel when we try to produce a vacuum. According to his calculations there is at least enough air to balance out the weight of the mercury.

Then comes that fine metaphor: *Noi viviamo sommersi nel fondo d'un pelago d'aria elementare*—"we live submerged at the bottom of an ocean of elementary air." Galileo must have known this too, but he thought that the air exerted no pressure.

Galileo's experimental value for the weights of equal volumes of water and air was far from correct, but Torricelli used it, his respect outweighing his confidence in estimates of the height of the atmosphere derived from observations on the twilight. He knew, as had Baliani before him,[41] that the air aloft was lighter, but he ascribed its lightness to its greater purity, as Aristotle would have done.[42]

Then comes the familiar experiment, including the brilliant idea of using water to show that the space above the quicksilver was empty of air, a demonstration that made a great impression on all who saw the experiment in the following years, and the clever use of two glass vessels of very different volume. We cannot know whether it was Torricelli or Viviani who thought of these devices.

[41] Galileo, *Le opere*, ed. naz., Vol. XIV, p. 160.

[42] Probably as a result of his early education. Indeed his reference to "elementary" air may have come from this source.

The discussion of the experiment is so clear that it needs little comment, except perhaps about the curious phrase "either from the vacuum, or from that extremely rarefied stuff" (*o di vacuo, o di quella roba sommamente rarefatta*), which suggests that Torricelli had the Cartesian "subtle matter" in mind. But at the very end of the letter there is an interesting confession of failure. The instrument ought not to have been sensitive enough to heat and cold to hide the variations in the atmospheric pressure, and it is possible that it was set up again with the tube that had been used for the experiment with the water, but without drying it first.

Ricci replied almost at once, on June 18, 1644,[43] expressing his admiration for the experiment but raising three objections, the last two of which were simply due to his lack of knowledge of hydrostatics. But the first was more interesting: what would happen if the bowl of mercury had a cover with only a hole for the tube, and everything sealed to exclude the air? In this event, argues Ricci, the air above the vessel would not succeed in pressing on the surface of the quicksilver, but would press on the cover; and if then the quicksilver remains up in the tube as at first, the effect can no longer be attributed to the weight of the air holding it up as if in equilibrium.

On June 28, 1644, Torricelli dealt with Ricci's "objections" in a letter[44] that discloses a good deal more of his thought about the atmosphere. To the above criticism, he replied:

> . . . if you infer that the top, cemented so that it covers the surface of the basin, does it in such a way that it touches the quicksilver in the basin, then that which is raised in the neck of the vessel will remain raised as before, not by the weight of the atmosphere [*sfera aerea*], but because there will be no room for it in the basin.
>
> But if you would have the cover put on so that it still takes some air inside, then I ask whether you will admit that the enclosed air will have the same degree of condensation as the external air? In this case the quicksilver will be held up as before (as in the example of the wool that I shall now give); but if the air that you include is more rarefied than the external air, then the suspended metal will descend by the proper amount; if now it were infinitely rarefied, i.e., a vacuum, then the metal would descend all the way, provided that the enclosed space were able to take it.
>
> The vessel *ABCD* [Fig. 1.4] is a cylinder filled with wool or other compressible material—let us say air. This vessel has two ends, *BC* fixed and *AD* movable, adapted to it. Let *AD* be loaded above with a lead weight *E* that weighs 10,000,000 lb. I know you will understand how much force the bottom *BC* will feel. Now if we forcibly push in a plane or sharp knife *FG* so that it goes in and cuts the compressed wool, then I say that if the wool *BCFG* is compressed as before, the bottom *BC* will suffer the same pressure as it felt at first, even if it no longer feels any of the superimposed weight of the lead *E*.

[43] Torricelli, *Opere*, Vol. 3, pp. 193–95. Part of the letter is in Dati, *Lettere a Filaleti*, pp. 21–22.

[44] Torricelli, *Opere*, Vol. 3, pp. 198–201. Dati, pp. 22–23. All the pertinent part is translated in *Hist. Bar.*, pp. 26–28.

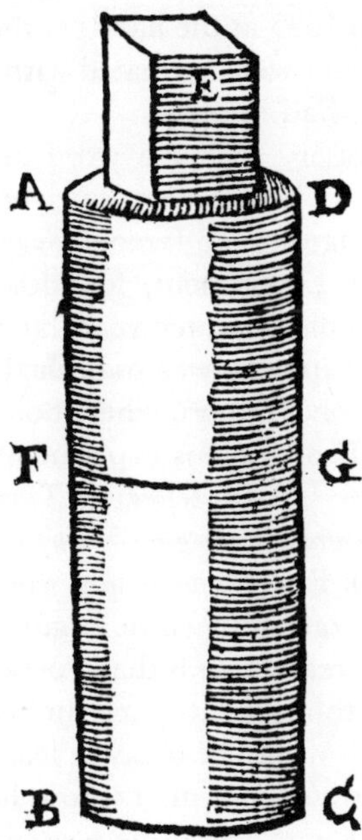

Fig. 1.4 To answer Ricci's first objection.

It will be seen that in June, 1644, Torricelli understood the elasticity of the air as well as its weight; he realized the essential difference between aerostatics and hydrostatics. We do not know how he had decided that the weight of the column of air is variable, but he had, and he announced in the second sentence of the first letter that he was trying to make an instrument to show this variation. *This is why the invention of the barometer should be ascribed to Torricelli.* It is true that he did not entirely succeed, but there is little doubt that success would have followed if he and Viviani had made further experiments. Dati, writing long after the event, claimed that further experiments were made by Torricelli,[45] but this is not elsewhere confirmed.

This brings us to the consideration that there is remarkably little information about the Torricellian experiment. As far as Italy is concerned, it was hidden as if it had been the ritual of some secret society, and it was first revealed to the world on the other side of the Alps, as we shall see. Torricelli himself wrote nothing further about it, as far as we know; and in the decade following his death it seems to have been mentioned in only two works published in Italy.[46] There is little doubt that the reason was prudence in the face of the attitude of the Church, to which the idea of a vacuum was anathema; it is natural that the Italian scientists would remember what had happened

[45] *Lettere a Filaleti*, p. 20.
[46] *Cf.* De Waard, *L'expérience barométrique*, p. 130. In *Hist. Bar.*, pp. 30–32, I have assembled a good deal of evidence of the concealment of the experiment.

to Galileo in 1633 at the hands of the Holy Office, and would imagine that they might well be treated with much less consideration than that famous man had received.

It is necessary to say a word about the date of the Torricellian experiment. The date usually given is 1643, and this probably results from a passage in the famous *Saggi* of the *Accademia del Cimento*, or Academy of Experiment, founded in Florence in 1657 by Prince Leopold and dissolved ten years later. The *Saggi*, a collective account of their experiments, was published in 1667, and at the beginning of the chapter on "Experiments about the natural pressure of the air" we read: "That famous experiment with the quicksilver which rose up before the great intellect of Torricelli in the year 1643 (*che l'anno 1643 si paró davanti al grande intelletto del Torricelli*) is now known in every part of Europe, as is also the high and wonderful thought that he derived from it, when he began to speculate on the reason."[47]

What this really says is that Torricelli thought about the experiment in 1643. I think it extremely unlikely that it was performed in that year, for why would he delay at least six months before reporting such a beautiful result to a man whom he called the best friend he had in Rome,[48] and with whom he was in regular correspondence? Torricelli's letter of June 11, 1644, looks to me like an enthusiastic report of a result just obtained. Our confidence in the date given in the *Saggi* is further diminished by the fact that Carlo Dati, a member of the Academy, says that Torricelli *imagined* what would happen and told Viviani, who had the apparatus made, procured the mercury, made the experiment, and saw the effect *forecast* by Torricelli.[49] The *Saggi* puts the experiment first, the thinking second. Torricelli was dead and twenty years had gone by. I prefer to believe that the famous experiment was made in the spring of 1644.

In Italy outside of Florence, there is no doubt that the experiment was performed privately in Rome by, or for, the Cardinal Giovanni Carlo de' Medici, the Grand Duke's brother, in February, 1645. Kircher,[50] Zucchi,[51] and most probably Maignan[52] were there. Kircher and Zucchi agree about the date, but Maignan's account is by far the most circumstantial. He states definitely that the experiment was made in tubes of widely different shapes and sizes, and that he saw all this in Rome.[53] Figure 1.5 is his illustration, and it will be

[47] *Saggi di naturale esperienze fatte nell'Accademia del Cimento* (Florence, 1667), p. 35.
[48] Ghinassi, *Lettere*, p. lix.
[49] Dati, *Lettere a Filaleti*, p. 20.
[50] Kircher, *Musurgia*, p. 11.
[51] Zucchi, *Experimenta vulgata*, p. 4 (cited by De Waard, *L'expérience barométrique*, pp. 177–78).
[52] Maignan, *Cursus*, p. 1897, in which he says that he saw the experiment "in Rome six years ago," without further identification. The "approbation" in the *Cursus* is dated Feb. 28, 1650, but the work was published in 1653. It is never safe to assume that no part of such works postdates the "approbation," and it was very probably the Cardinal's experiment that he saw.
[53] *Cursus*, p. 1888.

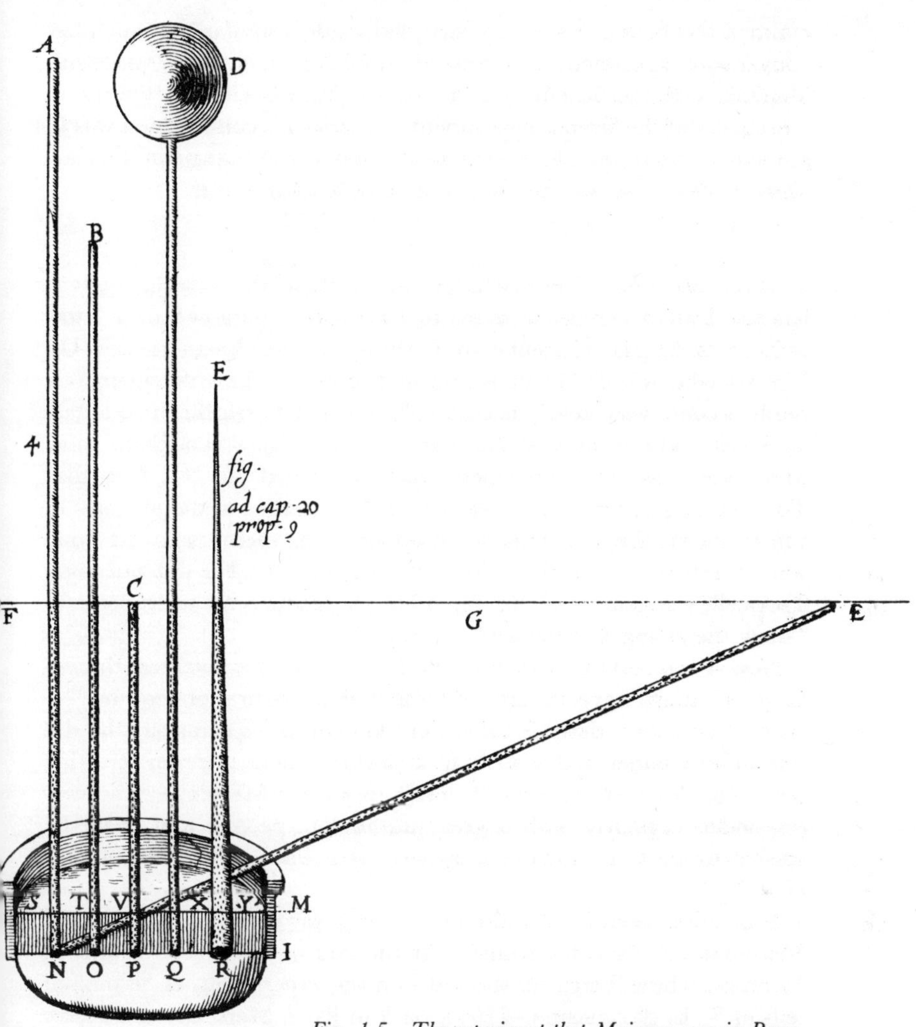

Fig. 1.5 The experiment that Maignan saw in Rome.

seen that the experiment goes beyond that reported by Torricelli, especially in the use of the sloping tube. Certainly the Cardinal or someone in his suite was thinking hard about the experiment. Then they kept "the five tubes" under observation, and it was found that the changes in the level of the mercury could not be correlated with the observed changes in temperature and humidity, so that "the height of the mercury is not changed by the alteration of the air around it, but by that of all the air in a broad region."[54] This may well be considered the beginning of the scientific study of meteorology.

Torricelli's real position in the history of the barometer should now be clear. Beeckman and Baliani before him had understood that the pressure of the air acts in every direction, and Torricelli nowhere

[54] *Ibid.*, p. 1898.

claimed this idea as his own. Berti had made a similar but much less convenient experiment, to which Magiotti had suggested a variation. Viviani, with the benefit of a broad hint from Galileo on the use of mercury, did the famous experiment. Torricelli's contribution was to see that it might provide an *instrument* to show "the changes of the air." Only in this sense was he the inventor of the barometer.

3. Across the Alps. The first information about the experiment that left the Italian peninsula seems to have been contained in a letter written to Marin Mersenne in Paris by a Frenchman named Du Verdus who was living in Rome at the time. This document[55] is really a copy, very nearly accurate, of parts of Torricelli's two letters to Ricci. The parts that Du Verdus copied were enough to show Mersenne how the experiment was performed, to tell him that Torricelli explained it by reference to the pressure of the air, and to communicate the idea that if the air over the mercury in the bowl were rarefied, the level in the tube would fall. He did not copy Torricelli's statement of his intention to produce an instrument to "show the changes of the air."

Now if a scientific worker in the first half of the seventeenth century—he might have thought of himself as a "natural philosopher"—wanted to ensure that his ideas were known to his compeers, he did not write a paper and send it to a journal; there were no scientific journals. Instead, he wrote a letter to Father Mersenne, who corresponded regularly with a great number of people, and could be trusted to see to it that the ideas were redirected to the best advantage.[56]

Soon after receiving this document, or at any rate in October, 1644, Mersenne left Paris for Rome. At the end of December he was in Florence, where Torricelli showed him the experiment, as he himself tells us.[57] In the autumn of 1645, back in Paris, Mersenne and Pierre Chanut tried to repeat the experiment, but could not get adequate glass tubes,[58] so Mersenne told Pierre Petit about it. Petit was at Rouen, where there was a good glass works. In November, 1646, Petit wrote a long letter to Chanut in Stockholm describing the experiment, and this was published in 1647.[59] It is of much interest that this letter brings the name of Blaise Pascal into the subject: Pascal, the

[55] BN, *fonds lat., nouv. acq.* 2338, fol. 45^r–46^v.

[56] This enormous correspondence has been collected by Mme. Paul Tannery and Cornelis de Waard, *Correspondance du P. Marin Mersenne, réligieux minime* (7 vols., Paris, 1933–65).

[57] Mersenne, *Novarum observationum physico-mathematicarum* (Paris, 1647), Vol. III, p. 216.

[58] *Ibid.*, p. 217.

[59] Pierre Petit, *Observation touchant le vuide faite pour la première fois en France* [etc.] (Paris, 1647).

famous mathematician, physicist, and religious philosopher, whose life and works have led to untold reams of controversy.[60] Pascal's connection with the barometer has produced some of the most violent of the polemics.

Petit told Pascal what he had learned from Mersenne about what Torricelli had done, and Pascal "was delighted to hear about such an experiment, as much because of its novelty as because he has believed in the vacuum for a long time, as you know."[61]

They performed the experiment on rather a grand scale, using a tube 12 or 15 mm in bore ("as big as the little finger inside") and buying "40 or 50 pounds" of mercury. They repeated Torricelli's inspired use of a layer of water over the mercury in the bowl. The detailed description given by Petit shows him, or Pascal, or both, to have had great experimental ability. In particular, Petit emphasizes the necessity of drying the inside of the tube very carefully if the experiment is to be set up again, a point that Viviani and Torricelli appear to have missed, as we have seen.

From his discussion of the experiment two things are clear. The first is that they were convinced that there was a vacuum, that is to say complete emptiness, in the tube above the mercury. The second is that neither of them seems ever to have considered at this time that the elevation of the mercury might be due to the pressure of the air. Petit's explanation was extremely vague: ". . . it seems to me that we might say that the vacuum, if there is one, or the rarefaction, is in no way limited, but follows the force of the agent that causes it, in such a way that the weight of the mercury [has] only the force to descend to a certain height and consequently to produce vacuity or rarefaction in all the remainder of the tube. . . ."[62] Mersenne, who died in 1648 still unconvinced of the pressure of the air, can scarcely have transmitted Torricelli's ideas on the subject to Petit.

Before Petit's letter was printed, an entirely independent account of the experiment with the tube and the mercury appeared in Poland,[63] the work of the Capuchin friar Valeriano Magni, who wrote of it with no reference to Torricelli, and indeed later claimed that he had thought of it independently.[64] The title of Magni's book shows that it was the vacuum, and that alone, which interested him; it was a frontal attack on the followers of Aristotle. Whether or not he had really arrived at the experiment by himself, his interest to us is that

[60] The best edition of Pascal's works is probably the *Oeuvres de Blaise Pascal, publiées suivant l'ordre chronologique, avec documents complémentaires, introductions et notes, par Léon Brunschvicg, Pierre Boutroux, et Félix Gazier* (14 vols., Paris, 1904–14). Reference will be made to this edition as Brunschvicg.

[61] Petit, *Observation*, etc., p. 4.

[62] *Ibid.*, p. 17.

[63] *Demonstratio ocularis loci sine locato: corporis successive moti in vacuo: luminis nulli corpori inhaerentis. A Valeriano Magno fratre Capuccino exhibita* [etc.] (Warsaw [1647]). The permission to print is dated July 12.

[64] Letter to Roberval, in Brunschvicg, Vol. II, pp. 503–6.

his little book was the very first *printed* account of it, and was reprinted at the end of Petit's published letter to Chanut.[65] The latter publication is interesting in another way. It is introduced by a dedication, signed "Dominicy,"[66] to Seguier, the Chancellor of France, and by a preface that is not signed at all, containing the following passage: "It will be seen that the glory of the invention belongs to Italy, and in my opinion to that admirable philosopher and mathematician Galileo, not to Torricelli at all. Also, that of having first observed it in France cannot be denied to Mr. Petit."[67] Now the historian J. Thirion came to the conclusion that this preface was written by Pascal himself, and that the latter had the ambition at the time to be considered the successor of Galileo.[68]

This brings Pascal into our story again, and indeed the history of the barometer in the years 1647 and 1648 revolves around that extraordinary man, even though the exact nature of his contribution is in some doubt. It is certain that he performed the experiment in public at Rouen in the winter of 1646/47, not only with mercury, but also in great tubes filled with water and with wine. He invited the local Aristotelians, as well as some other people, including Adrien Auzout and Florin Perier (Pascal's brother-in-law), of whom we shall hear later. These experiments had nothing to do with the pressure of the air, but were intended only to show that the thesis of the *horror vacui* is untenable. Those who held this theory asserted, of course, that the space at the top of the tube is filled with volatile "spirits" from the liquid, so Pascal asked them to predict whether the water or the wine would stand higher in its tube. They forecast that the water would be the higher, for they knew that wine is more "spirituous" than water, and the young Pascal took great delight[69] in demonstrating that they were wrong.

Later in 1647 Pascal wrote an account of his experiments,[70] and it is clear from this little work that at this time he was convinced that the space above the top of the mercury column was really a vacuum, for as his "conclusion" he writes: "After having demonstrated that none of the substances that can be perceived by the senses, or of which we have knowledge, fill this apparently empty space, it will be my opinion, until someone has demonstrated to me the existence of some matter that fills it, that it is truly empty, and destitute of all matter."[71] It is interesting that he examines and rejects the possibility (among others)

[65] Note 59 above.
[66] Marc Antoine Dominicy was a historian.
[67] Petit, *Observation*, etc., fol. a5^{v}.
[68] Thirion, *Rev. des questions scientifiques* (Brussels), 3e sér., Vol. 12 (1907), pp. 428–31.
[69] See Brunschwicg, Vol. II, p. 5.
[70] Pascal, *Expériences nouvelles touchant le vuide* (Paris, 1647) (Braunschvicg, II, 55–76). The *permission* is dated October 8, 1647.
[71] Brunschvicg, II, 73.

that the space may contain a very small amount of air, enormously rarefied, or of quicksilver or water "that is rarefied and converted into vapors."[72] He will have it really empty—a view much more extreme than that taken by any other seventeenth-century writer.

Nowhere in the book is there any indication that Pascal knew or cared about the pressure of the air. He ascribed the elevation of the liquid to a force which tends to prevent the occurrence of an empty space between bodies, but is limited to that with which water about thirty feet deep tends to flow downward. A force only slightly greater can produce a vacuum as large as desired.

These were his views in October, 1647. On November 15 of that year, he is supposed to have written a famous letter to his brother-in-law Florin Perier at Clermont, which was published in the autumn of 1648.[73] Almost at the beginning of this letter, Pascal says that he is inclined to attribute to the pressure of the air the effects that the philosophers impute to the *horror vacui*. After the experiment that he performed in Perier's presence "a few days ago" (*ces jours passez*) with two tubes, one inside the other, he is convinced of the pressure of the air, but it would be possible to explain this experiment by a limited force. There is therefore only one decisive test: "It is to make the ordinary experiment with the vacuum several times during the same day, in the same tube, with the same quicksilver, both at the bottom and at the summit of a mountain at least five or six hundred fathoms high, in order to find out whether the height of the quicksilver, suspended in the tube, will be the same or different in the two situations."[74] Pascal hopes that Perier, who lives near the foot of a high mountain, the Puy-de-Dôme, will make this experiment, and in the belief that he will, Pascal has told all his learned friends in Paris of the idea, especially Father Mersenne, who has written about it to Poland, Sweden, Holland, and other countries. Will Perier please do it as soon as possible, to satisfy the impatience of so many people who are waiting to hear the result.

For a long time this letter was accepted at its face value, until in 1906 Félix Matthieu roundly declared it to be a falsification, written for publication rather than for Perier, and in the summer of 1648 rather than in November, 1647.[75] This naturally caused a first-class intellectual row, but it is impossible to give an account of this controversy in the space available.[76] I shall note only that in a letter dated December 13, 1647, Descartes wrote to Mersenne that he had "advised Pascal to make an experiment to see whether the quick-

[72] *Ibid.*

[73] Blaise Pascal, *Récit de la grande expérience de l'équilibre des liqueurs* [etc.] (Paris, 1648) (Brunschvicg, II, 153–62).

[74] Brunschvicg, II, 159–60.

[75] Matthieu, *La Revue de Paris*, Vol. 13, part III (1906), p. 206.

[76] The contributions of Thirion and Matthieu alone come to about 420 pages. I have dealt with the controversy in *Hist. Bar.*, pp. 40–52.

silver rises as high when one is on a mountain as when one is right at the bottom," but that he did not know whether he had done it.[77] Mersenne then wrote about it to Huygens, and then to Le Tenneur, whom he believed to be at Clermont, inviting him to make the experiment on the Puy-de-Dôme.[78] But Le Tenneur had moved to Tours and replied from there that he could not do it, and at any rate he thought that it would be useless.[79] It is inconceivable that Mersenne would have asked Le Tenneur to make the experiment if he had known that Pascal had already requested Perier to do so. There are several other scarcely less cogent reasons for doubting Pascal's good faith in this matter. It seems highly probable that Descartes suggested the experiment to Pascal at a time when the latter did not believe in the pressure of the air, and that the suggestion remained in some back room of his capacious mind until the summer of 1648, when his opinion had changed.

A few words should be written about the experiment with the two tubes one inside the other, which Pascal claimed to have performed for Perier. This experiment, which is simply a way of setting up one barometer with its cistern in the vacuum of another, was certainly done during 1648 by either Adrien Auzout or Gilles Personne de Roberval.[80] The point of the experiment was that the mercury of the inner barometer stood at the same level in tube and cistern until air was let into the outer one, when the mercury rose to the usual height in the inner tube. What would happen had indeed been predicted by Torricelli in his second letter to Ricci.[81]

Let us return to the experiment on the mountain. Whenever it was that Pascal asked him to do it, it was on September 19, 1648, that Perier took the Torricellian tube up the Puy-de-Dôme. He was accompanied by five friends, and he left a second tube as a control at the monastery in Clermont, to be observed by one of the monks. At the monastery the mercury stood at 26 inches, 3½ lines,[82] remaining at this height all day, but at the top of the mountain the height of the column, measured several times, was only 23 inches, 2 lines. A careful experimenter, Perier compared the two tubes at the monastery after his return in the afternoon.

His description of the ascent was published by Pascal in the autumn of 1648, together with the controversial letter, in the pamphlet already

[77] Brunschvicg, II, 165. Descartes, *Oeuvres*, ed. cit., V, 99.
[78] Descartes, *Oeuvres*, V, 102.
[79] *Ibid.*, p. 103.
[80] See Brunschvicg, II, 291–93. Both men were distinguished devotees of the new science.
[81] See p. 14 above. Various forms of the experiment are described: by Jean Pecquet, *Experimenta nova anatomica* [etc.] (Paris, 1651), pp. 56–58; in the *Saggi* of the Accademia del Cimento (1667), pp. XLVI–XLVIII; by Jacques Rohault, *Traité de physique* (Paris, 1671), pp. 92–94.
[82] Twelve lines = one inch. These were Paris inches, about 6.5 per cent longer than English inches.

referred to.[83] Very few people seem to have seen this; it is evident that Descartes was not one of them, for on June 11, 1649, he asked Carcavi to let him know whether the experiment had succeeded.[84]

Pascal's other contribution to the history of the barometer is less controversial; he was the inventor of the siphon barometer. It is possible that he had this idea during the experiments at Rouen in 1646, for in September, 1647, Roberval wrote in a letter to Des Noyers that "the observation of the mercury was made by . . . Mr. Pascal in the recurved tubes that we commonly call siphons."[85] It is certain that the invention dates from before about 1650, the epoch at which he wrote his posthumously published *Traictez de l'equilibre des liqueurs et de la pesanteur de la masse de l'air* [etc.],[86] one of the great classics of physical science. We read:

> The most suitable instrument for observing all these variations is a glass tube three or four feet high, sealed at the top and curved round at the bottom, to which is glued a band of paper, divided in inches and lines. For if it is filled with quicksilver, we shall see that this will fall part of the way, and will remain partly suspended, and we shall be able to note exactly the degree to which it is suspended; and it will be easy to observe the variations in the weight of the air which happen because of the changes of weather, and those which take place in carrying it to a more elevated place.[87]

The reader will note the clear reference to a scale. But the honor of attaching a scale to the Torricellian tube for the first time belongs, as far as we know for certain, to Descartes, who was thus in a sense the inventor of a barometer. In a letter to Mersenne dated December 13, 1647, Descartes wrote: "So that we may also know if changes of weather and of location make any difference to it, I am sending you a paper scale two and a half feet long, in which the third and fourth inches above two feet are divided into lines; and I am keeping an exactly similar one here, so that we may see whether our observations agree."[88]

In this short account it is not possible to enter into a discussion of the philosophical effects of the invention of the barometer. We can pause to note only that it led to the gradual abandonment of the Peripatetic dogma of the *horror vacui*, though the doctrine had its adherents for the rest of the century. It is interesting that the main stumbling-block for many people—even celebrated men like Boyle—was the fact that light could pass through the supposed vacuum.

[83] *Récit de la grande expérience* [etc.].
[84] Descartes, *Oeuvres*, ed. cit., Vol. 5, p. 365.
[85] Brunschvicg, II, 32.
[86] Paris, 1663. This was translated into English by I. H. B. and A. G. H. Spiers (New York, 1937). It is in Brunschvicg, III, 156–292.
[87] *Traictez*, p. 100. (Brunschvicg, III, 233).
[88] Descartes, *Oeuvres*, ed. cit., Vol. 5, p. 99.

4. Experiments in England. Some time before April, 1648, the experiment was being made in England, to which news had evidently been sent by Mersenne in a letter to Theodore Haak,[89] who was associated with the group of men in London that eventually developed into The Royal Society. By July, Mersenne had told them about Auzout's experiment of the vacuum within a vacuum, though they did not quite succeed in making it work.[90] It is noteworthy that nothing about this experiment had been published at that time.

In England during the Commonwealth little experimenting of this sort was done except by Robert Boyle at Oxford, but after the Restoration the Torricellian tube became one of the most frequent concerns of the new Royal Society, and for the rest of the century more attention was paid to it in England than in any other country. It was there, probably in 1663, that the barometer received its name.[91]

The experiments made by The Royal Society were many and various, but must take second place to the really remarkable *New Experiments Physico-Mechanicall touching the Spring of the Air, and its Effects* [etc.] of Robert Boyle, to quote the title of the first of his physical treatises, published at Oxford in 1660.[92] Boyle, assisted by Hooke, improved the air pump of Otto Guericke, and used it to establish many of the properties of the vacuum and to investigate the behavior of things placed in it. I shall mention only his seventeenth experiment,[93] in which he set up the barometer in a bell-jar and evacuated the space, watching the mercury fall in the tube. This would seem to establish that the pressure of the air held the mercury, but it did not satisfy the Jesuit Francis Line or Linus, alias Hall, who managed to find a completely original explanation.[94] This was that a *funiculus*, an invisible elastic cord, extended from the top of the mercury to the end of the tube and held the mercury suspended. Linus also denied that the air had enough "spring" to sustain the mercury. This led Boyle to make further experiments, in which he showed that a much taller column of mercury than that in the barometer could be held up by air under pressure, and in doing so he discovered "Boyle's law"—that the volume and the pressure are inversely proportional.[95]

And yet, in spite of all these and many further pneumatic experiments,[96] Boyle refused to make up his mind that a vacuum could

[89] See Miss R. H. Syfret, *Notes & Records Roy. Soc.*, Vol. 5 (1948), p. 135.

[90] See Brunschvicg, II, 307.

[91] Robert Boyle, *New Experiments and Observations touching Cold* (London, 1665), p. 27 (*Works*, ed. Thomas Birch, 6 vols., London, 1772, Vol. 2, p. 487).

[92] *Works*, ed. cit., I, 1–117.

[93] *Ibid.*, I, 33–39.

[94] Franciscus Linus, *Tractatus de corporum inseparabilitate, in quo experimenta de vacuo . . . examinantur* [etc.] (London, 1661).

[95] Boyle, *A Defence of the Doctrine touching the Spring and weight of the Air . . . against the Objections of Franciscus Linus* (London, 1662), Works, ed. cit., I, 123–85.

[96] Boyle, *A Continuation of New Experiments Physico-Mechanical touching the Spring and Weight of the Air* (Oxford, 1669). *Works*, ed. cit., III, 173–276.

exist: ". . . I have formerly made, and now renew a solemn profession, that I do not in this treatise intend to declare either for or against the being of a vacuum. . . ."[97]

5. Early mercury barometers. When we consider the early development of the mercury barometer, we find that three sorts of improvement were gradually made to the Torricellian tube, for three purposes: to expand the scale, to render the instrument more portable, and to increase the real accuracy of reading the instrument. It would be quite wrong to assume that the solution of the first problem necessarily carries with it that of the third.

In the seventeenth century people were not used to making fine measurements, and the smallness and slowness of the day-to-day variations of the barometer must have been a disappointment. No less than six different schemes for magnifying the scale of the instrument were devised before the year 1700, some of them in more than one form. In the order of their first appearance they were: barometers with one or more liquids in addition to mercury; the wheel barometer; the balance barometer; the diagonal barometer; the "square" or L-shaped barometer; and the conical barometer.

We have the word of Pierre Chanut[98] that René Descartes produced a barometer such as that shown diagrammatically in Figure 1.6, with

Fig. 1.6 Descartes' two-liquid barometer.

water above the mercury. Chanut tried to get such an instrument made, but it defeated the Stockholm glass blowers.

Water was a very poor choice for the second liquid, because the water vapor in the upper tube made the instrument very sensitive to temperature. With something like vacuum-pump oil, it would have worked.[99] The magnification of such a barometer is something less than the ratio of the specific gravities of the liquids.

A better scheme (Fig. 1.7) was found by Hooke, who demonstrated his "double barometer" to The Royal Society on June 18, 1668.[100]

The cylindrical bulbs *A* and *B* are long enough to accommodate pressure changes, and the reading is taken from the water surface at *D*. In 1672 Christian Huygens re-invented both the barometers,[101] noting that the second is the better form because the water is not in

[97] *Ibid.*, p. 29 (*Works*, III, 198).
[98] Letter to Perier from Stockholm, September 24, 1650, in Pascal, *Traitez de l'équilibre des liqueurs* (2nd ed., Paris, 1698), pp. 207–08 (Brunschvicg, II, 437–38).
[99] As was suggested by F. Guthrie, *Phil. Mag.*, Vol. 3 (1877), pp. 139–41.
[100] Thomas Birch, *History of The Royal Society* (4 vols., London, 1756–57), II, 298.
[101] *Hist. & Mém. Acad. r. Sci.* (10 vols., Paris, 1733), X, 540–44. Also in Huygens, *Oeuvres complètes, publiés par la Société Hollandaise des Sciences* (22 vols., La Haye, 1888–1950), VII (1919), 242.

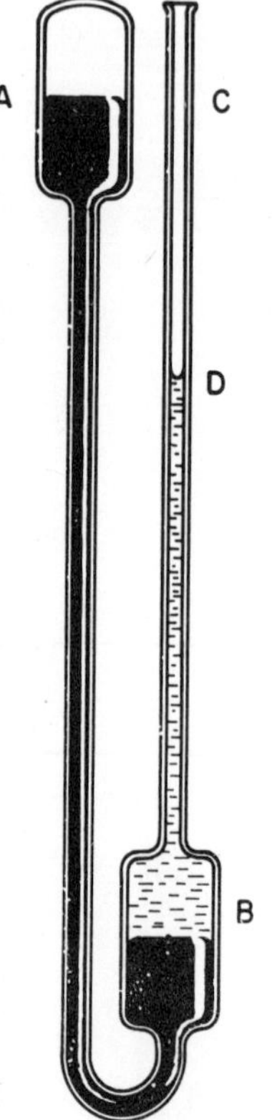

Fig 1.7. Hooke's two-liquid barometer.

the vacuum. In spite of its imperfections this barometer was immensely popular for two centuries, often combined with the ordinary barometer on the same board, because the two-liquid barometer suffered from evaporation. The water also became dirty and soiled the tube. Aware of these defects, Hooke invented a three-liquid barometer (Fig. 1.8), by adding a third bulb, at *C*, having the same diameter as *A* and *B*.[102] *BC* contained two immiscible liquids, such as alcohol and turpentine, the reading being taken at their common surface *D*. This worked much better, but the alcohol in *BD* still acted as a thermometer.

About 1685 this type of barometer was suggested by Guillaume Amontons to the Paris instrumentmaker Hubin, who informed him to his surprise that Hooke had proposed exactly the same instrument.[103] This is only one of many examples of the failure of the French academicians of the time to read the works of their English contemporaries.

An entirely different means of expanding the scale was devised, probably in December, 1663, by Robert Hooke; this was the well-known "wheel barometer," which became immensely popular as an article of furniture in England. It was first described by Hooke in his influential work *Micrographia*,[104] from which Figure 1.9 is taken;[105] it is simply a siphon barometer in which a "steel bullet" floats on the mercury in the open limb. This float is attached to a thread passing around a pulley to a small weight, and on the shaft of the pulley is a pointer. The bend of the tube was made as shown in the smaller drawing, for ease of filling. The closed end of the tube was blown out into a large spherical bulb, with the mercury surface in a diametral plane, in order to throw nearly all the motion into the short limb.

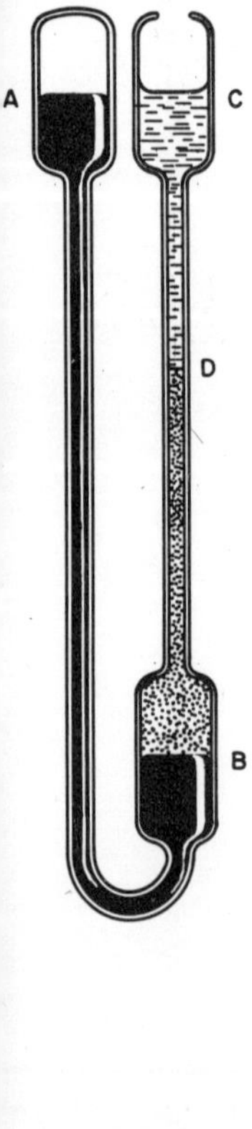

Fig. 1.8 Hooke's three-liquid barometer.

Later[106] Hooke, realizing that this was not important, lengthened the tube so that "more room is allowed for any remainder of air, to expand the better."[107]

The usefulness of the wheel barometer depended on the degree to which friction could be reduced, and this was affected not only by the

[102] *Phil. Trans.*, Vol. 16 (1686), pp. 241–44.

[103] Amontons, *Remarques et expériences phisiques sur la construction d'une nouvelle clepsidre, sur les baromètres, termomètres, & higromètres* (Paris, 1695), pp. 147–48. Hubin had been in England.

[104] Hooke, *Micrographia, or some Physiological Descriptions of Minute Bodies, made by Magnifying Glasses, with Observations and Inquiries thereupon* (London, 1665). Preface, sig. c2–c3.

[105] There are two other versions, which I have discussed in *Hist. Bar.*, pp. 94–98.

[106] *Phil. Trans.*, Vol. 1 (1666), pp. 218–19.

[107] *Ibid.*, p. 219.

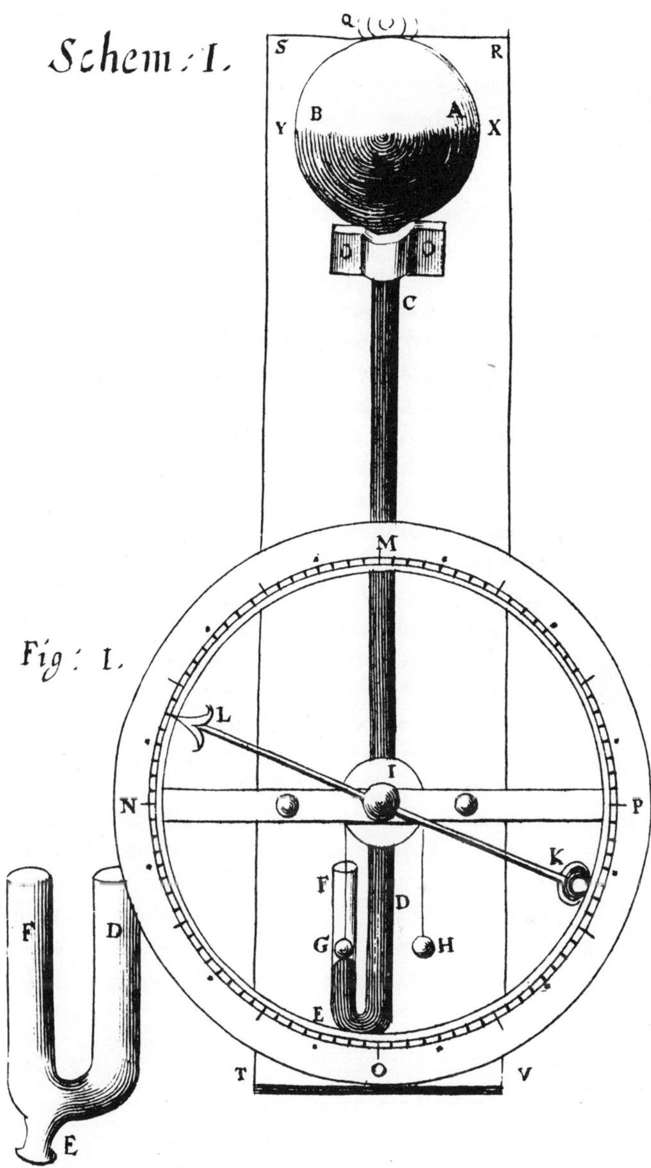

Fig. 1.9 Hooke's wheel barometer, from the Micrographia.

mechanical design and workmanship, but also by the cleanness of the exposed surface of the mercury and the shape of the float. I have recently published a holograph of Hooke's[108] which states quite clearly that the recognition of this friction led him to invent the "double barometer" described above. Hooke knew the value of tapping the wheel barometer with the finger, "for thereupon you will always find

[108] Royal Society, *Classified Papers*, xx, item 58 (fol. 124 and 125). See W. E. K. Middleton, *Notes & Records Roy. Soc. London*, Vol. 20 (1965), pp. 145–51.

that the hand will remoue a little either forwards or backwards according as the tendency of the pressure of the air at that time operates, which though it be a defect in it self yet upon this account is of good use."[109]

Another sort of barometer that could have an expanded scale might be constructed by hanging either the tube or the cistern of an ordinary barometer from one arm of a balance. Marin Mersenne had weighed a Torricellian tube hanging in its cistern,[110] and on August 13, 1662, Dr. Jonathan Goddard performed similar experiments before The Royal Society.[111] The weight seemed to be that of the glass tube, plus all the mercury in it above the level in the cistern. This result was puzzling, because it seemed that the quicksilver must in some way be attached to the tube that hung from the balance. It was hard to see that the closed end of the barometer tube represented a horizontal cross-section with the air pressing down on it and nothing at all pressing upward beneath; indeed, an eminent judge, Sir Matthew Hale, published an anonymous book[112] in which he used this very experiment to demonstrate that it cannot be the pressure of the air that sustains the mercury.

It was not until some time later, however, that the effect was used to produce a barometer. This was done by Sir Samuel Morland, an engineer,[113] probably in 1679,[114] and we have Hooke's word for it that two schemes were tried, hanging either the tube or the cistern from a balance.[115] The magnification in the first type, as Francis North clearly saw,[116] ought to be equal to the ratio of the specific gravities of mercury and glass, if the tube is of uniform cross-section and is allowed to rise and fall, counterpoised by a constant weight, and if the cistern is very wide. Unfortunately none of Sir Samuel's barometers have survived, and we can do no more than guess at their design. In fact the oldest surviving barometer of this kind, part of a meteorograph in the Museo Copernicano at Rome, dates from about 1865.[117]

Sir Samuel Morland is also given credit for the invention of the diagonal barometer, a very simple instrument made by bending the Torricellian tube through a little less than a right angle at a point about 69 cm above the point at which the mercury will stand in the

[109] *Ibid.*, p. 146. Note that the word "tendency" is still used to express the direction and rate of the change in pressure.

[110] Mersenne, *Novarum observationum physico-mathematicarum Tomus III* (Paris, 1647), first preface.

[111] Royal Society, *Register Book*, I, 185–86.

[112] *Difficiles nugae: or, Observations touching the Torricellian Experiment* [etc.] (London, 1674).

[113] *Cf.* Roger North, *Life of the Rt. Hon. Francis North, Baron Guilford* [etc.] (London, 1742), p. 293. Morland's own account seems to have been lost.

[114] Middleton, *Notes & Records Roy. Soc. London*, Vol. 20 (1965), p. 150.

[115] *Ibid.*, p. 147.

[116] *Life of . . . Francis North*, pp. 293–94.

[117] *Cf.* Angelo Secchi, *Collegio Romano, Bull. Meteorol.*, Vol. 5 (1866); also *Hist. Bar.*, pp. 304–5; and below, p. 258.

cistern. The magnification is the cosecant of the angle that the upper part of the tube makes with the horizontal. Morland's claim to the invention rests on a passage in a pamphlet published in 1688 by John Smith, a London clockmaker,[118] who wrote of "[the baroscope] with a tube, whose top inclines, devised by Sir Samuel Morland," as if it were a familiar instrument at that time. It was independently invented about 1695 by Bernardo Ramazzini of Modena.[119] In spite of its shape (to which John Smith had objected), it quickly became a common household instrument. John Patrick mounted it at the top and left-hand side of a rectangular frame, with a large thermometer at the right-hand side, "and a looking glass commodiously plac'd on the same frame, between the barometer and the thermometer, whereby gentlemen and ladies at the same time they dress, may accommodate their habit to the weather."[120] As a scientific instrument it suffered from excessive friction and from the curious shape of the mercury surface in the inclined tube.

Another attempt to magnify the motion of the mercury resulted in the "square" barometer, adequately explained in Figure 1.10. The

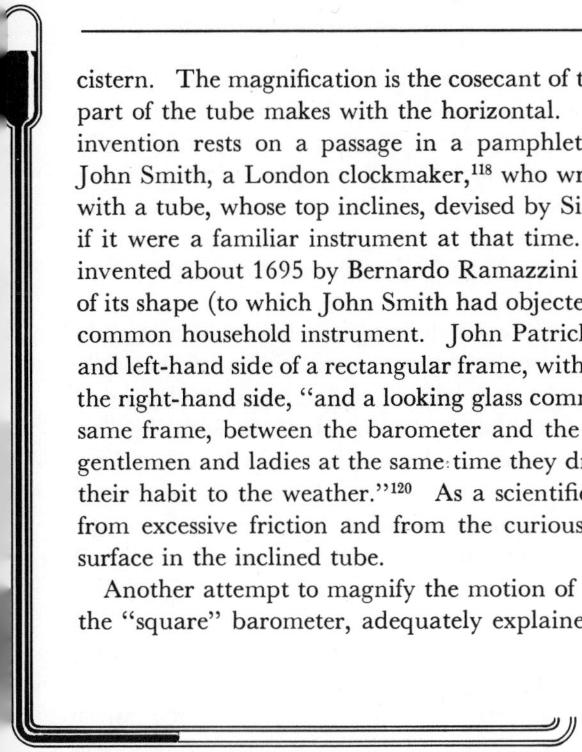

Fig. 1.10 The "square" barometer.

invention of this is usually ascribed to Johann Bernouilli,[121] but I have found a manuscript[122] that shows quite clearly that Jean Dominique Cassini installed one at the Observatory of Paris on January 3, 1673,[123] when Bernouilli was five years old.

I shall mention one more seventeenth-century attempt to extend the scale; this was the "conical barometer" invented by Guillaume Amontons[124] and, perhaps independently, by John Patrick.[125] It was simply a slightly conical tube, closed at the smaller end and containing a column of mercury of suitable length. When the tube was hung with the closed end upward, the column would fall in the tube until

[118] Smith, *A Compleat Discourse of the Nature, Use and right Managing of that Wonderful Instrument the Baroscope or Quicksilver Weather Glass* (London, 1688), pp. 1–2.

[119] Ramazzini, *Ephemerides barometricae mutinenses anni M.DC.XCIV. una cum disquisitione causae ascensus ac descensus mercurii in Torricelliana fistula, juxta diversam aëris statum* (Modena, 1695), pp. viii–ix.

[120] *A New Improvement of the Quicksilver Barometer, made by John Patrick, in the Old-Baily, London* (broadsheet, n.d., probably 1700 on internal evidence). There is an instrument of this sort, and of about this date (but without the mirror) in the History of Science Museum, Oxford.

[121] *Johannis Bernouilli opera omnia*, ed. Cramer (4 vols., Lausanne & Geneva, 1742), II, 204–7.

[122] Observatoire de Paris, ms. AD1, Tom. 4, fol. 3v.

[123] See also *Hist. Bar.*, pp. 115–17.

[124] *Remarques et expériences phisiques* [etc.] (Paris, 1695), pp. 123–35.

[125] Patrick, *A New Improvement*, etc.

its length, decreasing as its diameter became greater, equaled the barometric height. The magnification would depend on the rate of change of the diameter of the tube. Amontons intended it for a marine barometer, thinking that it would not be affected by the motion of the ship, but of course it was not practical, because the inside of the tube became dirty in a short time. There is one of these barometers, about 1.8 meters long, in the Conservatoire National des Arts et Métiers.[126]

We must now return to better designs, and first to seventeenth-century improvements on the original tube and cistern of Torricelli. Naturally the cistern ought to be closed, both for portability and to keep the mercury clean, but the air must be given access to the mercury so that it may press upon it. A small hole may be closed by a screw or plug for transport, but at least by 1688 it was known that boxwood[127] is previous to air along the grain, but impervious to mercury.[128] In addition it is strong and eminently machinable, so that it quickly became the favorite material for barometer cisterns, which were usually made of two pieces of boxwood glued together, with a plugged hole for filling, or screwed together, with a thin leather gasket between them. In the former construction the plug was also used, probably before 1700, to adjust the amount of mercury in the cistern by overflow. Referring to Figure 1.11,[129] the barometer tube was first filled and glued into the piece *BCD*. The cistern was then filled with mercury to the level *ef*, and the piece *A* was glued on. When the glue had dried, the instrument was carefully turned right side up, and the plug *E* was removed in order to let the mercury run out to its final level, the plug being then replaced. Such a barometer was made portable by laying it horizontal with *E* on top, and filling the cistern full of mercury. The possibility of adjusting the level in the cistern by overflow permitted a great advance in accuracy.

The outstanding portable barometer of the seventeenth century, however, was that made and patented[130] by Daniel Quare of London in 1695. Many of Quare's beautiful barometers survive,[131] easily recognizable by their turned cases of wood or ivory, their scales engraved on silver or brass and enclosed in a box with a glass front, and their three little knobs at the top (Fig. 1.12).

The cistern is a hollow cylinder of boxwood with the tube cemented into the top. The botton of the cistern is a leather bag, resting on the

[126] Inventory no. 1580.

[127] From the trees *Buxus sempervirens* and *B. balearica*.

[128] [Joachim Dalencé]. *Traittez des baromètres, thermomètres, et notiomètres, ou hygromètres*. Par Mr. D*** (Amsterdam, 1688), p. 35.

[129] Adapted from Jacob Leupold, *Theatrum machinarum Pars III, Theatri statici universalis, sive theatrum aerostaticum* [etc.] (Leipzig, 1726).

[130] A manuscript copy of the patent is at Oxford, Bodleian Library, Rawlinson MS. A241, fol. 90r–91r. But N. Goodison, *Ann. Sci.*, Vol. 23 (1967), pp. 287–93, argues that Quare was not really the inventor.

[131] Some are undoubtedly copies by John Patrick and others.

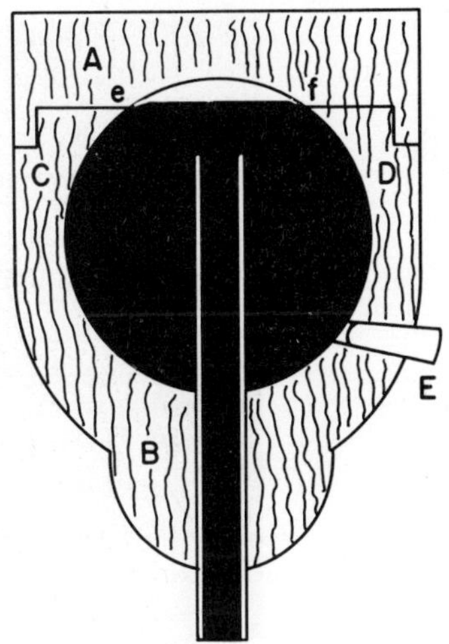

Fig. 1.11 Simple barometer cistern, inverted as in process of filling.

padded end of a screw that works in a nut set into the bottom of the outer wooden or ivory case of the cistern. To make the instrument portable the screw was advanced until both the tube and the cistern were full of mercury, and the barometer was carefully inverted and carried upside down. On reinstallation it was put right way up and the screw was simply lowered as far as possible. A further safeguard introduced by Quare was a constriction near the top of the tube that left a small bulb into which the mercury could not flow very fast, so that it could never strike the top of the tube hard enough to break it.

Seventeenth-century barometers were almost always provided with simple triangular pointers for reading the level of the mercury in the tube, usually to be set against the edge, not the top, of the meniscus. Because there was no provision for avoiding parallax, no great accuracy or even precision could be attained, so that William Derham's scheme[132] of a rack-and-pinion drive for the pointer, the pinion having a pitch circumference of exactly one inch and carrying a graduated dial on its shaft, was really no improvement. A much better idea was that of Stephen Gray,[133] whose sketch (Fig. 1.13) will be quite clear to the reader, with the note that there was a cross-hair at the focus of the objective of the compound microscope, and a spring at the top of the screw to avoid one of the possible sources of backlash. But in 1698 this excellent design was made superfluous by the errors resulting from the primitive state of other parts of the barometer. The vernier as a means of reading the level of the mercury surface was probably introduced in England in the second half of the eighteenth century; the earliest description of a barometer with a vernier that I

[132] Derham, *Phil. Trans.*, Vol. 20 (1698), pp. 45–48.
[133] Gray, *Ibid.*, pp. 176–78.

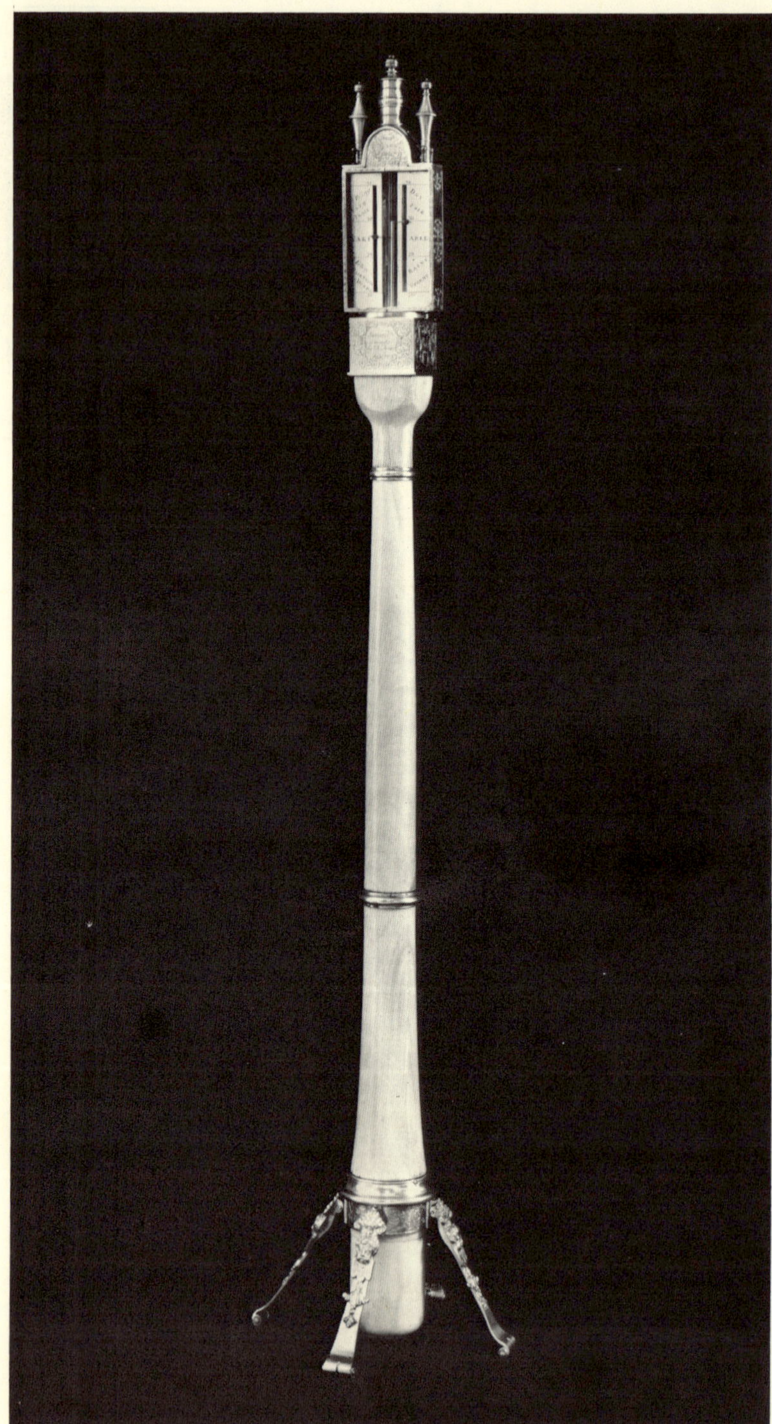

Fig. 1.12 A barometer by Quare, in a private collection (copyright by N. Goodison; reproduced by permission).

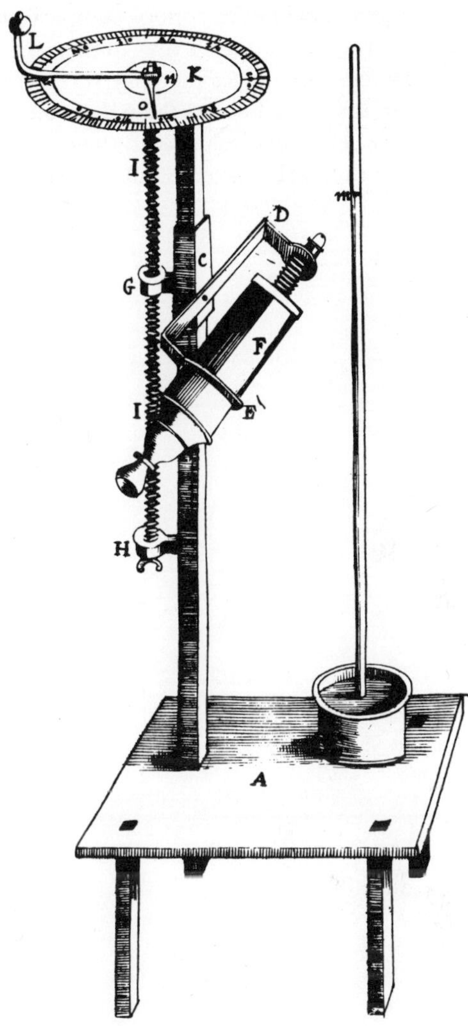

Fig. 1.13 Gray's micrometer, 1698.

have been able to find dates from 1772[134] and refers to an instrument by the famous London maker Jesse Ramsden, who is also credited[135] with making the first barometer in which the summit of the mercury meniscus could be lined up with the plane lower end of a tube attached to the vernier, light coming from the back through a slot, as is universal in the mercury barometers used at meteorological stations today. By about 1800 this construction was becoming common, especially for portable instruments, and must have increased the precision of reading almost tenfold.

[134] *Obs. sur la phys.*, Vol. 1 (1772), pp. 509–12.
[135] By the Paris instrumentmaker Megnié, quoted by Truchot, *Ann. Chim. et Phys.*, Vol. 18 (1879), p. 305.

Meanwhile ever more complex cisterns were being devised for the purpose of ensuring that the zero of the scale should coincide with the level of the lower mercury surface. It is quite out of the question to go into detail about these developments, and I shall confine myself to the celebrated solution to the problem that was arrived at about 1800 by the Paris maker Nicolas Fortin.

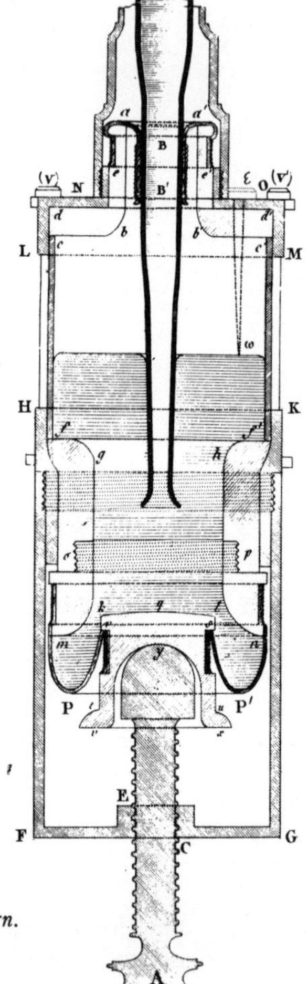

Fig. 1.14 Fortin's barometer cistern.

Figure 1.14[136] shows Fortin's cistern, in which the mercury can be seen through the glass cylinder *ee'f'f* and brought up to touch the ivory point ω by means of the screw *A*, which raises the leather bag *PP'*. Fortin introduced neither the leather bag nor the ivory point nor the glass cylinder; but he combined them, with the result that

[136] From J. N. P. Hachette, *Programmes d'un cours de physique* [etc.] (Paris, 1809). The barometer is described on pp. 221–25.

people tend to apply the term "Fortin barometer" to any barometer in which the mercury is adjusted to touch a point.

One alternative to adjusting the level of the mercury in the cistern is to leave it alone and correct for its change of level as the atmospheric pressure rises and falls. In the seventeenth century it was believed that it was sufficient to make the cistern large enough so that this change could be neglected, though Hooke saw clearly that an appropriate contraction of the scale could be used.[137] This idea was reintroduced in 1792 by Hugh Hamilton of Armagh in Ireland.[138] Fixed-cistern barometers with contracted scales first became common in marine use, and the tubes of such instruments were furnished with a section of narrow bore to prevent the violent oscillations of the mercury column that would otherwise result from the motion of the ship. This idea was also described by Robert Hooke,[139] in the form of "a small stop cock placed just in the bending of the two pipes" of his wheel barometer, "the key of the stopcock [being] just soe farr turn'd as to leave the least imaginable passage for the mercury between the two stemms." Nobody remembered this, however, and in the form of a capillary tube it was reinvented by Edward Nairne about 1770.[140]

We have noted that Pascal invented the siphon barometer;[141] Boyle[142] described one at great length, calling it a "portable barometer," and providing an illustration. The motion of the mercury in each limb of the tube was of course only half that in the tube of a cistern barometer, and this made it unpopular at the time, until makers found that they could save mercury, and restore the movement in the vacuum space, by blowing a little glass bottle, generally pear-shaped, at the open end of the tube. It is safe to say that the vast majority of the barometers made for private use between 1670 and 1850 (at least) were such "bottle barometers." They made no pretence to accuracy, and their only advantage was simplicity and cheapness.

The first man to make a serious effort to perfect the siphon barometer was Jean André Deluc.[143] At some time shortly after 1750 he began experiments that finally resulted in the celebrated instrument shown at the right-hand side of our Figure 1.15. After attempts, shown at the upper left, to replace the bulb of the "bottle barometer"

[137] This was in a description of the wheel barometer, *Phil. Trans.*, Vol. 1 (1666), p. 219.

[138] Hamilton, *Trans. R. Irish Acad.*, Vol. 5 (1792), pp. 117–27.

[139] To The Royal Society, on January 2, 1667/8. See Royal Society, *Classified Papers* XX, item 48 (Hooke's autograph).

[140] See Constantine Phipps, Baron Mulgrave, *A Voyage towards the North Pole undertaken by His Majesty's Command* (London, 1773), pp. 123–24.

[141] See p. 23 above.

[142] Boyle, *Continuation of New Experiments*, etc. (1669), pp. 68–73. (*Works*, ed. cit., III, 219–23).

[143] Deluc, *Recherches sur les modifications de l'atmosphère* (2 vols., 4°; Geneva, 1772. Also 4 vols., 8°; Paris 1784), ¶¶459–507. The paragraphs are numbered identically in the two editions.

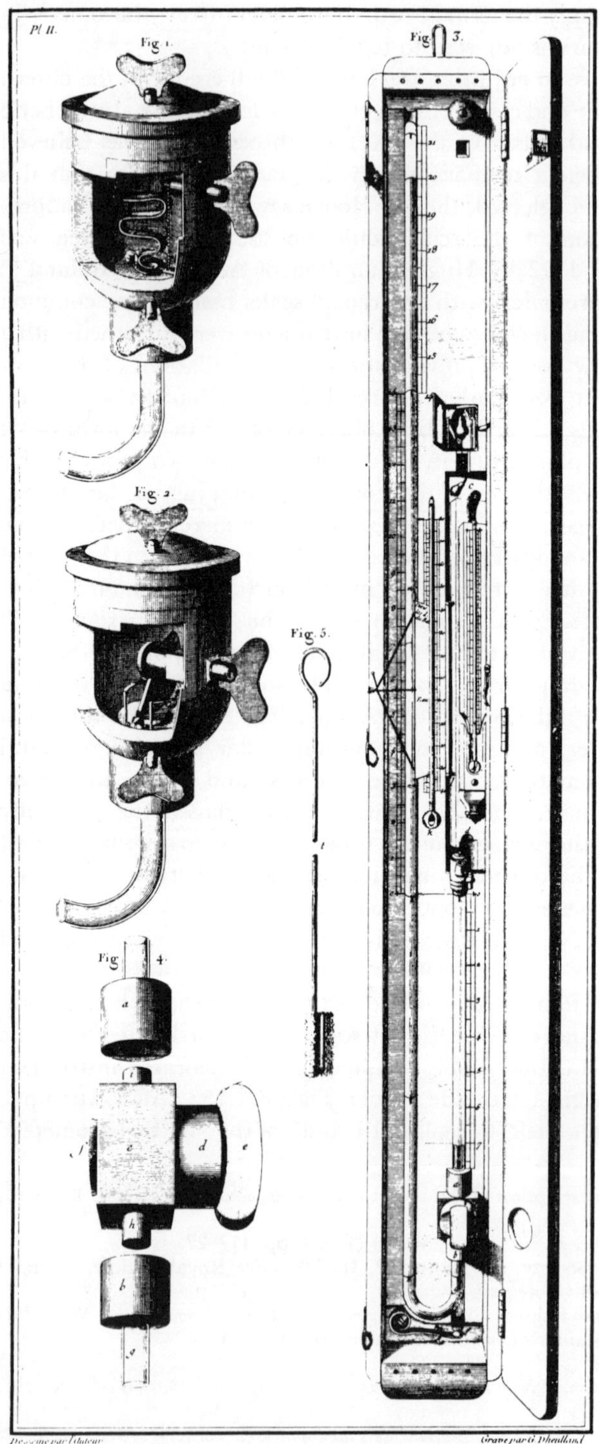

Fig. 1.15 Deluc's portable barometer.

by an elaborate cistern, he went over to the siphon barometer, and produced an instrument that was about as easily portable as a mercury barometer can be, largely as a result of the well-designed valve of ivory and cork used to isolate the mercury in the longer tube after inclining the instrument. This valve is shown enlarged at the lower left of Figure 1.15.

The accuracy of this barometer was probably limited by the grade of instrument making available to Deluc—apparently he made it himself. There were no verniers, not even any pointers; the tubes were sunk in the wood up to their diametral planes, and the paper scales came right up to the walls of the tubes. In principle the accuracy was limited by the erratic variations in the capillary depression of a mercury surface exposed to the air, a phenomenon not suspected in the eighteenth century. Although improved siphon barometers were later constructed according to the ideas of Joseph Louis Gay-Lussac[144] and others, this limitation continued to operate.

6. *Water barometers and air barometers.* I have already referred to the experiments of Berti and Pascal. Water barometers were set up many times in the seventeenth century, for example by Otto Guericke[145] about 1654, by Richard Townley and others in 1661,[146] in 1663 by The Royal Society,[147] and about 1683 by Edme Mariotte in a shaft twenty-eight meters deep dug beneath the Observatory of Paris.[148] This was an excellent location for a water barometer because of the almost unvarying temperature. In the eighteenth century it seems that almost no water barometers were made. I shall say nothing further about these instruments, none of which advanced the science of meteorology in any way.

Air barometers, by which term I refer to barometers making use of the relations between the pressure and volume, or pressure and density, of some gas, are scarcely of more interest, but one or two must be mentioned. The earliest was described independently by Boyle[149] and by Guericke;[150] it consisted of a light hollow vessel hung on a balance and counterpoised by a small weight. The importance of this simple instrument for the understanding of aerostatics is shown by its continuing use with the air pump in the teaching of elementary physics.

A very different instrument was described to The Royal Society

144 Gay-Lussac, *Ann. Chim. et Phys.*, Vol. 1 (1816), pp. 113–19.

145 Guericke, *Experimenta nova (ut vocantur) Magdeburgica* [etc.] (Amsterdam, 1672), p. 100.

146 Henry Power, *Experimental Philosophy in Three Books* (London, 1664), p. 131.

147 Birch, *History of The Royal Society* (4 vols., London, 1756–57), I, 255.

148 Mariotte, *Traité du mouvement des eaux* (Paris, 1686), pp. 90–92.

149 *New Experiments and Observations Touching Cold* [etc.] (London, 1665), pp. 19–20. *Works*, ed. cit., II, 485.

150 *Experimenta nova*, p. 101. Guericke had written to Kaspar Schott about it on Dec. 30, 1661. See Schott, *Technica curiosa* [etc.] (Nürnberg, 1664), pp. 52–53.

by Robert Hooke on January 2, 1667/8,[151] as a means of measuring the pressure of the air at sea. It was simply an open air thermometer[152] with mercury as the liquid, and a sealed spirit thermometer. The latter was to determine the effect of the ambient temperature on the air in the air thermometer, and Hooke explained clearly how it was to be calibrated on land before the voyage. Halley took one on his famous scientific expendition in the Atlantic (1698–1700), and was most enthusiastic.

Hooke's "marine barometer" has been reinvented or "improved" many times, most importantly, perhaps, by Alexander Adie of Edinburgh, in 1818, who gave it the learned name sympiesometer (*sympiezo*, compress; *metron*, a measure).[153] This instrument (Fig. 1.16)

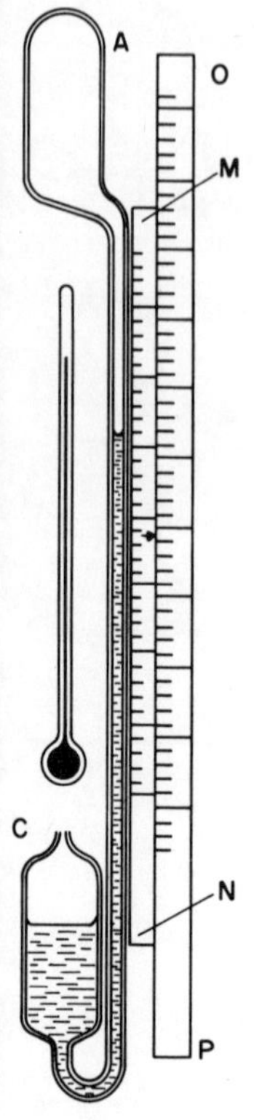

Fig. 1.16 Adie's sympiesometer, 1818.

has a bulb *A*, filled with hydrogen, and another bulb *C*, which, with part of the connecting tube, contains colored almond oil. A thermometer is mounted on the same board. The scale of pressures *MN* slides against a fixed scale of temperatures *OP*, both of which are graduated empirically. To use the instrument, an index mark on *MN* is set opposite the temperature (as read from the thermometer) on *OP*. The pressure is then read from *MN* opposite the level of the oil in the tube.

The sympiesometer found some acceptance as a marine barometer for two or three decades, after which it was displaced by the improved mercury barometer with the constricted tube.

7. *The Aneroid Barometer*. We come finally to barometers that depend on neither hydrostatic equilibrium nor the gas laws, but on the measurement of the force exerted by the atmosphere on some area in a mechanical system.

The first barometer of this sort was suggested by Blaise Pascal.[154] A "concertina" bellows was to be closed and sealed, one end fastened to a roof beam, and a heavy chain hung from the other. The chain was to reach the floor, and as the pressure of the air changed, various

[151] Royal Society, *Classified Papers*, XX, item 48. Autograph. Nothing was published about it until 1701, when Edmond Halley (*Phil. Trans.*, Vol. 22, pp. 791 ff.) praised it.
[152] See Chapter 2.
[153] Adie, British Patent 4323 (1818). See also *Edinburgh Phil. J.*, Vol. 1 (1819), pp. 54–60.
[154] *Traitez de l'équilibre des liqueurs et de la pesanteur de la masse de l'air* [etc.] (Paris, 1663), p. 144. *Oeuvres*, ed. Brunschvicg, II, 515–17. This was written before 1654.

amounts of it would be held suspended. In the seventeenth century it would certainly have been impossible to make such a bellows both flexible and airtight.

A few years later Robert Boyle realized that the pressure of the air might be measured by hanging a dead weight on the piston of an air pump, the piston having first been pushed to the end of the cylinder, and the valve closed.[155] The technical embodiment of this idea had to wait three centuries.[156]

These instruments used dead weight. The barometer that we now refer to as the aneroid (*a*, without; *neros*, a liquid) uses the elasticity of solid materials.

In a letter to Johann Bernouilli dated June 7, 1698,[157] the great Gottfried Wilhelm Leibniz put forward the idea of "a little closed bellows that would be compressed and dilate by itself, as the weight of the air increases or diminishes." In 1702 he thought that such a barometer, without mercury, could be carried like a watch.[158] Later he suggested a metallic bellows,[159] but we may suppose, and regret, that he had no instrumentmaker who could carry out his ideas.

Sixty years later I. E. Zeiher constructed a "marine barometer" in which a spring resisted the pressure of the air (Fig. 1.17).[160] It consisted of a cylinder *ABCD* with two pistons *E* and *H* carefully fitted into it. The upper one was attached to the top of the cylinder by a spring *G*. The lower one, at first in contact with the upper, could be pulled down by means of a special tool so as to evacuate the space, and clamped by a wing-nut *3* on to a leather washer *4*. After this had been done the piston *E*, and rod *EF* attached to it, would move with changes in atmospheric pressure. This motion was communicated to a pointer γ by a friction wheel β, moved by a spring ϵ. In case this "friction gearing" slipped, there was a scratch on the rod which coincided with the surface of the cover *AB* at a known pressure, permitting readjustment. If the vacuum was lost, it could easily be re-established by operating *H*. It is likely that this process would be needed very frequently, and at any rate nothing further seems to have been heard of this barometer.

Leibniz's idea was revived in 1798, probably independently, by Nicolas Jacques Conté,[161] who was not satisfied with it because it was too sensitive to temperature changes. The technical realization of the

[155] Boyle, *New Experiments Physico-Mechanicall* [etc.] (London, 1660), pp. 237–38. *Works* (1772 ed.), I, 71–73.

[156] See *Izmeritel'naya Tekhnika* (Measurement Techniques), Moscow, 1959, February, pp. 23–24 [I.S.A. Translation (1959), pp. 108–13].

[157] *Virorum celeberr. Got. Gul. Leibnitii et Johan. Bernouillii commercium philosophicum et mathematicum* (2 vols.; Lausanne and Geneva, 1745), I, 368.

[158] *Ibid.*, II, 70.

[159] *Ibid.*, II, 78.

[160] Zeiher, *Nov. Comm. Petrop.*, Vol. 8 (1763), pp. 274–78.

[161] *Bulletin des Sciences, par la Société philomathique*, 11 floréal an VI (April, 1798), p. 106.

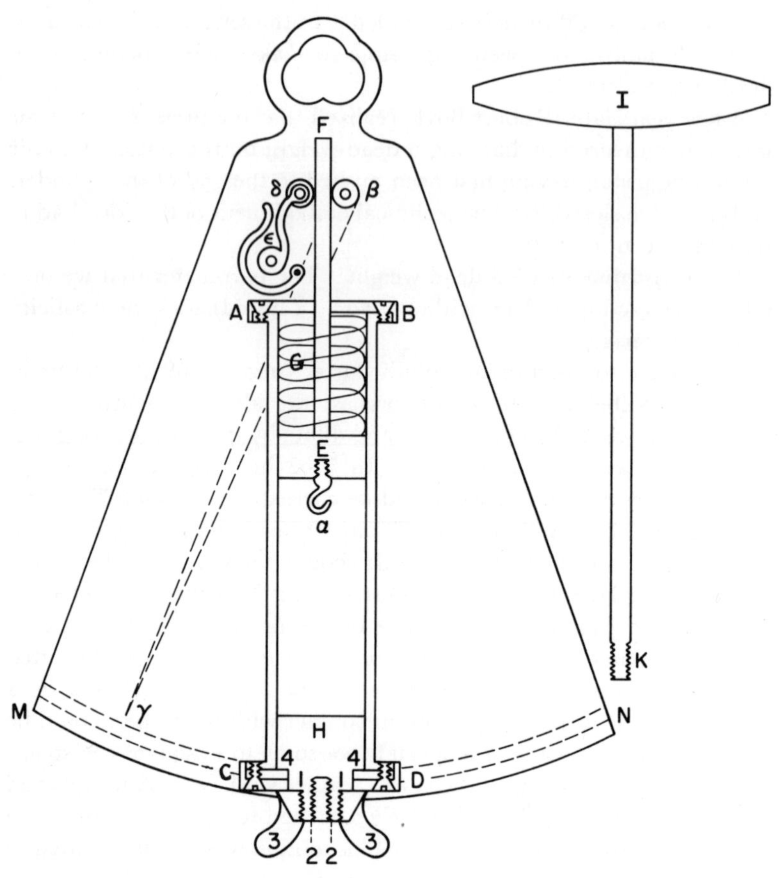

Fig. 1.17 Zeiher's marine barometer, 1763.

idea had to wait until 1843, when Lucien Vidie, an engineer of Nantes, produced a satisfactory metallic barometer,[162] in spite of a generally accepted belief that metals are not only slightly porous, but also have no real elastic limit, so that even a small force will produce a permanent deformation if it acts for a long time.

Figure 1.18 is taken from Vidie's first patent,[163] and shows a cross-section of his earliest barometer. We see a stong evacuated brass box *A*, closed by a corrugated diaphragm *B*, which is supported against the pressure of the atmosphere by thirty-three helical springs *C*, each fitted with a flat cap *E* and based in a shallow recess at the bottom of the box. A cup *G* is soldered into a hole in the center of the dia-

[162] Auguste Laurant, *Histoire des baromètres et manomètres anéroides. Biographie de Lucien Vidie* (Paris, 1867).

[163] British Patent 10,157 (1844). A French Patent, no. 12,473 (1844), and a U.S. Patent, no. 4,702 (1846), were also taken out.

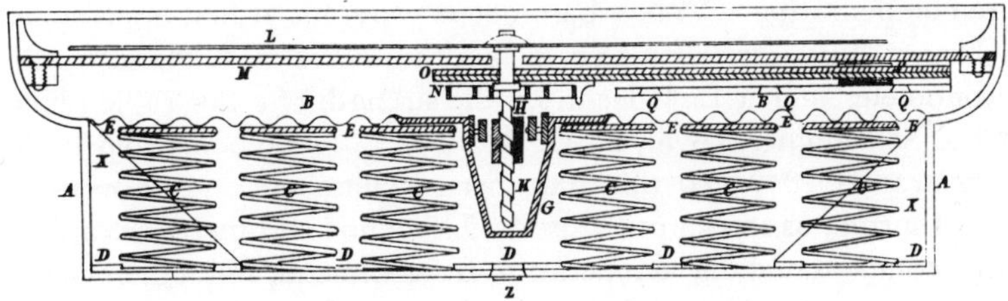

Fig. 1.18 Vidie's aneroid, first form, 1844.

phragm; in this cup, supported on gimbals, is a nut *H* which engages a steeply-threaded screw *K* to which the pointer *L* is attached. Before the upper part of the instrument was assembled, the chamber was exhausted through *D* and the plug *Z* was ingeniously soldered in. Temperature compensation is cleverly provided by a bimetallic strip *O* which supports the bearing in which *K* turns, and backlash is taken up by a spiral spring *N*. About five other possible constructions are illustrated, three of them using a metal bellows.

I have not space to recount the sad story of the long litigation begun by Vidie that in the end denied him the fruits of his labors. His opponent was the Paris instrumentmaker Bourdon, whose pressure gauge using the deformation of a curved tube of elliptical cross-section[164] was believed by Vidie to infringe on his patent.

In 1845 Vidie took out another patent in France[165] covering the use of the familiar external spring, as well as various methods of tempera-

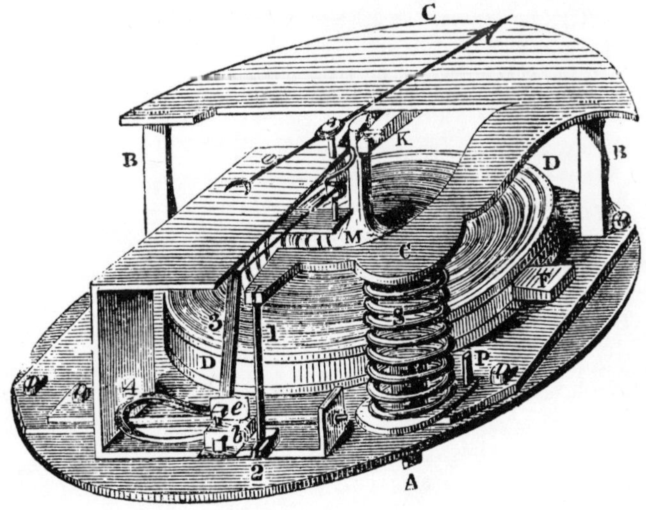

Fig. 1.19 Aneroid with external spring, about 1848.

[164] French Patent 4,408 (1849), June 19. The construction had been described by Schinz, a German railway engineer, in *Eisenbahn-Zeitung*, Vol. 7 (1849), March 5 and April 2.
[165] Vidie, French Patent 1,149 (1845), July 28.

ture compensation. Vidie's barometers were not at first appreciated in France, but were soon sold in large numbers in England by the chronometer maker E. J. Dent, on the strength of a favorable report by the Astronomer-Royal, Sir G. B. Airy.[166] Those being sold in the years around 1849 had thin flat aneroid chambers and an external helical spring, as shown in Figure 1.19.[167] The instrument was apparently compensated for temperature by a bimetal, but others had a certain amount of gas left in the chamber,[168] providing an exact compensation at one pressure and an approximate one at neighboring pressures. By about 1860 the well-known flat external spring had become the usual construction, and the compensation for temperature had again become bimetallic.

The Bourdon tube, which became ubiquitous in pressure gauges, and made a fortune for Bourdon, was not used very much in barometers, though it became familiar in French balloon meteorographs[169] after about 1890.

[166] Laurant, *Histoire*, pp. 391–92.
[167] From E. J. Dent, *A Treatise on the Aneroid, a Newly Invented Portable Barometer* [etc.] (London, 1848, pamphlet, 34 pp.).
[168] *Ibid.*, p. 16.
[169] See Chapter 9.

II

The Thermometer

1. Temperature and temperature scales. A thermometer is an instrument for measuring temperature, that is to say, for quantifying the sensations "hot" and "cold." Temperature is a quantity of a very special sort. In contrast to length, mass, and time, it is not additive; for example we should not expect to add a mass of water at a temperature of 60° to one at 40° and obtain 100°. Nor is it a property of a given substance, after the manner of refractive index or magnetic permeability. If we look into the matter closely we shall find that until recent times it has been necessary to specify the quantity "temperature" in terms of measurable changes in some chosen property of some chosen substance as it gets hotter or colder.[1] It is important to recognize that three independent and quite arbitrary choices have to be made in establishing a scale of temperature: the choice of a thermometric substance, the choice of the property whose changes are to be measured, and the choice of a system of numbering or naming the degrees of the scale.[2]

Because of the historical accident that the volume of a body is the property most easily seen to change as the body becomes hotter or colder, it is the property used in most thermometers and all the early ones. The history of the thermometer is largely an account of attempts to make and justify choices of the other two kinds. Thus the earliest choice of a substance was air, for the reason that its change in volume was large and easily observed. Later, when some objections to the use of air became apparent, various liquids were tried. For convenience in numbering the scale, fixed points were adopted; sometimes two were used, the scale between them being divided into some convenient number of degrees, or sometimes only one fixed point was used, the size of the degree being made to correspond to some arbitrary fractional increment in the volume of the thermometric substance at the fixed point.

[1] In the absolute thermodynamic scale of temperature the physical substance has been replaced by an ideal construct, the reversible heat engine.

[2] For a valuable discussion see Ernst Mach, *Die Prinzipien der Wärmelehre* (Leipzig, 1896); see also my *A History of the Thermometer and its Use in Meteorology* (Baltimore, 1966), pp. 48–50. This will be referred to as *Hist. Therm.*

The exact nature of these assumptions was not at first recognized. People tended to think that a thermometer was measuring the "amount of heat" in the medium by which it was surrounded, and the choice between various thermometric substances was often made on the basis of an unsupported belief that some particular substance would give the thermometer this desirable property. It should be remembered that until well into the nineteenth century heat was generally believed to be a material substance that could pass into bodies or out of them. The early experimenters were happily unaware of the arbitrariness of their thermometer scales.

To end this introduction I shall note that the idea of a scale of degrees of heat and cold is centuries older than the thermometer, dating at least from the time of the Greek physician Galen.[3]

2. Air Thermoscope and Air Thermometer. I take it as axiomatic that a "meter" must have a scale. An instrument that merely shows that the temperature has changed should therefore be called a thermoscope, and the word thermometer ought to be reserved for a thermoscope to which a scale has been added. The nonobservance of this distinction has greatly complicated the history of these instruments. I hope to demonstrate that the thermoscope and the thermometer may not have been invented by the same person.

There is no doubt whatever that the ancestor of the air thermoscope is to be found in a pneumatic experiment performed in antiquity by Philo of Byzantium and Hero of Alexandria. Philo's work was almost unknown in the sixteenth century, but a Latin translation of Hero's *Pneumatics* was printed in 1575,[4] catching the attention of a number of scholars, especially in Italy. The pertinent experiment is one in which water, partly filling a vessel that is closed except for an exit tube, is driven out through the tube by the expansion of the air in the vessel when it is heated. It was not a thermometer or even a thermoscope; it was "a fountain that drips in the sun."

One of those who made the experiment, almost certainly before 1600, was Galileo Galilei. Several lines of evidence converge to persuade us of this,[5] the most unambiguous being contained in a letter written on September 20, 1638, by Benedetto Castelli to Ferdinando Cesarini:

I remember an experiment shown me by our Signor Galileo more than thirty-five years ago. He took a small glass flask, about as large as a small hen's egg, with a neck about two spans long and as fine as a wheat straw, and warmed the flask well in his hands, then turned its mouth upside down

[3] See F. Sherwood Taylor, *Ann. Sci.*, Vol. 5 (1942), p. 129.
[4] *Hieronis Alexandrini spiritalium liber. A Federico Commandino Urbinate, ex graeco nuper in latinum conversus* [etc.] (Urbino, 1575). There were translations into Italian in 1589 and 1592.
[5] These are set out in detail in *Hist. Therm.*, Chap. 1.

into a vessel placed underneath, in which there was a little water. When he took away the heat of his hands from the flask, the water at once began to rise in the neck, and mounted to more than a span above the level of the water in the vessel. The same Sig. Galileo had then made use of this effect in order to construct an instrument for examining the degrees of heat and cold.[6]

This is the sort of vivid memory that any intelligent elderly man might have of an excellent demonstration lecture in his undergraduate days. Galileo's simple instrument was indeed a thermoscope. Whether or not Galileo provided it with a scale depends on the reliability of Castelli's last sentence, which was written at a time when the air thermometer was a common instrument. There is no doubt that by the year 1613 Galileo had *claimed* to be the inventor of the air thermometer, as is evident from a series of letters[7] written to him by Giovanfrancesco Sagredo in which the latter at first told Galileo about its invention by Santorio Santorre, the Professor of Medicine at Padua. It will seem at least possible to many twentieth-century scholars that Galileo, remembering the experiment described by Castelli, had persuaded himself that he had put a scale on the thermoscope, but however this may be, he published nothing about it, either at the time or later.

Santorio, famed as the first to apply quantitative methods to medicine, published a reference to his thermometer in 1612,[8] and described and illustrated several forms of it in 1625,[9] some of which are shown in Figures 2.1 and 2.2. By modern standards of publication, Santorio certainly has the priority. The earliest published figure of a thermoscope, however, is contained in the *Sphaera mundi* of Giuseppe Biancani or Blancanus,[10] who is apparently also responsible for the word "thermoscope" (*thermoscopium*). He credits Santorio with the invention.

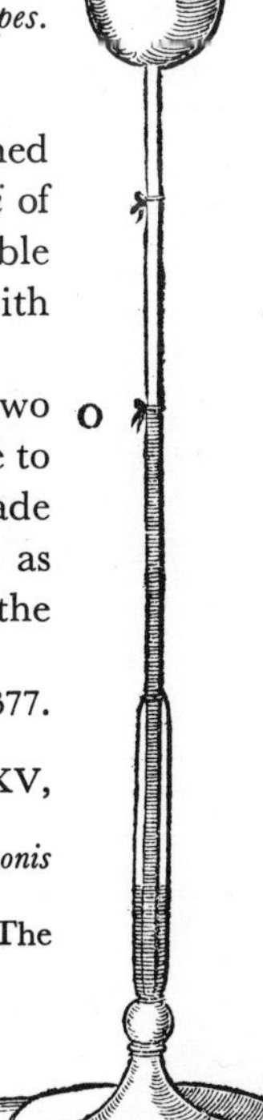

Fig. 2.1 One of Santorio's thermoscopes.

The instrument shown in Figure 2.1 has no scale, but only two threads round the tube, presumably movable. In his first reference to the instrument, Santorio tells us that the measurement was made "with the compasses."[11] The instrument was of meteorological as well as medical interest, for we are told that "the temperature of the

[6] *Le opere di Galileo Galilei*, ediz. naz. (20 vols., Florence, 1890–1909), XVII, 377.
[7] *Ibid.*, XI, 350–51; 506; 545; XII, 139; 157.
[8] Santorio, *Commentaria in artem medicinalem Galeni*, Part III, col. 62 (Cap. LXXXV, Particula X).
[9] *Sanctorii Sanctorii Iustinopolitani . . . commentaria in primam fen primi libri canonis Avicennae* [etc.] (Venice, 1625). There was a second edition in 1626.
[10] *Sphaera mundi, seu cosmographia demonstrativa* [etc.] (Bologna, 1620), p. iii. The dedication is dated Feb. 13, 1617.
[11] *Commentaria in artem . . . Galeni*, Part III, col. 62.

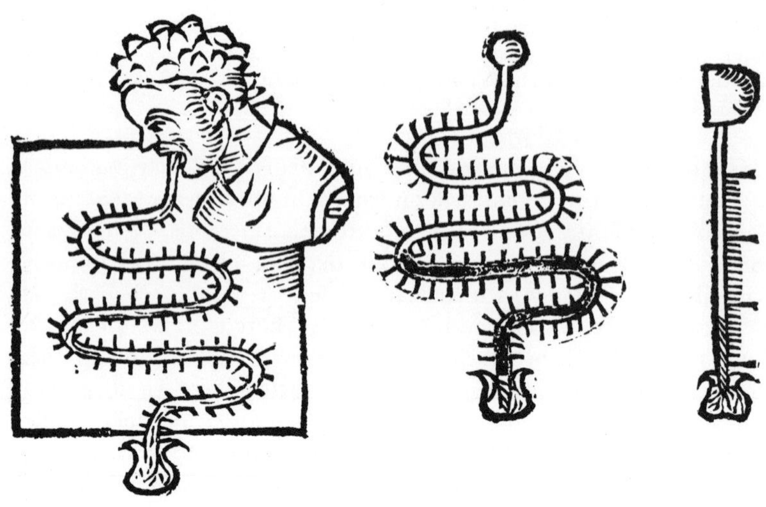

Fig. 2.2 Thermometers of Santorio.

air can be observed not only in so far as it belongs to the body, but also as a thing in itself; so that the mean between very hot and cold temperatures of the air can be exactly perceived. For we have an instrument with which not only the heat and cold of the air is measured, but all the degrees of heat and cold of all the parts of the body. . . ."[12] This does not indicate any fixed points, but in the second edition of the *Commentaries on Galen*, published in 1630, this lack is supplied in an added paragraph, in which we read that "we learn the extremes and the mean thus: we apply snow to the sphere of the glass instrument so that the water may ascend to its upper limit. Then with the flame of a candle we make the water descend as far as it will go. Knowing the extremes we shall at once find the mean and temperate. . . ."[13] As far as I know, this is the first published reference to fixed points, but on February 7, 1615, Sagredo had written to Galileo that he had buried the bulb of his thermometer in snow and in a mixture of salt and snow.[14]

Later instruments of Santorio's had scales (Fig. 2.2). Of various shapes, they all had the bulb at the top, and a tube dipping into an open vessel. It seems clear that Santorio invented the air thermometer before 1612; but how long before? The question is pointed up by the existence of a manuscript clearly dated Rome, 1611, by an unknown writer called Bartolomeo Telioux, who figures and de-

[12] *Ibid.*, col. 105.
[13] Santorio, *Commentaria in artem medicinalem Galeni* (2nd ed., Venice, 1630), col. 762.
[14] In Galileo, *Le opere*, ed. cit., XII, 140.

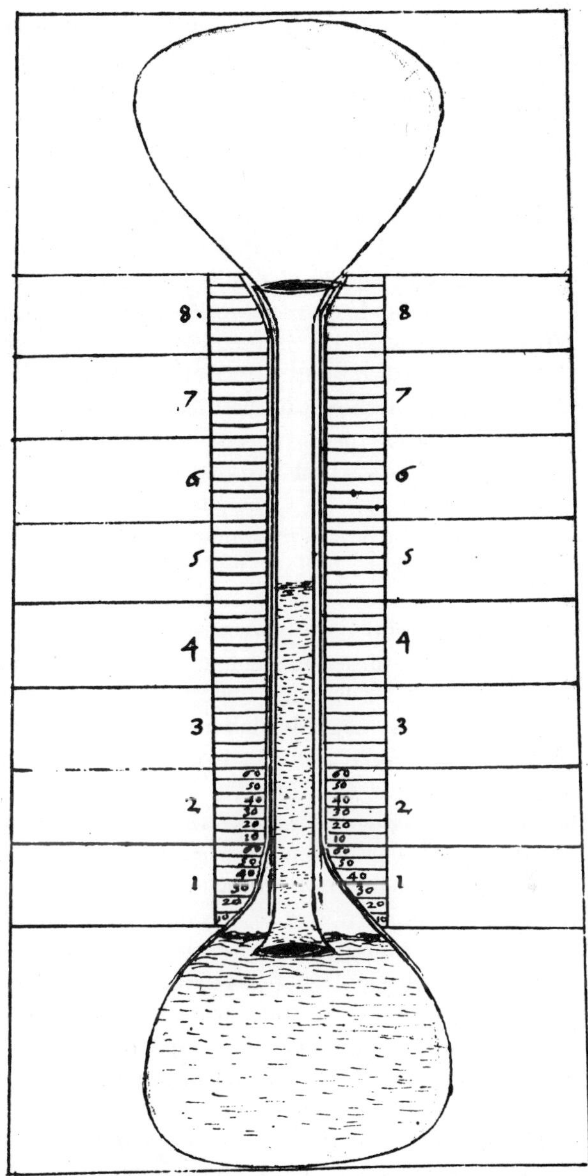

Fig. 2.3 Telioux's figure of an air thermometer, 1611.

scribes a thermometer with a scale,[15] albeit very confusedly. Telioux's figure of the thermometer[16] is reproduced in Figure 2.3. The rest of the manuscript is largely a pastiche of the "books of machines" that were popular at that period. Dr. A. G. Keller of Leicester, an

[15] Telioux, *Mathematica maravigliosa* [etc.]. Paris, Bibl. de l'Arsenal, ms. 8525. See also J. A. Chaldecott, *Ann. Sci.*, Vol. 8 (1952), pp. 195–201.
[16] *Ibid.*, p. 45.

authority on this literature, examined the manuscript at my request, and can find no reason to doubt its dating. We should be very much interested to know where Telioux got his rather garbled information.

There are two other men commonly named as inventors of the air thermometer, both from north of the Alps. These are Robert Fludd, a Welsh mystic, and Cornelius Drebbel, born at Alkmaar in Holland. I shall not here examine Fludd's qualifications for the honor, since F. Sherwood Taylor dealt with them in an excellent paper more than twenty-five years ago.[17] Not only did Fludd never actually claim the invention, but also his earliest work that contains a reference to Philo's experiment[18] was too late in the field, for by 1615 an air thermometer had been described from north of the Alps by Salomon de Caus, "Engineer and Architect to the Elector Palatine," who is not usually numbered among its inventors.[19] This instrument, mechanically similar to Hooke's wheel barometer,[20] is shown in Figure 2.4. "It may be used," wrote De Caus, "to mark the coldest or hottest days; for if the said machine is in some part of the room that the sun never shines on, the copper ball will rise according to the temperature of the day."[21]

Cornelius Drebbel is a more serious contender. He was trained as an engraver but turned to mechanical invention and in 1598 took out a patent on a self-winding timepiece.[22] About 1604 he came to England and later made an elaborate clock for James I, described and figured by Thomas Tymme in 1612[23] as a "perpetual motion." Part of it, Tymme was told, was "a ring of cristall glasse, which being hollow, hath in it water, representing the sea, which water riseth and falleth . . . twice in 24 houres, according to the course of the tides in those parts, where the instrument shall be placed. [Drebbel] extracted a fierie spirit, out of the mineral matter, ioyning the same with his proper aire, which encluded in the axle-tree, being hollow, carrieth the wheeles."[24] The reference to the tides fooled Tymme, but several others were well aware that the motive power was the expansion and contraction of the air contained in the instrument.[25] There is no doubt that Drebbel understood the principle. But did he invent an air thermometer?

The ascription of the instrument to Drebbel was quite casually made in a chapter heading of a book published in 1628 by Gaspar Ens

[17] Taylor, *Ann Sci.*, Vol. 5 (1942), pp. 129–56. See also *Hist. Therm.*, pp. 15–19.
[18] Fludd, *Utriusque cosmi historia* (Oppenheim, 1617).
[19] De Caus, *Les raisons des forces mouvantes* [etc.] (Frankfurt, 1615), pp. 18–19.
[20] See above, p. 26.
[21] De Caus, *Les raisons*, p. 19.
[22] *Register Acten States General 1589–1602*, no. 3328. Cited by G. Tierie, *Cornelius Drebbel* (Amsterdam, 1932), p. 41.
[23] Tymme, *A Dialogue Philosophicall* [etc.] (London, 1612), pp. 60–61.
[24] *Ibid.*
[25] Tierie, *Cornelius Drebbel*, pp. 38–42.

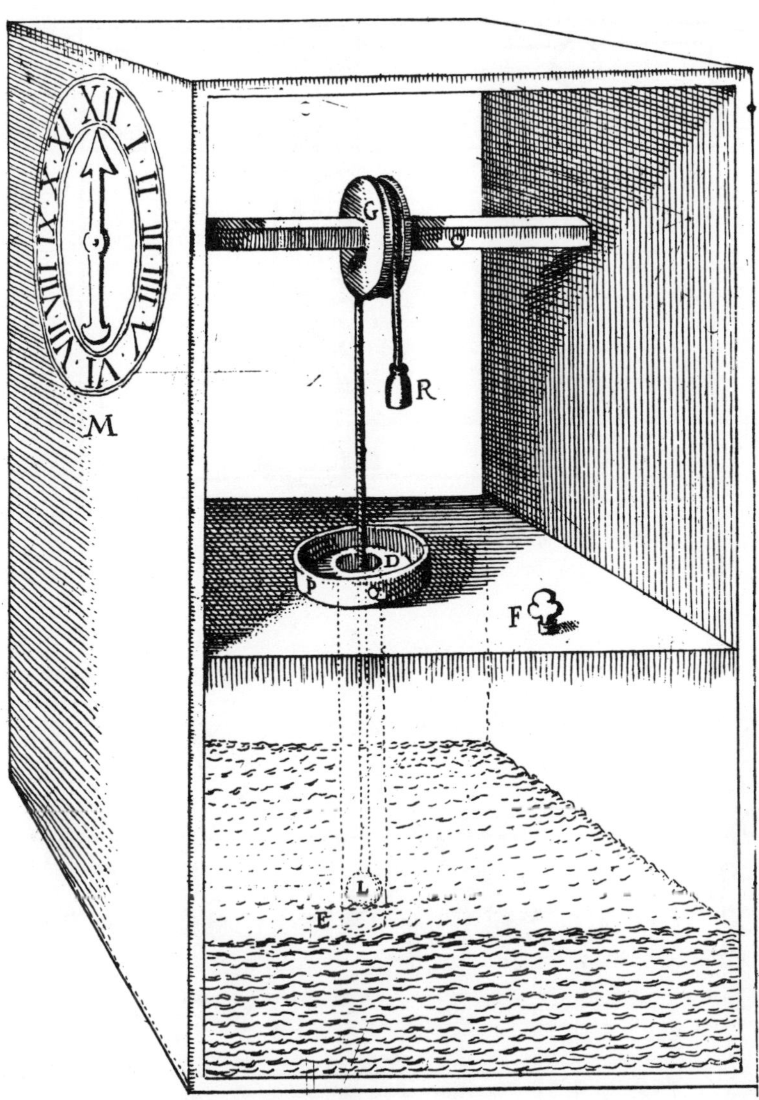

Fig. 2.4 The air thermometer of Salomon de Caus, 1615.

of Cologne.[26] He referred to it as "the thermometer or Drebbel's instrument." This sounds as if Drebbel was widely thought of as the inventor, and indeed several authors in the seventeenth century supported him,[27] but probably none of them was in a position to know. Nevertheless it is certain that Drebbel could have made a thermometer, or at any rate a thermoscope, if it had occurred to him to do so.

[26] Ens, *Thaumaturgus mathematicus* (Cologne, 1628).
[27] See Taylor, *Ann. Sci.*, Vol. 5 (1942), p. 154.

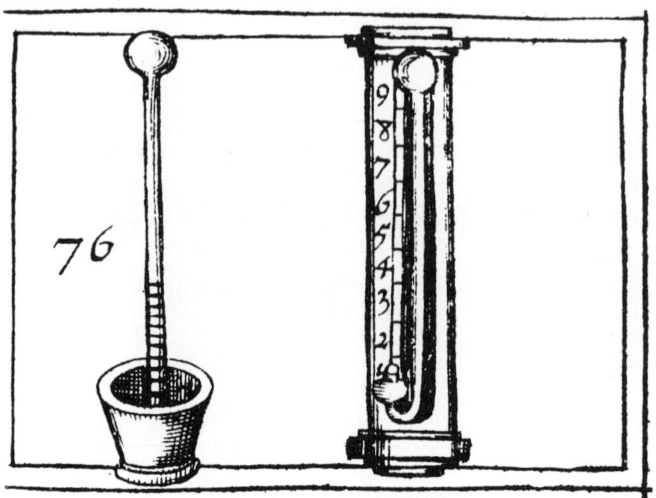

Fig. 2.5 The two sorts of air thermometer, after Leurechon.

Fig. 2.6 Guericke's air thermometer.

Some time before 1625 a distinctive type of air thermometer came into use in the Low Countries. It consisted of a J-shaped tube with a closed bulb at the end of the long leg and an open bulb at the end of the short one, as shown at the right in Figure 2.5, taken from a work by a Jesuit, Jean Leurechon, who wrote under the pseudonym H. Van Etten.[28] This work interests us because it contains the first use of the word "thermometer."

I shall end this section with a reference to the famous air thermometer constructed by the mayor of Magdeburg, Otto von Guericke[29] (Fig. 2.6). It was made entirely of copper and brass, about ten feet high, and fastened to a north wall, so that Guericke might see which was the warmest and which the coldest day of the year. By this time, Guericke should have realized that its indications would be influenced by the changes in atmospheric pressure, and this brings us to our next subject.

3. The invention of the liquid-in-glass thermometer. About 1630 a country doctor named Jean Rey thought of making a thermometer out of a small phial with a very long slender neck, almost filled with water, but not sealed. He told Marin Mersenne of this in the course of some correspondence[30] about a remarkable book that Rey had published in that year.[31] It transpired that Rey had not even heard of the air thermoscope. Mersenne does not seem to have recognized the possibilities of Rey's instrument, and it was forgotten.

The invention of the barometer revealed the variability of the pressure of the air, and thus showed up a very serious defect of the air thermometer. Some better thermometer was needed, and before very long the sealed liquid-in-glass thermometer was invented by no less a personage than the Grand Duke of Tuscany, Ferdinand II, one of the celebrated Medici family. In fact two sorts of thermometer, shown in Figures 2.7 and 2.8, were invented in Florence before 1660. The first consists of a phial of spirit of wine containing a number of glass balls having slightly different ratios of weight to volume. If the temperature is very low these will all float, but as it rises and the spirit becomes less dense they will sink, one after another, so that the

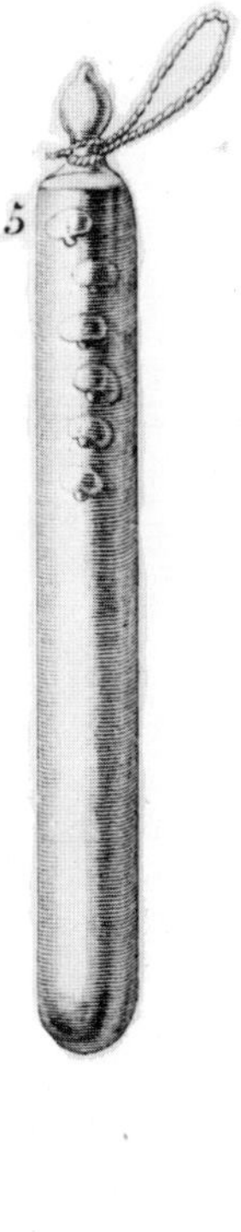

Fig. 2.7 Sealed thermometer with immersed glass balls.

[28] H. Van Etten (pseud.), *Récréation mathématique* [etc.] (Pont-à-Mousson, 1626; 2nd ed., Paris, 1626), fig. 76.

[29] First described by Gaspar Schott in *Technica curiosa, sive miriabilia artis* [etc.] (Nuremberg, 1664), p. 871. Our figure is from Gureicke, *Ottonis de Guericke experimenta nova (ut vocantur) Magdeburgica de vacuo spatio* [etc.] (Amsterdam, 1672), p. 124.

[30] *Correspondance du P. Marin Mersenne, religieux minime. Commencée par Mme. Paul Tannery, publiée & annotée par Cornelis de Waard* (7 vols., Paris, 1933–65), III, 244.

[31] Rey, *Essays . . . sur la recherche de la cause pour laquelle l'estain et le plomb augmentent de poids quand on les calcine* (Bazas, 1630).

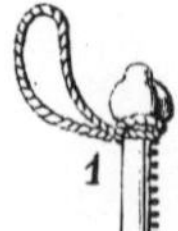

temperature can be estimated by noting the number that have sunk. On November 7, 1646, Evangelista Torricelli told Balthasar de Monconys about such thermometers,[32] but they must have been invented by 1641, because there is an entry in the original diary of the Accademia del Cimento, dated June 20, 1657, which reads: "A thermometer made sixteen years ago was cut open with a diamond."[33] The idea may have been to find out whether the spirit had changed.

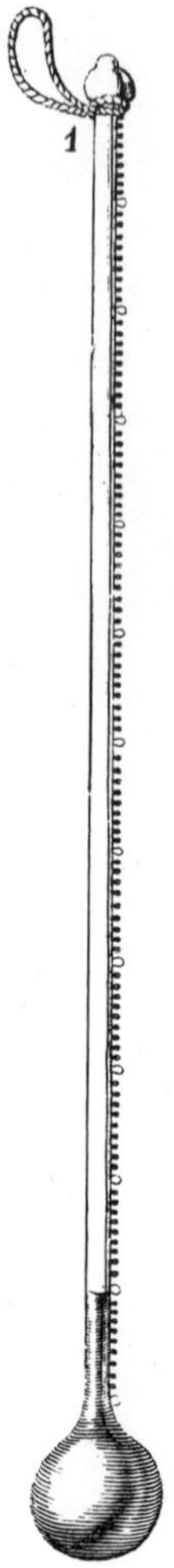

Fig. 2.8 The "100-degree" Florentine thermometer.

The second thermometer has a more familiar appearance. This had a scale of 100 degrees, and there were others with 50-degree scales, and even some with 300 degrees. The degrees are marked with tiny glass beads fused onto the tube, every tenth one white, the rest black. All these thermometers are figured in the *Saggi* of the Accademia del Cimento[34] and many still exist in the Museo di Storia della Scienza at Florence. I cannot say exactly when this type of thermometer was invented. It was probably after 1646, for Monconys did not know of it. There is an unmistakable reference to it in a manuscript at the National Library in Florence, apparently in the hand of Vincenzio Viviani. On the evening of February 23, 1653, five different liquids—pure water, salt water, pure red wine, salted red wine, and pure Malmsey wine (*malvagia pura*)—were exposed to the air in pottery dishes. "The following morning when the thermometer in the air went down to $7\frac{2}{3}$ degrees the various liquids were found as follows:"[35] (some of them had frozen). As we shall see shortly, $7\frac{2}{3}$ degrees on the 50-degree Florentine thermometer would be a reasonable temperature to expect on a cold winter night in Florence.

It is quite certain that a number of comparable 50-degree thermometers had been made by December, 1654, and sent from Florence to observers at various Italian cities, for on December 22 of that year Antonio Terillo wrote a letter[36] from Parma to Luigi Antinori in Florence (Fig. 2.9). It begins: "I have received the letter from your Reverence, with the box, in which were the two ampoules for measuring the variation of heat and cold; and keeping them for some time in the same place in a room I reflected that they both moved at the same rate without any considerable difference occurring. I made the

[32] *Journal des voyages de M. de Monconys* (3 vols., Lyon, 1665–66), I, 130.

[33] Quoted in G. Targioni-Tozzetti, *Notizie degli aggrandimenti delle scienze fisiche accaduti in Toscana nel corso di anni LX del secolo XVII* (3 vols. in 4; Florence, 1780), I, 150.

[34] *Saggi di naturali esperienze fatti nell'Accademia del Cimento* (Florence, 1667). There have been several editions, the most useful being the third Florentine one of 1841, edited by Vincenzio Antinori, because of its long preface and appendix.

[35] Florence, *Bibl. Naz.*, ms. Gal. 259, fol. 9^{r}.

[36] Florence, *Bibl. Naz.*, ms. Gal. 307, fol. 87. Terillo was an English priest (Anthony Terill).

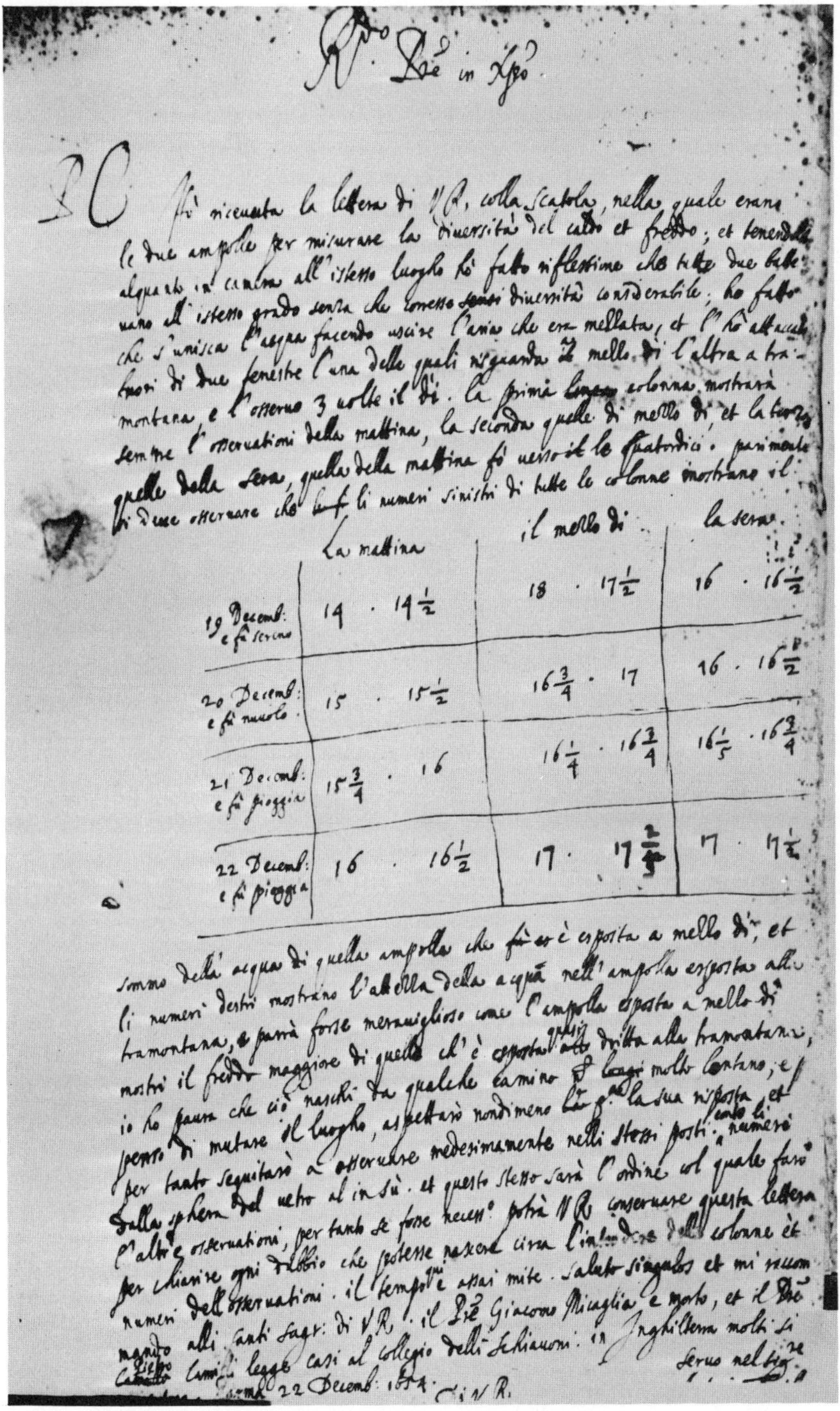

	La mattina	il mezzo di	la sera
19 Decemb. e fù sereno	14 · $14\frac{1}{2}$	18 · $17\frac{1}{2}$	16 · $16\frac{1}{2}$
20 Decemb. e fù nuvolo	15 · $15\frac{1}{2}$	$16\frac{3}{4}$ · 17	16 · $16\frac{1}{2}$
21 Decemb. e fù pioggia	$15\frac{3}{4}$ · 16	$16\frac{1}{4}$ · $16\frac{3}{4}$	$16\frac{1}{5}$ · $16\frac{3}{4}$
22 Decemb. e fù pioggia	16 · $16\frac{1}{2}$	17 · $17\frac{2}{4}$	17 · $17\frac{1}{2}$

Fig. 2.9 Antonio Terillo's letter (courtesy of the Biblioteca Nazionale, Florence).

liquid unite by getting out the air that was mixed with it; and I have fastened them outside two windows, one facing south, the other north, and I am observing them three times a day."

Other letters exist acknowledging the receipt of similar instruments in Milan and Bologna. Some sample observations in Terillo's letter make it clear that the thermometers in question had the 50-degree scales. The stations established by Luigi Antinori at the command of Ferdinand II constituted the world's first meteorological network.

The Accademia del Cimento (Academy of Experiments) was founded in 1657 by the Grand Duke's brother, Prince Leopold, and disbanded in 1667. Although the sealed thermometer was invented before the Academy came into being, there is a long account of its construction in the first chapter of the *Saggi*.[37] The 50-degree thermometer was made in very large numbers, and they appear to have been remarkably uniform. In 1828 in Florence a chest was discovered. It contained a great many of these, besides other thermometers and various glass instruments that had belonged to the Academy.[38] Libri made more than 200 comparisons between the 50-degree thermometers and found a really astonishing agreement. In melting ice they stood very near 13½°; their zero corresponded to −15°R. (−18.75°C.), and their 50° mark to 44°R. (55°C.).

The *Saggi* tells us how it was done. The comparability of the thermometers seems to have depended entirely on the extraordinary skill of the Duke's glass blower, Mariani, who "was accustomed to say that if the 50-degree thermometers were desired he could very well manage to make two or three or any number which, surrounded by the same atmosphere, would always move equally . . ."[39] but he would not guarantee this with the 100-degree thermometers, and still less with the 300-degree ones.

There seems to have been an attempt to set the 100-degree thermometers at 20 degrees in melting ice or snow. This is not entirely clear in the *Saggi*, but is placed beyond doubt by a passage in one of the earlier drafts of that book, still preserved.[40] The 50-degree thermometer was considered merely a small version of the other, and the value 13½ degrees at the ice point was fortuitous. What is really remarkable is the uniformity with which these beautiful little instruments were constructed.

Most of the Florentine thermometers have uncolored spirit in them, but at first it was dyed red, until this was found to soil the tube. There is a group of these thermometers with red spirit in Rome. In their experiments the Academy tried water and found that its volume passes through a minimum value as it is cooled.[41] Of course they

[37] 1841 edition, pp. 12–13. I have translated this in *Hist. Therm.*, pp. 33–34.
[38] G. Libri, *Ann. Chim.*, Vol. 45 (1830), pp. 354–61.
[39] *Saggi*, ed. cit., p. 14.
[40] Florence, *Bibl. Naz.*, ms. Gal. 264, fol. 83^{v}–84^{r}.
[41] *Ibid.*, ms. Gal. 263, fol. 50^{r}.

tried mercury, and were disappointed by the smallness of its expansion. This was in September, 1657.[42]

The earliest published description of the spirit-in-glass thermometer came from Rome, clearly ascribing its invention to Ferdinand II, and referring, if only by implication, to his meteorological network.[43] The thermometer reached Poland in 1657, Paris in 1658, and England in 1661.[44]

4. Later Seventeenth-century Thermometers. In the 1660's the Fellows of the newly founded Royal Society naturally experimented with thermometers, at first with some mechanical devices connected with a meteorograph of Christopher Wren's.[45] Wren seems to have devised two such thermometers; the first was an air thermometer in which the expansion of the air in a large bulb changes the level of a mercury surface in a vertical cylinder, the motion of a float on the mercury being recorded on a chart.[46] This instrument was seen by Balthasar de Monconys on June 10, 1663.[47]

Wren's second thermometer evidently had two glass bulbs connected by a pipe in the form of an arc of a circle. The bulbs apparently were partly full of mercury; the temperature was indicated by the rotation of this assembly as its center of gravity shifted.[48] This also was an air thermometer, but Hooke applied the same principle to a spirit-in-glass thermometer which he exhibited to the Society on January 4, 1664/5.[49] In addition to these, there was a similar instrument in the form of "a spiral turning on an axis."[50]

These mechanical schemes are unimportant in comparison to the success of Robert Hooke in constructing thermometers that would agree with one another without the necessity of making their dimensions exactly similar.[51] To do this, Hooke adopted the principle of using one fixed point, the temperature of distilled water when it begins to freeze. From this point, marked zero, his degrees were made to correspond with measured increases in the volume of the thermometric liquid, spirit of wine, each degree representing 0.001 of the volume at the freezing point.

[42] Targioni-Tozzetti, *Notizie*, II, 179.
[43] *Trattato della sfera di Galileo Galilei, con alcune prattiche intorno a quella . . . di Buonardo Savi* [pseudonym of Urbano Daviso] (Rome, 1656), pp. 191–93.
[44] The details are in *Hist. Therm.*, pp. 37–38.
[45] See also Chap. 7 below.
[46] See Chap. 7.
[47] Monconys, *Journal des voyages*, II, 53.
[48] Wren to Lord Brouncker, July 30, 1663; *BM*, Sloane ms. 2903, fol. 104^{r}–105^{v}. See also Chapter 7, p. 245.
[49] Birch, *The History of The Royal Society of London* (4 vols., London, 1756), II, 1–2.
[50] Sir Robert Moray to Christiaan Huygens, January 27, 1664/5; in *Oeuvres complètes de Christiaan Huygens* (22 vols., 1888–1950), V, 228.
[51] Hooke, *Micrographia: or some physiological Descriptions of minute Bodies made by Magnifying Glasses with Observations and Inquiries thereon* (London, 1665), pp. 38–39.

Hooke's method of carrying out the graduation of his thermometers was characteristically ingenious. He made a cylindrical vessel of metal (Fig. 2.10) with a glass tube sealed into the top, the internal diameter of the tube being exactly one tenth that of the vessel. Then if a scratch *GH* was made at a height *GE* from the top of the vessel equal to its depth *BC*, the volume of the tube between *EF* and *GH* was 0.01 that of the metal cylinder. This distance was then divided into ten parts. He next put both his thermometer and the vessel into

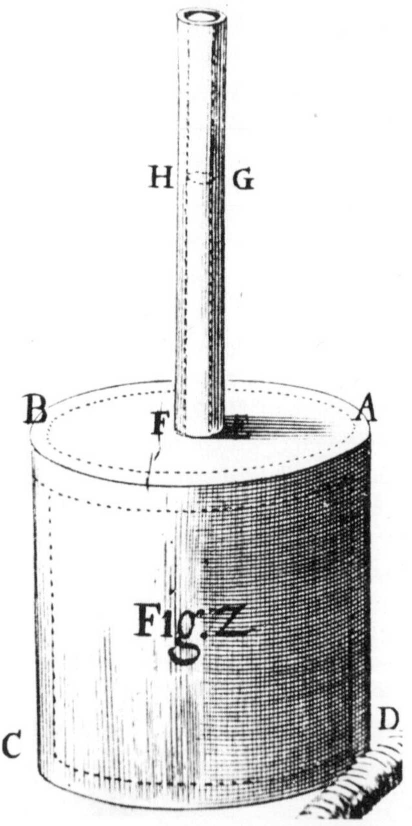

Fig. 2.10 Hooke's apparatus for calibrating his thermometers.

freezing water, and adjusted the amount of spirit in the vessel until it was just level with *EF*. Then, letting the whole apparatus warm up gradually, he marked the tube of the thermometer with degrees by observing the rise of the spirit in *EFGH*.

Like so many of Hooke's innovations, this was forgotten, and in the next century Réaumur described his rather similar method with no reference to Hooke at all. Nevertheless, thermometers were made on this principle in England for about thirty or forty years, and fairly widely used,[52] so that it is astonishing that in 1688 the great astronomer Edmond Halley complained that no way had ever been found to make comparable thermometers except by having them all constructed by the same workman.[53]

It must also be recorded that at about the time when Hooke was making his thermometers, Christiaan Huygens suggested in a letter to Sir Robert Moray that *either* the freezing point *or* the boiling point of water should be used, "the capacity of the bulb [having] a certain proportion to that of the tube."[54] Moray replied[55] that Hooke had done this.

An entirely different sort of "fixed" point, the introduction of which had far-reaching and rather unfortunate results, was provided by Edme Mariotte's discovery, at some time before 1679, of the approximately constant temperature in the deep cellars under the Paris Observatory.[56]

I shall now consider the earliest attempts to base thermometer scales on two fixed points. As far as I know, the first suggestion to use snow and boiling water as fixed points was published posthumously in 1679 by Sebastiano Bartolo, a professor at Naples.[57] Bartolo divided the interval into eighteen degrees. He went out of his way to emphasize that his two points are "fixed, and everywhere immutable," but few people seem to have read his book on hot springs.

In the following year Francesco Eschinardi published another method, also using ice and boiling water, but with a means of dividing the interval, as he thought, into uniform "degrees of heat."[58] This was rather confused, and I have not space to deal with it, so I shall pass on to Carlo Renaldini or Rinaldini, to whom nearly all the textbooks ascribe the idea of using ice and boiling water as fixed points. Renaldini, who was seventy-nine years old at the time, described his method in 1694.[59]

He provided six vessels, each able to hold twelve ounces of water, and a dipper holding one ounce. He first noted where the spirit stood in his thermometer when the bulb was surrounded with ice. Then into one of his vessels he put eleven ounces of ice-cold water, immersed

[52] See Louise Diehl Patterson, *Amer. J. Phys.*, Vol. 19 (1951), pp. 523–35; *Isis*, Vol. 44 (1953), pp. 51–64.

[53] Halley, *Phil. Trans.*, Vol. 17 (1693), pp. 655–56. This paper was read in 1688.

[54] Huygens to Moray, January 2, 1664/5, in *Oeuvres complètes*, V, 188.

[55] *Ibid.*, V, 228.

[56] Mariotte, *Essay du chaud et du froid* (Paris, 1679), pp. 38–44.

[57] Bartolo, *Thermologia Aragonia, sive historia naturalis thermarum* [etc.] (2 vols., Naples, 1679). The passage about the thermometer is on some unnumbered pages (sig. L2r–L4r) near the end of Volume 2. The dedication of Volume 2 to P. V. Diazio is dated October 9, 1672, but the additional pages may well have been written later. Bartolo died in 1676.

[58] Eschinardi, *Raguagli . . . sopra alcuni pensieri sperimentabili* [etc.] (Rome, 1680), pp. 62–63.

[59] *Caroli Renaldini . . . naturalis philosophia . . . corrigente J. B. Sanctio auctoris amanuensi* (3 vols., Padua, 1693, 1694), Vol. 3, pp. 275–56.

the thermometer in this, added one ounce of boiling water, and marked on the tube the height to which the spirit ascended. The process was repeated with ten ounces of ice-cold water and two ounces of boiling water, and so on, giving twelve degrees of heat between ice-cold and boiling. The idea of having six vessels was so that he could start with a cold one each time.

The reason for this elaborate procedure was that Renaldini believed that the resistance of the air included in the tube would prevent the spirit from rising in a linear relationship with the amount of heat. This supposed effect of pressure in hindering the expansion of liquids was greatly overestimated through most of the succeeding century.

Other pairs of fixed points were suggested, for example: the temperature at which water freezes, and the melting point of butter; or a mixture of ice and salt, and the temperature of a very deep cellar.[60]

Isaac Newton devised a thermometer scale, which was communicated to The Royal Society on May 28, 1701, and published anonymously.[61] He chose linseed oil as the thermometric liquid, and for the lower fixed point (0°) "the heat of the air in winter when water begins to freeze," which, he adds, can be found by putting the thermometer in melting snow. The upper fixed point, 12°, was blood heat, "the maximum heat that the thermometer can attain by contact with the human body." On this scale, rapidly boiling water turned out to be at 34°.

There is one further scale which may have been invented just at the end of the seventeenth century, the so-called "Royal Society" scale.[62] This scale is inverted, the zero representing the greatest heat, and the numbers increasing as the temperature decreases. The freezing point was supposed to be 65°. The connection of the scale with The Royal Society is that thermometers of this sort were recommended by the Secretary of the Society, James Jurin, who issued a general invitation to the learned in all parts of the world to make meteorological observations ar d send the results to London.[63] Jurin recommended that for the sake of comparability everyone should obtain his instruments from Francis Hauksbee the younger, the Society's instrument-maker. Unfortunately we have much evidence that Hauksbee's thermometers with their inverted scale did not agree among themselves, even approximately.[64]

The origin of this peculiar scale is debatable. I have given reasons

[60] Our source for these is [Joachim Dalencé], *Traittez des baromètres, thermomètres, et notiomètres, ou hygromètres. Par Mr. D**** (Amsterdam, 1688).

[61] *Phil. Trans.*, Vol. 22 (1701), pp. 824–29.

[62] Discussed at length by Louise Diehl Patterson, *Amer. J. Phys.*, Vol. 19 (1951), pp. 523–35; *Isis*, Vol. 44 (1953), pp. 51–64.

[63] Jurin, *Phil. Trans.*, Vol. 32 (1723), pp. 422–27.

[64] J. H. van Swinden, *Dissertation sur la comparaison des thermomètres* (Leiden, 1792), p. 230. George Martine, *Essays Medical and Philosophical* (London, 1740), pp. 226–27.

elsewhere[65] for thinking that it dates from just before 1700 and that it was invented by John Patrick. If this is so, it is likely that Francis Hauksbee the younger, going into business for himself about ten years later, took it as a conveniently available scale and copied it, though not very successfully.

5. Rømer and Fahrenheit. With the coming of the eighteenth century we begin to encounter serious attempts to produce accurate thermometers with scales that could be reproduced from first principles. Many scales were suggested, but only three survived, characterized by intervals of 180°, 100°, and 80° between the freezing and the boiling points of water, and popularly associated with the names Fahrenheit, Celsius, and Réaumur. The earliest of these was the Fahrenheit scale, which nowadays is defined by having its ice point at 32°F. and its steam point at 212°. The history of this scale has led to much controversy, with strong national overtones, and the last word on the subject may not be said for a long time, especially about the extent of Fahrenheit's indebtedness to the great Danish astronomer Ole Rømer.

Daniel Gabriel Fahrenheit, a native of Danzig, spent most of his life in the Netherlands.[66] In 1708 he visited Ole Rømer. Rømer left a large notebook entitled *Adversaria*,[67] which is now in the Royal Library at Copenhagen,[68] bearing a great number of marginal remarks by Rømer's successor Peter Horrebow. It was published, with Horrebow's remarks, in 1910,[69] and the part of it dealing with thermometry was discussed very fully by one of its editors elsewhere.[70]

In the *Adversaria* there is a long passage on thermometer-making, beginning with several pages about the investigation, by means of a drop of mercury, of the uniformity of the bore of a glass tube. Finally Rømer gives formal instructions for the construction of a standard thermometer, ending with the following sentence: "When the thermometer has been made, filled, and sealed, the point of division 7½ is fixed by means of snow or crushed ice, the point 60 by boiling."[71] Five of Rømer's thermometers had survived, and Horrebow, with the

[65] Middleton, *Hist. Therm.*, pp. 58–62.

[66] For biographical details see Ernst Cohen and W. A. T. Cohen-De Meester, *Kon. Akad. Wet., Verhand., Afd. Natuurkunde*, Section 1, Vol. 16, no. 2 (1936), pp. 1–37.

[67] Latin for "notebook."

[68] Ms. E don. var. 16.

[69] *Ole Rømers Adversaria . . . udgivne af det Kgl. Danske Videnskabernes Selskab, ved Thyra Eibe og Kirstine Meyer* (Copenhagen, 1910).

[70] Kirstine Meyer, *Arch. Gesch. Naturw. Techn.*, Vol. 2 (1910), pp. 323–49. See also Meyer, *Die Entwicklung des Temperaturbegriffs im Laufe der Zeiten* (Brunswick, 1913; no. 48 in the collection "Die Wissenschaft"). Also *Nature*, Vol. 82 (1910), pp. 296–98.

[71] *Adversaria*, ed. cit. p. 210. "Confecto impleto et sigillato thermometro per nivem vel glaciem contusam constituatur punctum divisionis 7½ per ebullitionem punctum 60."

help of Rømer's widow, deduced that these had been made in 1702. In April, 1741, Horrebow recalibrated them in snow and then in boiling water, finding "precisely the same marks that Rømer had scratched on them."[72] It therefore seems certain that Rømer was the first to make reproducible thermometers with two fixed points, dividing the scale into equal increments of volume—precisely the method still employed.

The strange number 7½ used for the ice point has caused some people[73] to doubt that Rømer really meant what he said. But I do not think that the choice was entirely arbitrary. I suspect that he chose 60—a number very familiar to an astronomer—for the boiling point and then put ⅞ of his entire scale above the freezing point and ⅛ below, to leave space, as he thought, for all meteorological temperatures without having to use negative numbers. This interpretation is confirmed by two of Horrebow's notes in the *Adversaria*. Horrebow had been considering a centigrade thermometer, and he writes: "for 60 I take 100 and for 7½ I take 20, that is, ⅕ of the whole scale, in order to leave space for a greater cold than is observed in Denmark. This I do in Rømer's name, as he advised it."[74] And again: "It seems to me more convenient as follows: between boiling point and the snow four equal parts are taken, and one below; divide each of these five parts into 20, and you have 100 parts, a round number, below which the thermometer has never been observed to fall at Copenhagen. But nevertheless the invention is to be attributed to Rømer, for it is really his. P. Horrebow."[75] I feel that the number 7½ is satisfactorily accounted for.

We now come to the question of how much, if anything, Fahrenheit learned from Rømer. Nothing was known about this until Ernst Cohen found among Herman Boerhaave's correspondence at Leningrad a letter from Fahrenheit to Boerhaave, dated April 17, 1729.[76] The most important part of the letter is as follows:

Now concerning the way in which I came to begin improving thermometers, I am glad to inform you that I obtained the first incitement to it in the year 1708 through conversation with the excellent Rømer in Copenhagen. Once, when I went to see him on a fine morning, I found that he had stood several thermometers in water and ice, and later he dipped these in warm water, which was at blood-heat [*welches blutwarm war*]. And after he had marked these two limits on all the thermometers, half the distance found between them was added below the point in the vessel with ice, and the whole distance was divided into 22½ parts, beginning with 0 at the bottom, then 7½ for the

[72] *Ibid.*, p. 213.

[73] E.g., J. Newton Friend, *Nature*, Vol. 139 (1937), p. 586, who could not believe that Rømer "could be so inartistic" as to choose such a number.

[74] *Adversaria*, ed. cit. p. 207.

[75] *Ibid.*, p. 211.

[76] First published in its original Dutch in an article about Fahrenheit: *Chem. Weekblad*, Vol. 33 (1936), pp. 374–93; and later in German in *Verh. K. Akad. Wetensch. Amsterdam, Natuurkunde* (first section), Vol. 16, no. 2 (1936), pp. 1–37.

point in the vessel with ice and 22½ degrees for that at blood-heat. I also used this graduation until the year 1717, but with the difference that I divided each degree into 4 smaller ones. And in this manner were also divided the two thermometers, about which Professor Wolf[77] wrote a report in the *Acta Lipsiana* for August 1714. As this graduation is inconvenient and awkward because of the fractions, I decided to alter the scale, and to use 96 instead of 22½ or 90; this I have always used since then.[78]

If Fahrenheit was remembering correctly, it would seem that Rømer must have altered his technique between 1702 and 1708. Nothing was said in the *Adversaria* about blood heat; his upper fixed point, 60°, was the temperature of boiling water. This discrepancy has supported some doubts about the extent to which Fahrenheit was indebted to Rømer.[79] In the absence of further documents my interpretation is as follows: Rømer had noticed that the temperature of the air never went above about 20° on his scale, so that most of the length of the thermometer was superfluous. He had therefore calibrated some thermometers of shorter range by comparison with his earlier instruments, making the top of the scale 22½°, three-eighths of the way from his zero to the boiling point. Now there would be no means of holding a vessel of water at blood heat or anything near it without the aid of a thermometer, and therefore one of the thermometers that were "dipped" into the warm water must have been calibrated. The water was probably not actually at blood heat, but in view of the non-linear scale of a spirit-of-wine thermometer it may not have been more than 2° or 3°C. below it, and it would be very natural for Rømer to refer to this point as "blood heat."

It is also of interest that at some time between Fahrenheit's visit and the end of 1708, Rømer had changed his scale so that the freezing point became 8° instead of 7½°, as is shown in a graph of daily temperatures taken during the very cold winter of 1708–09 and preserved in the manuscript of the *Adversaria* at Copenhagen.[80] Fahrenheit does not seem to have known that Rømer had made this change.

Fahrenheit had begun to make mercury thermometers by about 1617. He later[81] ascribed his use of mercury to learning of Amontons' demonstration[82] that the readings of the mercury barometer should be corrected for its temperature.

His own description of his thermometers and of their calibration was published in 1724, the year of his election to The Royal Society.[83] They were of two kinds, those filled with mercury and those filled with spirit of wine. The length of the tube, he is careful to state, does not

[77] [Christian Wolff], *Acta Eruditorum* (1714), pp. 380–81. The article is not signed.
[78] *Verh. K. Akad. Wetensch.*, pp. 9–10. I have translated from the German version, which is stated to be "möglichst wort- und stilgetreu."
[79] Cf., N. Ernest Dorsey, *J. Washington Acad. Sci.*, Vol. 36 (1946), pp. 361–72.
[80] It is reproduced in the printed edition, p. 214.
[81] Fahrenheit, *Phil. Trans.*, Vol. 33 (1724), pp. 1–3.
[82] G. Amontons, *Mém. Acad. r. Sci. Paris* (1704), pp. 164–72.
[83] Fahrenheit, *Phil. Trans.*, Vol. 33 (1724), pp. 78–84.

matter, because they are graduated between fixed points. Those intended only for meteorological observations were graduated from 0° to 96°. We are told how the fixed points (*termini fixi*) were found:

The division of their scales is based on three fixed points, which can be produced accurately as follows: The first is placed at the lowest part or beginning of the scale, and is attained with a mixture of ice, water, and sal-ammoniac or sea-salt; if the thermometer is placed in this mixture, its fluid descends to a point that is marked zero. This experiment succeeds better in winter than in summer. The second fixed point is obtained if water and ice are mixed together without the above-mentioned salts. If the thermometer is placed in this mixture its fluid takes up the thirty-second degree, which I call the point of the beginning of congelation, for in winter stagnant waters are already covered with a very thin layer of ice when the liquid in the thermometer reaches this degree. The third fixed point is found at the ninety-sixth degree, and the spirit expands to this degree when the thermometer is held in the mouth, or under the armpit, of a living man in good health, for long enough to acquire perfectly the heat of the body . . . The scale of thermometers for determining the heat of boiling liquids also begins at zero, and contains 600 degrees; for the mercury that fills the thermometer begins to boil at about that point.[84]

It is quite impossible for us to believe that Fahrenheit's lowest "fixed point" was really constant, if only because either of two salts is permitted; and his remark that "the experiment succeeds better in winter than in summer" makes it likely that he himself had no confidence in it. The real fixed points were the ice point and the temperature of the healthy human body. The boiling point of water was

[84] *Phil. Trans.*, Vol. 33 (1724), pp. 78–79.

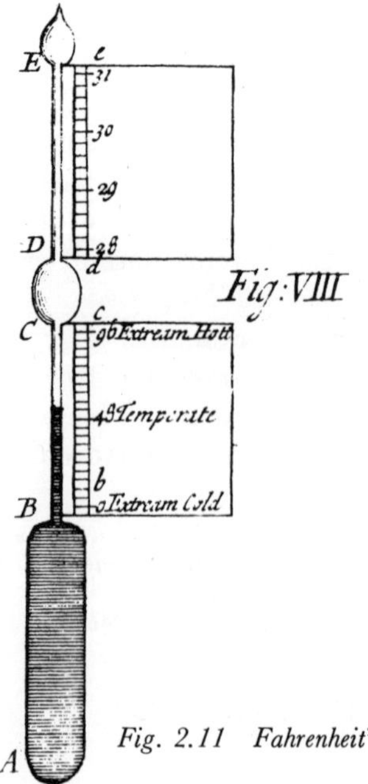

Fig. 2.11 Fahrenheit's "new barometer."

definitely not used by Fahrenheit as a fixed point; indeed, he discovered its variation with changes of atmospheric pressure, and devised a "new barometer," a thermometer graduated in terms of barometric readings in inches of mercury (Fig. 2.11).[85] But only a few years after Fahrenheit's death in 1736 it became common practice to consider the boiling point as a fixed point, with the value 212°, to the exclusion of blood heat.

6. Réaumur and Delisle. We now come to two thermometer scales based, like that of Hooke, on one fixed point, together with numerical values of the dilatation of the thermometric substance relative to its containing vessel. The scale devised by René-Antoine Ferchault de Réaumur[86] was among the least satisfactory productions of that eminent scientist. His actual methods, much inferior in some respects to those of Hooke, were soon modified or neglected by the makers of "Réaumur thermometers," but his scale, with the freezing point at 0° and the 80° point moved to agree with the boiling point of water, survived in France and Central Europe for well over a century.

We are told about Réaumur's methods at very great length in two memoirs,[87] which I shall attempt to summarize briefly. After complaining of the chaotic state of thermometer-making in France at that period, he says that he chose spirit of wine because of its large dilatation, although he knew that the amount of this depends on the purity of the spirit. So he started with spirit of wine of known properties, made a thermometer containing some known volume of spirit at the temperature of water beginning to freeze, and divided the tube into degrees, each containing a known fraction of this volume. This was achieved volumetrically by the use of small pipettes made specially for this purpose. In order for it to be done with accuracy, the thermometers were made very large, with bulbs about 4½ inches in diameter and tubes up to ¼ inch in bore, and several feet long.

The freezing point was established by putting the bulb into a vessel of water surrounded by a larger vessel containing a freezing mixture. When most of the water had frozen and the spirit had descended as far as it would go, the level of the spirit was adjusted to a mark made in advance. The tube was sealed after warming it so that the air left inside would be somewhat rarefied.

There was a tradition, well known to Réaumur, that the freezing point of water was not constant, but was influnced by the number of "saline particles" in the air.[88] Réaumur took pains to show that his

[85] *Ibid.*, pp. 179–80.
[86] See Centre International de Synthèse, *La vie et l'oeuvre de Réaumur* (*1683–1757*) (Paris, 1962).
[87] Réaumur, *Mém. Acad. r. Sci. Paris* (1730), pp. 452–507; *ibid.* (1731), pp. 250–96.
[88] See for example J. J. d'Ortous de Mairan, *Dissertation sur la glace* (Bordeaux, 1716), pp. 29ff.

artificial ice made the thermometer go down to the same point, no matter in what season or in what weather it was made. The reason that he succeeded so well was that the enormous size of his thermometer bulbs and his way of doing the experiment ensured that there was plenty of ice and water in thermal equilibrium by the time his huge instrument had steadied down. Nevertheless it appears that Réaumur, probably on the advice of the Abbé Jean Antoine Nollet, soon realized that melting ice was a more convenient standard than freezing water.[89]

So far, Réaumur's method was theoretically sound, even if unhandy. But it was necessary for him to choose a standard dilution of spirit of wine, for he had ignored the advice of Carlo Taglini to use the purest spirit available.[90] So he calibrated a narrow-necked flask and determined, for various dilutions, the dilatation between the freezing point of water and the point where the spirit just ceases to boil after it has come to the boil and cooled down below its boiling point several times, presumably so that any air it contains—and unfortunately some of the alcohol as well—will have been driven off. He seems to have used a water bath to heat the spirit. He summarized his very long-winded description in a sentence that confused many people: "What I was trying to do was to find, in parts of [the volume at the freezing point], its difference from the volume of this same quantity dilated by the heat of boiling water (*par la chaleur de l'eau bouillante.*)"[91]

He finally chose a dilution of which the expansion, measured in this way, was 80 parts in 1,000, mainly because 80 is "a number convenient for dividing into parts."[92] He then made confusion inevitable by referring to the 80th degree as "the boiling-water point" (*le terme de l'eau bouillante*).[93] The predictable result was that everyone came to believe that 80°R. is the boiling point of water.

It is therefore not surprising that in the decades after 1730 Réaumur's principles were gradually lost sight of, and it became the general practice to calibrate mercury thermometers at the melting point of ice and the boiling point of water, to divide the interval into 80 degrees, and to call such instruments "Réaumur thermometers." The history of this development is too complicated to permit a brief summary.[94] It culminated in the detailed researches of Jean André Deluc, carried out in the 1760's, and published in 1772 in a book of undoubted importance but unexampled verbosity.[95] Deluc set out to discover the precise relationship between the original Réaumur scale

[89] See Nollet, *L'art d'expériences* [etc.] (3 vols., Paris, 1770), III, 147–48.

[90] Taglini, *De thermometro disputatio* [etc.] (Andreas Aloysius, *praeses*) (Pisa, 1725), p. 31.

[91] *Mém. Acad. r. Sci. Paris* (1730), pp. 482–83.

[92] *Ibid.*, p. 489.

[93] *Ibid.*

[94] See *Hist. Therm.*, pp. 84–87 and 115–21.

[95] J. A. Deluc, *Recherches sur les modifications de l'atmosphère* [etc.] (2 vols., 4°; Geneva, 1772. Reprinted in 4 vols., 8°; Paris, 1784).

and that of the 80° mercury thermometer. One interesting result was that, because of the lack of proportion between the expansions of mercury and of spirit, the scales agree fairly well at ordinary meteorological temperatures, though 80° on the original scale corresponds to only 66.6° on the mercury thermometer.

Even after the publication of Deluc's book, proposals for the "reform" of the Réaumur scale continued to be made in France,[96,97] though they were given short shrift by the Académie Royale des Sciences, still jealous of the reputation of the great Réaumur.

A thermometer scale analogous to that of Réaumur was devised in 1732 by Joseph Nicolas Delisle, a French astronomer working in St. Petersburg. A description of this scale was taken to Paris by Daniel Bernouilli in 1733, and read to the Academy on February 3 and 10, 1734.[98] It was never published by the Paris Academy,[99] and appeared in print only in 1738.[100]

Delisle's thermometer was filled with mercury. The zero of the scale was at the boiling point of water, and the scale was inverted, lower temperatures being expressed by higher numbers. One degree represented a contraction of 1/100,000 (later 1/10,000) of the volume at the boiling point of water, and his thermometers were usually graduated down to 2,400 (240) or 2,700 (270) degrees, sufficient for the temperatures in St. Petersburg.

Delisle had said nothing about the freezing point of water, but in the winter of 1737–38 his colleague Josias Weitbrecht found that the temperature of the water flowing under the ice on the river Neva remained constant at 149.5° Delisle all winter.[101] After *ad hoc* experiments Weitbrecht[102] then proposed to make mercury thermometers by using two fixed points, marking 0° at the boiling point, and 150° at the freezing point, and extrapolating to 200°. Thermometers made in this way remained in use for more than half a century in Russia.

7. The centigrade thermometers. In our decimal system of counting, 100 is a "round" number, and it was natural that someone should propose a thermometer scale with 100 degrees between fixed points— a "centigrade" thermometer. Nowadays this means a scale with the ice

[96] E.g., L. A.[bbé] B.[ossier] *Mémoires sur la réforme des thermomètres, avec des avis particuliers, & des notes justificatives, critiques & instructives* (Tours, 1779).
[97] *Jean Gaussen, Dissertation sur la thermomètre de Réaumur* (Béziers, 1789).
[98] According to the *Registres.*
[99] A copy of the manuscript given to Bernouilli is preserved at the Observatoire de Paris, ms. A.7.4, item 35.
[100] In *Mémoires pour servir à l'histoire & au progrès de l'astronomie & de la géographie physique* [etc.] (St. Petersburg, 1738), pp. 267–84.
[101] Observatoire de Paris, ms. A.7.3, item 29.
[102] Weitbrecht, *Comm. Acad. Petrop.*, Vol. 8 (1736), pp. 310–33. This volume was published in 1741; it was common for the transactions of eighteenth-century learned societies to be left open for papers long after the nominal date of publication.

point at 0° and the steam point at 100°, but the earliest such scales were different. We shall merely mention the first two of these. In 1724 the Delisle brothers made four spirit thermometers that showed 0° in boiling water and 100° in the cellars of the Paris Observatory.[103] I know of no references to the later use of this scale.

The next was exactly the inverse of this, the cellars being at 0° and the boiling point, under a pressure of 27¾ Paris inches of mercury, at 100°. Its author, who published it anonymously in 1741,[104] was a Genevan, Jacques Barthélemi Micheli du Crest.

I now come to the centigrade scale in the modern sense, and must deal with the contributions of the Swedish astronomers Anders Celsius and Märten Strömer, the great botanist Linnaeus, and Jean Pierre Christin of Lyon. Before doing so, I should record the existence of manuscript evidence that about 1741 either Réaumur or Horrebow—it is not clear which—devised a scale with the boiling point at 0° and the freezing point at 100°, and then did nothing further about it.[105] The same inverted scale, and not the present-day one, as most people believe, was described in 1742 by Anders Celsius,[106] who had his thermometer ready for use on December 25, 1741, on which date a column headed "Cels. Th." was added to a table of observations taken with a Réaumur thermometer,[107] and it is clear from the comparison of the two that the new thermometer was the one that I am discussing. But he must have been interested in the making of thermometers at the time of his visit to Italy in 1733, for on May 9, 1734, one Giovanni Bianchi wrote from Rimini begging Celsius to "write . . . about that way of making a mercury thermometer, as you promised me when you were here."[108] We do not know what sort of thermometer this was, but we do know that Hauksbee's thermometer had been used in Uppsala since 1726.[109] By 1740, Celsius was also observing a Réaumur thermometer, made by Nollet, and one of Delisle's.[110]

The last-named instrument, one of two sent by Delisle from St. Petersburg in December, 1737,[111] survives (Fig. 2.12), the treasured possession of the Meteorological Institute of Uppsala University, and probably the most interesting eighteenth-century thermometer to have come down to us. The very slender tube is tied over a paper

[103] J. N. Delisle and L. Delisle de la Croyère, *Mém. Acad. r. Sci. Paris* (1724), pp. 316–19.
[104] *Description d'un thermomètre universel* (Paris, 1741), pamphlet.
[105] I have dealt with this in *Hist. Therm.*, pp. 91–95.
[106] Celsius, *Kongl. Svensk. Wet. Akad., Handlingar*, Vol. 3 (1742), pp. 171–80. Also in German in *Abh. Schwed. Akad.*, trans. A. G. Kästner (Hamburg and Leipzig, 1750), Vol. 4 (1742), pp. 197–205.
[107] Uppsala, Univ. Lib., ms. A530.
[108] Uppsala, Univ. Lib., ms. A533, item 19.
[109] M. Strömer, *Abh. Schwed. Akad.*, Vol. 7 (1745), p. 168.
[110] *Observationes Meteorologicae, habitae Upsaliae Anno 1740 ab And. Celsio.* Paris, Acad. Sci., dossier "Réaumur, thermomètre et baromètre," 8 small leaves.
[111] The covering letter, dated December 11, 1737, new style, is in Uppsala, Univ. Lib., ms. A533, item 47 (one folded sheet).

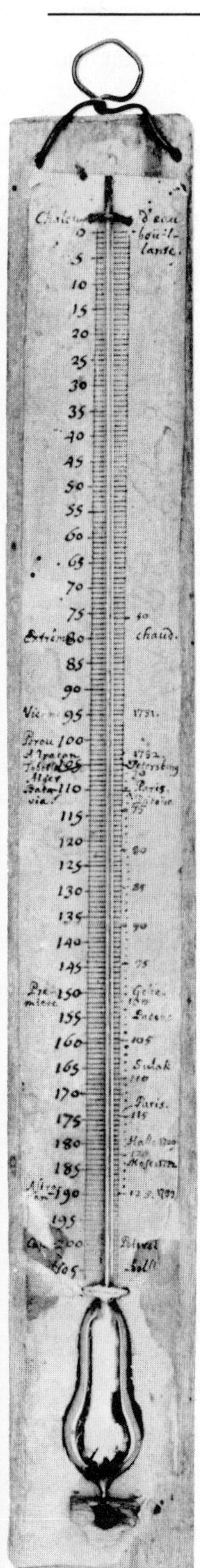

Fig. 2.12 A thermometer sent by Delisle to Celsius (courtesy of Dr. G. H. Liljequist).

scale to a thick wooden base. Delisle's scale runs from 0° at the top (*Chaleur d'eau bouillante*) past 150° (*Premiere Gelée*) to 205°. On the right is part of Celsius' scale which is probably the result of a recalibration, for 100° Celsius corresponds to 151.8° Delisle, and not to 150°, as one might expect.

Celsius' paper of 1742[112] shows that he had read extensively and thought carefully about thermometry. He used melting snow for his ice point. The boiling point was to be determined when the barometer stood at 25¼ Swedish inches,[113] and he gave a rule for marking the point 0° if the water had to be boiled at other atmospheric pressures.

Celsius died in 1744, but his thermometer was used in the official observations at Uppsala until April 13, 1750,[114] being then replaced by one ascribed to Märten Strömer, his successor. We are told that "The thermometer of the late Professor Celsius has 0 at the boiling point of water and 100 at the freezing point; that of Professor Strömer, on the contrary, has 0 at the freezing point and 100 at the boiling point."[115] This statement has led to the standard textbook assertion that Strömer "inverted" Celsius' scale in 1750; but both Strömer and the Stockholm instrumentmaker Daniel Ekström appear to have had such thermometers in 1746, whether independently or not we do not know.[116] Strömer's thermometer had been added to those at the official observatory on September 1, 1747.[117]

But earlier than this, in 1745, the important botanist Linnaeus had such a thermometer, which he showed to the Senate of the University on December 2, 1745.[118] A fortnight later one of his students described the botanical garden of the University in a dissertation, and in his discussion of the *calidarium* or hothouse stated that "our thermometer is 0 at the freezing point and counts 100 to the degree of boiling water."[119]

This does not provide a clear answer to the question of who in Uppsala first thought of the centigrade thermometer with the ice point at zero (Ekström or Linnaeus, or perhaps Strömer). In later life Linnaeus claimed to be the inventor, for on October 30, 1758, he wrote in Latin to Boissier de la Croix de Sauvages, a botanist at Montpellier: "I was the first who decided to construct our thermometers in which the freezing point is 0, and the heat of boiling

[112] *Abh. Schwed. Akad.*, Vol. 4 (1742), pp. 197–205.

[113] Hilding Köhler has pointed out (*Uppsala, Kungl. Vetenskapssocietatens Årsbok*, 1944, p. 111) that on the present-day definition Celsius' upper fixed point was at 99.67°C.

[114] *Kon. Svensk. Wet. Akad., Handlingar*, Vol. 14 (1735), pp. 254–55.

[115] *Ibid.*, p. 253.

[116] See N. V. E. Nordenmark, *Svenska Linné-Sällskapets Årsskrift* (1935), p. 131. See also Nordenmark, *Anders Celsius* (Stockholm, 1936).

[117] Nordenmark, *Anders Celsius*, p. 162.

[118] *Ibid.*, p. 163.

[119] Samuel Naucler, *Hortus Upsaliensis* [etc.] (Diss. Uppsala, Dec. 16, 1745), p. 23n.

water 100, and this for the greenhouses of our garden. I am sure that if you were accustomed to these, they would please you."[120] We may conclude that Celsius made a 100-degree thermometer and that Linnaeus, most probably, inverted its scale not later than the autumn of 1745. But there was a centigrade thermometer with the ice point at zero more than two years before this at Lyon in France, the invention of Jean Pierre Christin.[121]

In 1740 we find Christin reading papers to the Academy which show him making a decision in favor of mercury as a thermometric liquid. But he was still far from thinking of a centigrade thermometer, and in 1741 he was no nearer it, for we find him writing a letter to one Joannon on February 15 of that year in which he states that "Mr. de Réaumur used in the construction of his first [thermometers] a spirit of wine the dilatability of which was found to be 80 degrees from the freezing point to that of boiling water; I see that this is a division into a fixed number, which should no more be changed than that which has been fixed at 360 for the circle."[122] Later on in the same letter he says that ". . . whatever liquids . . . thermometers are filled with, they must always be made on Mr. de Réaumur's principles to make them comparable."[123]

On May 2, 1743, the centigrade thermometer arrived. I shall translate part of the summary of Christin's paper in the Academy's manuscript *Journal des Conférences* for that date:[124] "An experiment . . . had shown [Christin] that a quantity of mercury condensed by the cold of pounded ice, and then dilated by the heat of boiling water, formed, in these two states, volumes that were to each other as 66 to 67, and that a volume of 6,600 parts, condensed, became one of 6,700 parts by dilatation. The difference, 100, . . . is the number of degrees that he gives to the scale of a new mercury thermometer between these two points." This seems to be a poor reason for the choice of a centigrade scale, but the summary continues: "This number is found to be advantageous for the precision of observations, and each degree represents one of 6,600 condensed parts taken from zero, the freezing-point . . . Mr. Christin has remarked that several advantages can be derived from this discovery; among others, that of being able to construct mercury thermometers by using boiling water, without the help of freezing, and conversely with ice, without the heat of boiling water." This shows that Christin thought he was simply doing what Réaumur

[120] Louis Augustin d'Hombres-Firmas, ed., *Lettres inédites de Linné à Boissier de la Croix de Sauvages* [etc.] (Alais, 1860), p. 226.

[121] The relevant manuscripts in the libary of the Académie des Sciences, Belles-Lettres et Arts de Lyon were graciously placed at my disposal by its secretary, M. Marcel Chamaraud. I have dealt more fully with the development of Christin's ideas in *Hist. Therm.*, pp. 101–05.

[122] Lyon, Palais des Arts, ms. 199, fol. 131^v.

[123] *Ibid.*, fol. 133^v.

[124] Also published in *Mém. de Trévoux* (July, 1743), pp. 2126–28. I have not found the paper itself.

would have done, had the great man used mercury; in fact one is impressed, reading these manuscripts at Lyon, with the way that the huge shadow of Réaumur hung over the subject of thermometry in France.

Not everyone was quite as inhibited as this, for on July 31, 1743, one De Moronval wrote to Christin from Paris, eliciting a twelve-page reply[125] which makes it evident that De Moronval had divided a mercury thermometer into 100 parts between the ice point and the boiling point without bothering to determine the capacity of the bulb.

There is an original "thermomètre de Lyon," made by Pierre Casati, in the Science Museum, London.[126]

8. Further Development of the Air Thermometer. It was natural to think of protecting the air thermometer from the variations of the atmospheric pressure, and a simple way of doing this was found by the Paris instrumentmaker Hubin in 1672.[127] Hubin modified Hooke's (or Huygens') two-liquid barometer[128] by adding the bulb shown in broken lines in Figure 2.13. The liquid, instead of water, was *l'eau seconde*, an acid solution of copper nitrate produced in the refining of gold, which was green and easily visible. Such a thermometer would respond to the vapor pressure of the *eau seconde* as well as the dilatation of the air in the bulb.

Hooke's three-liquid barometer[129] was similarly modified by closing the open bulb. This was done by Guillaume Amontons,[130] who is much more important to us as the first to make a serious study of the principles of the air thermometer, finding experimentally in 1699 that different amounts of air have their "spring" increased equally by the

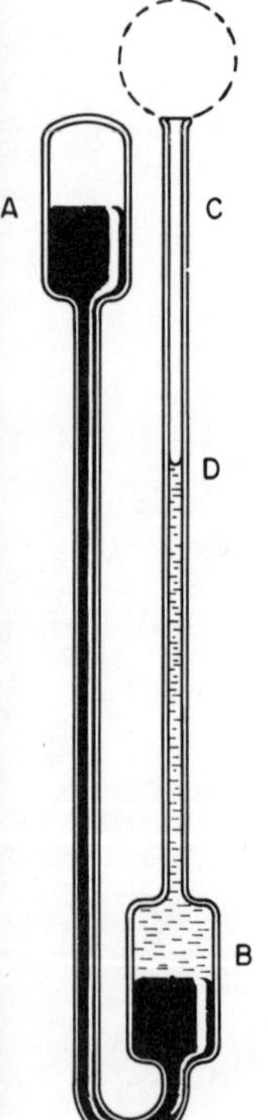

Fig. 2.13 Hubin's air thermometer.

same rise in temperature.[131] He deduced from Mariotte's (*i.e.*, Boyle's) law that any volume of air would expand by the same fraction between, for example, room temperature and the boiling point if it were at liberty to expand without changing its "spring."

[125] Lyon, Palais des Arts, ms. 199, fol. 171r–176v. De Moronval's letter seems to have been lost.

[126] Inventory no. 1951–581. See J. A. Chaldecott, *Handbook of the Collection illustrating Temperature Measurement and Control*, Part II (London, 1955), pp. 7–8.

[127] Hubin, *Machines nouvellement executées et en partie inventées par le Sieur Hubin* [etc.]. Pamphlet (Paris, 1673).

[128] See above, p. 26 and Fig. 1.7.

[129] See above, p. 26, and Fig. 1.8.

[130] Amontons, *Remarques et expériences phisiques sur la construction d'une nouvelle clepsidre, sur les baromètres, termomètres, & higromètres* (Paris, 1695), pp. 146–57.

[131] Amontons, *Mém. Acad. r. Sci. Paris* (1699), pp. 112–26.

Later[132] he showed that the form of the relation between the temperature and the pressure of an enclosed mass of air is independent of the initial pressure, and made a very simple air thermometer out of a J-shaped tube with a bulb on the short end (Fig. 2.14). This had the advantage over the earlier air thermometers that the only liquid involved was mercury, but the disadvantage that it was sensitive to atmospheric pressure. All that was necessary to obviate this was to close the long tube, filling the instrument like a siphon barometer

Fig. 2.14 Amontons' quantitative air thermometer.

before sealing the bulb with some air in it; this seems to have been done first by Vittorio Francesco Stancari of Bologna about 1708.[133]

J. H. Lambert believed that the air thermometer was the only one that shows "the actual degree of heat," and introduced a scale of his own, based on experiments with it, in which the freezing point is 1,000° and the boiling point 1,370°.[134] In the nineteenth century the air thermometer, or the hydrogen thermometer, did become the standard to which others were referred, but this is outside the scope of this book.

9. *Reforms in calibration.* We have already seen that by about 1740 the lower fixed point was commonly obtained by surrounding the bulb of the thermometer with crushed ice that was melting. The Abbé Nollet recognized that the melt-water should be drained away "when it is seen to be getting too abundant."[135] For the upper fixed point, the bulb was immersed in rapidly boiling water, and we have noted that Celsius adopted a standard barometric pressure at which the water was to be boiled. In England the famous instrumentmaker John Bird calibrated his Fahrenheit thermometers at 30 inches of mercury.[136]

The suggestion that the bulb of the thermometer should be in the stream, rather than in the boiling water, was made in a paper about The Royal Society's instruments by Henry Cavendish,[137] who was at once made the chairman of a committee on the calibration and use of thermometers. The committee's report appeared in the following year.[138]

132 *Ibid.* (1702), pp. 155–74.
133 Cf. *Comment. Bonon.*, Vol. 1 (1731), pp. 209–10.
134 Lambert, *Pyrometrie oder vom Maase des Feuers und der Wärme* (Berlin, 1779), p. 78.
135 *L'art d'expériences*, III, 148.
136 Samuel Horsley, *Phil. Trans.*, Vol. 64 (1774), p. 224.
137 Cavendish, *Phil. Trans.*, Vol. 66 (1776), pp. 375–401.
138 *Phil. Trans.*, Vol. 67 (1777), pp. 816–57.

They found English thermometry in a bad state; thermometers by different makers varied over a range of 3¼° F. in steam. After many experiments, they recommended suitable apparatus and methods for determining the steam point, choosing 29.8 inches of mercury as the standard pressure. The entire thermometer, almost up to the 212° graduation, was to be in the steam, the bulb being "at least one or two inches" above the surface of the boiling water. The vessel had a close-fitting cover and a "chimney" with a small, loose piece of tin-plate on top of it to prevent the ingress of air without materially increasing the steam pressure. Although their apparatus has been much improved, the committee's general recommendations have been the basis for determinations of the steam point ever since.

10. Early Maximum and Minimum Thermometers. On July 5, 1698, Johann Bernouilli wrote to Leibniz, enclosing a sketch of three air thermometers (Fig. 2.15) having many appendices (*varices, seu tumores*) to measure the highest or lowest temperature or both, since the last visit by the observer.[139] This idea seems to have had no consequences except that about 1743 G. W. Krafft made a maximum thermometer of this sort[140] but sealed, like Stancari's air thermometer.

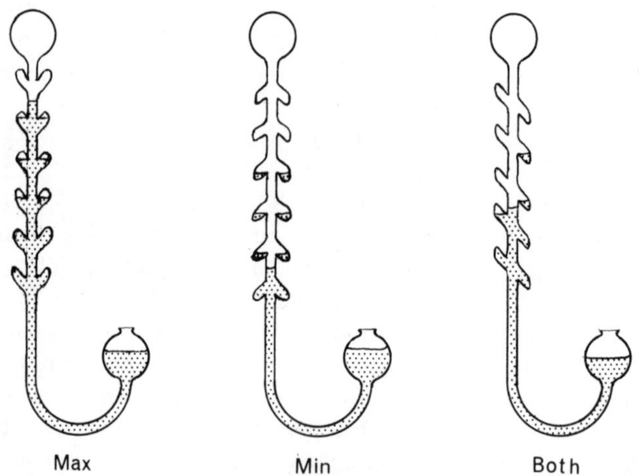

Fig. 2.15 Bernouilli's suggestion for maximum and minimum thermometers.

Apart from this unfruitful idea, the first maximum and minimum thermometers seem to have been invented by Lord Charles Cavendish,

[139] *Virorum celeberr. Got. Gul. Leibnitii et Johann Bernouillii commercium philosophicum et mathematicum* (2 vols., Geneva, 1745), Vol. 1, pp. 371–78.
[140] Krafft, *Comm. Petrop.*, Vol. 13 (1741–43), pp. 346–47.

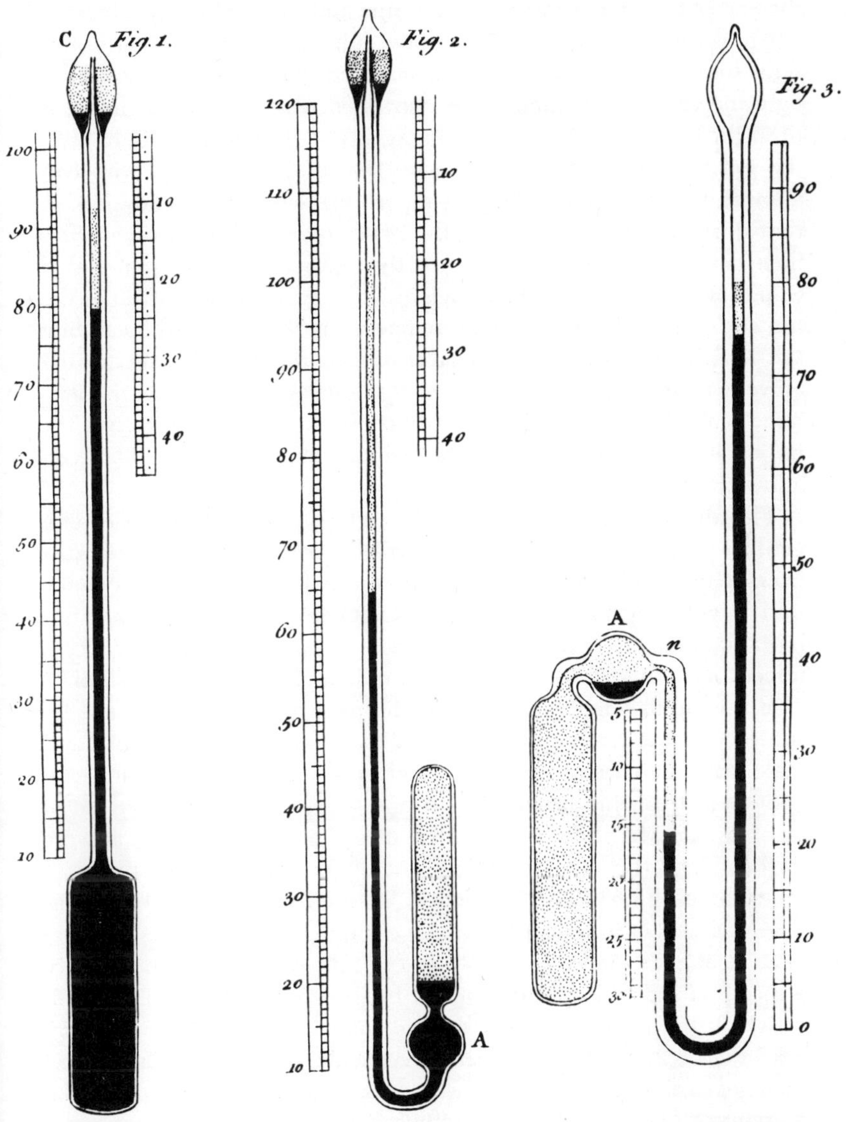

Fig. 2.16 Lord Charles Cavendish's thermometers.

the father of the famous chemist Henry Cavendish, and described in 1757.[141] Three of his thermometers are shown in Figure 2.16. They all depended on overflow.

The two on the left are maximum thermometers, differing in that the first depends on the expansion of mercury, the second on that of spirit. The former is like an ordinary thermometer with a large bulb,

[141] Cavendish, *Phil. Trans.*, Vol. 50 (1757), pp. 300–10.

the end of the tube being drawn out fine and surrounded by a bulb *C*. There is spirit over the mercury in the tube, and the bulb *C* also contains some mercury and some alcohol; the space remaining contains only the vapor of alcohol. The instrument was "set" by inclining it until the end of the tube was submerged and warming the bulb until the spirit in the tube joined up with that in *C*. The bulb was then allowed to cool to the ambient temperature while in this position, so that spirit was sucked in, filling the whole tube. The instrument was then put upright and left alone. If the temperature rose, some spirit would overflow into *C;* when it fell again, a space would be left at the top of the tube. The maximum temperature was obtained by adding the length of this space, measured on the right-hand scale, to the current temperature at the time of the observation, read on the left-hand scale at the top of the mercury column. The drawing gives the impression that the volume of the reduced part of the tube has been allowed for.

The minimum thermometer on the right is constructed on the same general principles. It should be noted that a fine thread of glass goes from *A* through the slight constriction *n* as far as the bottom of the instrument, which is set by tilting it until *n* is covered by the mercury in *A*, and then warming the bulb so that—we are told—the mercury is forced past *n* in drops until the shorter arm of the siphon is full of it. The minimum temperature before the next observation is found by subtracting the reading of the top of the mercury on the right-hand scale from the current temperature given by the left-hand scale.

Such instruments are hard to set and to transport, and were out of favor by the end of the eighteenth century, because better ideas were forthcoming, the earliest of these being the combined instrument invented by James Six,[142] which in its modern form is ubiquitous. The thermometric fluid is alcohol, contained in the long cylindrical bulb *ab* of Six's Figure 1 (Fig. 2.17 here), the mercury *defg* being only a means of indication, and confined between two volumes of alcohol, of which the second is given a little room to move in the bulb *hi*. Six's Figure 2 shows, on a much larger scale, one of the little indices, made of steel wire and covered with glass, that were pushed up by the mercury. These had long glass "tails" to act as springs.

Six's thermometer is no longer used as a serious meteorological instrument because the spirit wets the glass, and can at length pass between the glass and the mercury.

In 1790 Daniel Rutherford described a pair of thermometers (Fig. 2.18) invented by a namesake, John Rutherford.[143] The minimum is essentially the one used almost exclusively by meteorological services at the present day, and is simply a spirit thermometer with a conical glass index, its point toward the bulb, immersed in the spirit. The

[142] Six, *Phil. Trans.*, Vol. 72 (1782), pp. 72–81.
[143] D. Rutherford, *Trans. Roy. Soc. Edinburgh*, Vol. 3 (1794), pp. 247–49.

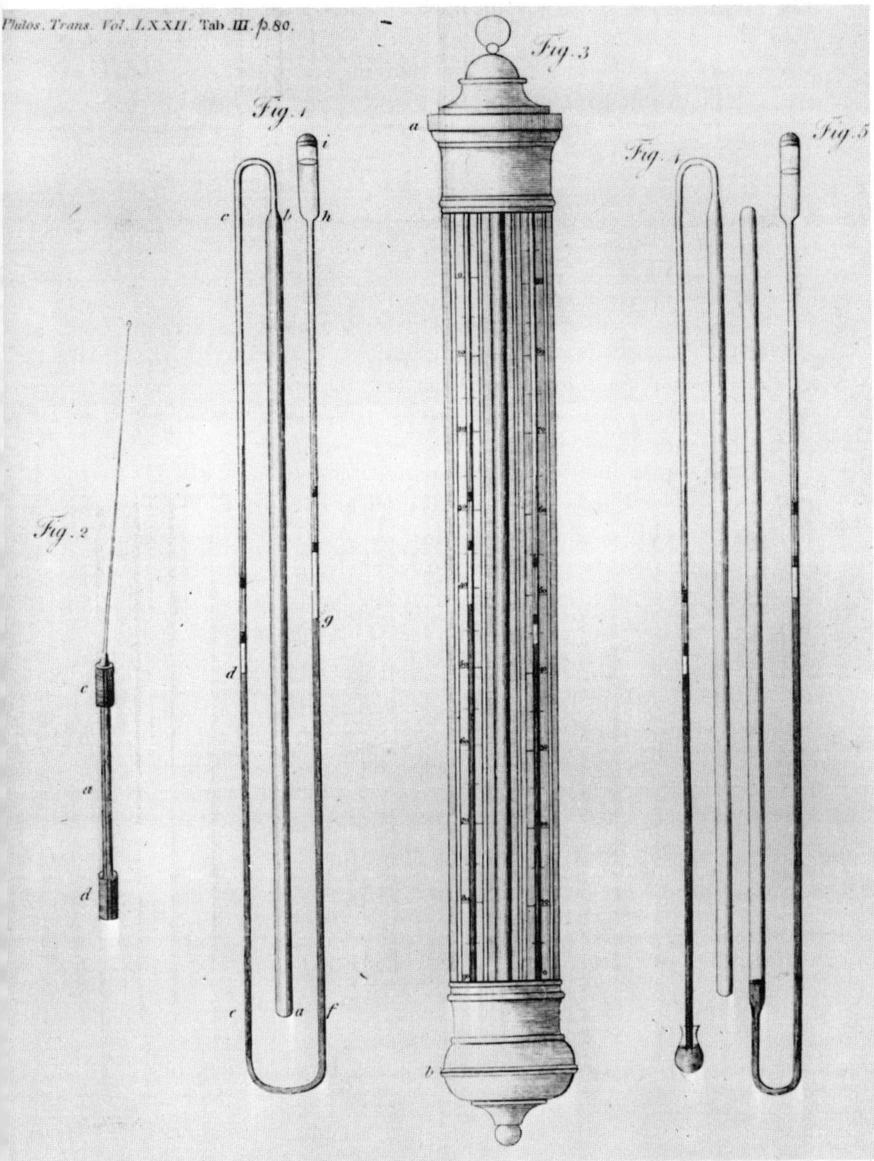

Fig. 2.17 Six's combined maximum and minimum thermometer.

maximum is a mercury thermometer with a conical index of ivory. The two instruments were placed horizontally on the same frame, one above the other, the bulb of the minimum and the tip of the maximum toward the right. Inclining the frame to the left was all that was needed to set both instruments.

The Rutherford maximum thermometer was finally found to be unsatisfactory, because the mercury tended to get past the index.

The minimum, its index of blue glass now in the shape of a dumbell, is used almost everywhere. This change in the form of the index had taken place by 1826, and the index of the maximum thermometer was also a little dumbell, made of blued steel.[144]

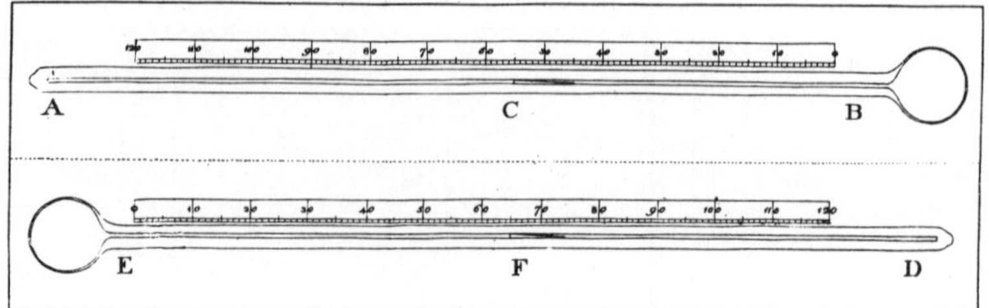

Fig. 2.18 Rutherford's maximum and minimum thermometers.

A better sort of maximum thermometer was devised in 1832 by the geologist John Phillips,[145] and reinvented in 1855 by F. H. Walferdin of Paris.[146] This was simply an ordinary mercury thermometer in which about one centimeter of the column, separated from the remainder by a very small bubble of air, acts as an index when the thermometer is installed horizontally. This sort of thermometer was still being made and sold in the 1870's, but was eventually superseded by the well-known instrument with a constriction in the tube near the bulb, patented by the London firm of Negretti & Zambra on March 8, 1852,[147] and now almost universally used at meteorological stations, and of course as a clinical thermometer. This instrument requires careful workmanship, and was not immediately adopted on the continent of Europe.[148]

11. Early Deformation Thermometers. By "deformation thermometer" I mean any thermometer which indicates a change of temperature by a change in the shape or configuration of some system of solid bodies, usually metallic. Practical thermometers of this sort, with one or two exceptions, are of two kinds: those using straight bars or tubes of different materials, and those which incorporate a compound bar made of two or more metals fastened together, which changes its curvature as the temperature varies. The earliest deformation thermometers were of the first kind, the very first probably being that made about 1735 by Cromwell Mortimer, who read a paper about thermometers on May 8 of that year,[149] only published twelve years later. But this large instrument was intended to measure high tem-

144 K. L. G. Winckler, *Ann. Phys.*, Vol. 6 (1826), pp. 127–32.
145 Phillips, B.A.A.S. Report, Oxford, 1832, *Sections*, pp. 580–81.
146 Walferdin, *Compt. Rend.*, Vol. 40 (1855), pp. 951–54.
147 British patent 14,002 of 1852.
148 There were a large number of less successful designs of maximum and minimum thermometers. I have dealt with many of them in *Hist. Therm.*, Chapter VI.
149 Mortimer, *Phil. Trans.*, Vol. 44 (1747), pp. 672–95.

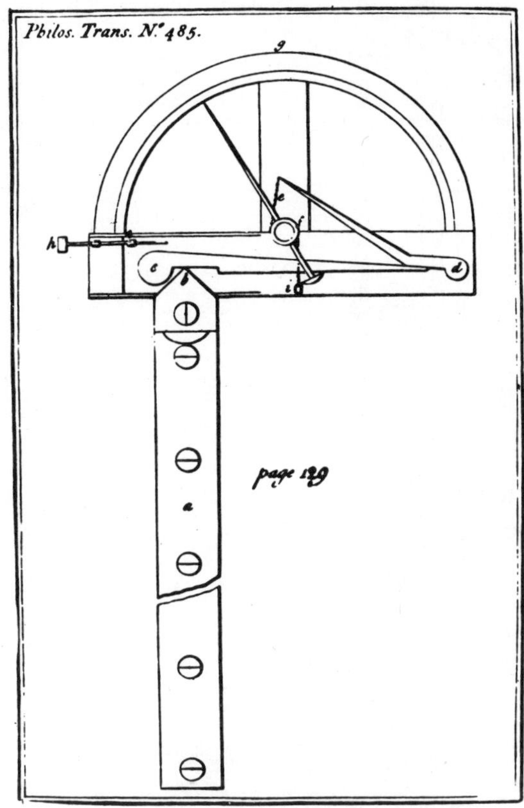

Fig. 2.19 Samuel Frotheringham's metallic thermometer.

peratures, and the first description of one for the meteorological range dates from 1748.[150] This instrument (Fig. 2.19) was made by John Ingram of Spalding, Lincolnshire, at the suggestion of Samuel Frotheringham. A rectangular brass bar four feet long was screwed to an iron bar at the lower end, and held to it at intervals by screws working in slots. At the top of the brass bar was a steel chisel-edge which communicated the relative motion of the top of the bar to the first of two magnifying levers and a chain and pulley. A pointer was fastened to the pulley, and its extremity must have had a magnification of about 500. This design is good in one respect that we may assume the maker never thought of, namely, in that the thermal lag of the two elements is about the same.

It is possible that some of the instruments invented by Count Hans von Löser, and made by the famous J. G. Zimmer in the Count's workshops at Reinhartz, near Wittenberg, were earlier than that of

[150] Maurice Johnson, *Phil.Trans.*, Vol. 45 (1748), pp. 128–30.

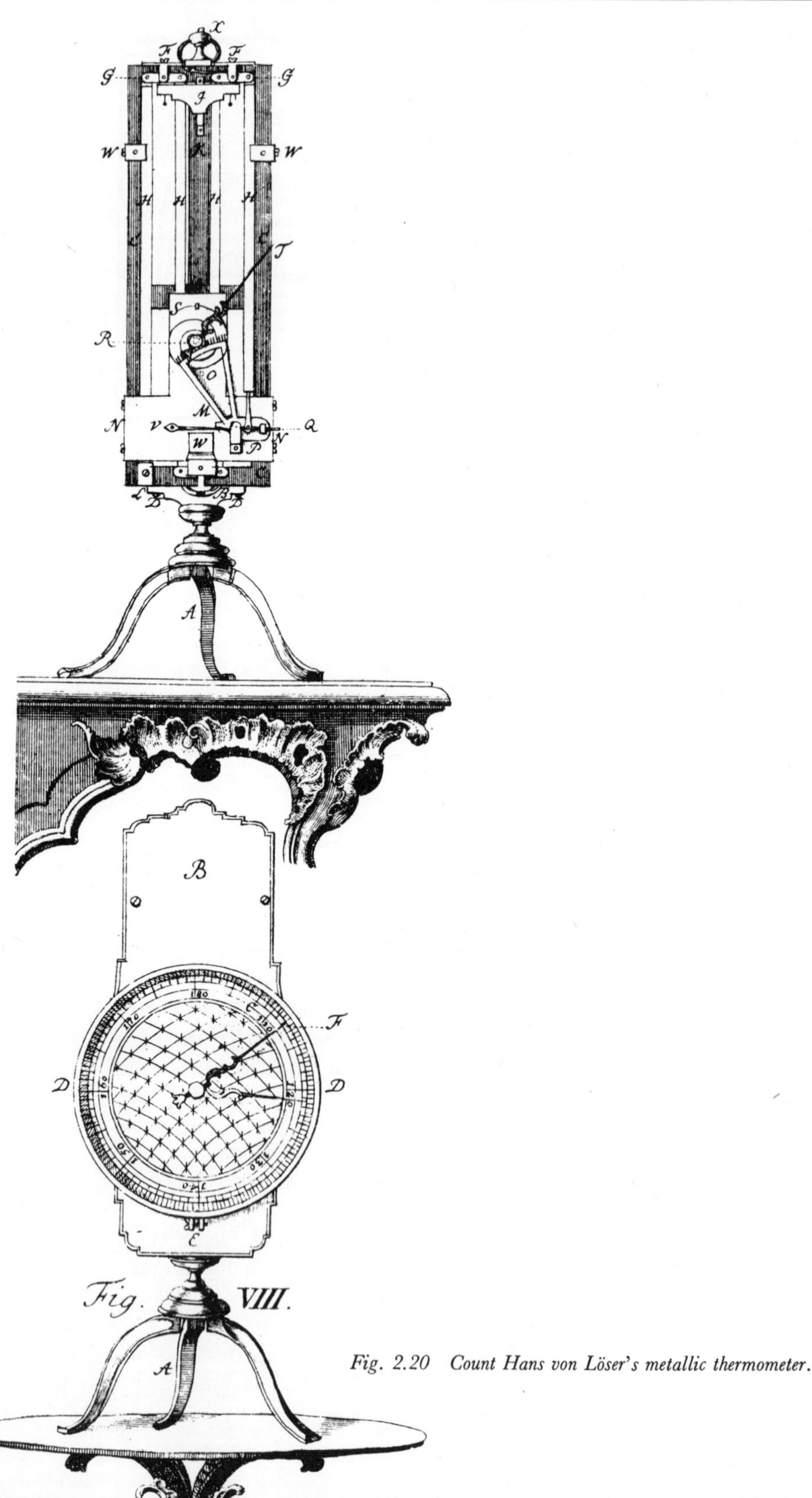

Fig. 2.20 Count Hans von Löser's metallic thermometer.

Frotheringham, but they were described only in 1764.[151] They each embodied a brass or steel frame and several lead bars, the latter arranged so that the relative motions of their ends were added by levers. The motion of the final free end was magnified by a sector and a train of gears, and there was an adjustment for the magnification. The earliest (Fig. 2.20) was made in 1746 according to Titius, and had four lead bars about nine inches long. It is plain that these instruments were not intended for use out of doors.

Apart from the choice between a pair of long bars or tubes and several shorter ones, it is not easy to be original in the design of such instruments. Keane Fitzgerald described one in 1760,[152] and an improved one in 1761,[153] the latter having two pointers that could be pushed by the main pointer to indicate the maximum and minimum temperatures.

I have not been able to establish when or by whom the compound strip that we know as a bimetal was invented and applied to thermometry. The earliest reference to it is by Lambert and dates from 1779.[154] He has been told, he says, that someone in England has had the idea of making bimetallic strips of copper and iron, and—which may surprise us—that this can be done by rolling, without soldering. "But such thermometers do not seem to be much in use."

In 1782 we find Jean Hyacinthe de Magellan, a Portuguese who had settled in England, writing of the bimetallic thermometer as if it were well known: "To construct it, a strip of steel and one of brass are solidly riveted or even soldered[155] together; for heat and cold make this double strip curve more or less, according to the reading of the thermometer. It is better to give this double strip a spiral form, so as to have it long enough to enlarge the movement . . . This strip, being fastened at one end, moves the pencil that is attached to the other end (or to a rod fastened to this) according to the different temperatures of the atmosphere."[156]

However, a pocket thermometer that must surely have contained a bimetal had been constructed in 1767 at Norriton in Pennsylvania by that remarkable man David Rittenhouse.[157] This was

an ingeniously contrived thermometer, constructed on the principle of the expansion and contraction of metals, by heat and cold, respectively. This instrument had, under glass, a face upon which was a graduated semi-circle;

[151] Johann Daniel Titius, *Descriptio thermometri metallici ab inventione Comitis ab Loeser* (Leipzig, 1764). There is a copy at Offenbach-am-Main; Poggendorf, *Biogr. Wörterbuch*, cites only an edition printed at Wittenberg in 1765.
[152] Fitzgerald, *Phil. Trans.*, Vol. 51 (1760), pp. 823–33.
[153] *Phil. Trans.*, Vol. 52 (1761), pp. 146–54.
[154] J. H. Lambert, *Pyrometrie oder vom Maase des Feuers und der Wärme* (Berlin, 1779), p. 124. Lambert had died in 1777.
[155] "Ou même on [les] soude ensemble." The verb *souder* can mean to solder, to braze, or to weld.
[156] Magellan, *Obs. sur la Phys.*, Vol. 19 (1782), p. 353.
[157] Brooke Hindle, *David Rittenhouse* (Princeton, 1964), p. 25.

the degrees of heat and cold corresponded with those of Fahrenheit's thermometer . . . Its square (or rather parallelogramical) form, its flatness and thinness, and its small size—together with its not being liable to the least sensible injury or irregularity, from any position in which it might be placed—rendered it safely portable; insomuch, that it could be conveniently carried in the pocket.[158]

In the absence of further evidence, we can conclude only that the bimetallic strip had been applied to thermometry by 1767. We need not assume that Rittenhouse was the first to do this, although with his astonishing ingenuity he might well have been the inventor of this device.

[158] William Barton, *Memoirs of the Life of David Rittenhouse* (Philadelphia, 1813), p. 155*n*.

III

Instruments for measuring humidity

---

1. Introduction. The measurement of atmospheric humidity is a process of giving numbers to the sensations "moist" and "dry," or more precisely to whatever properties of the air are responsible for these sensations. The first instruments for this purpose were invented long before anyone had a clear idea of what these properties are, and it will be necessary to preface this chapter with some remarks about the development of ideas regarding water vapor.[1]

The ancient Greeks and Hebrews both had a fairly clear idea of the hydrological cycle, the raising of water from the earth and the sea and its return as precipitation. The difficulty was that at one stage in the process the water becomes entirely invisible. What was more natural than to assume that it had changed into air? The doctrine of the interconvertibility of air and water was held by Aristotle, who considered the invisible vapor to be air formed from water, although he recognized that clouds were made of drops of water.[2]

It was only in the seventeenth century that it began to be accepted that water vapor is a distinct substance. This was largely a result of the great influence of René Descartes, who believed that all substances are composed of tiny particles, all of the same kind of matter, with the substances distinguished from one another entirely by the shapes and sizes of their constituent particles. He supposed water to be composed of long smooth eel-shaped atoms that can easily be separated, and compared the process of evaporation with the raising of dust from a field by the feet of a passer-by.[3] Even though the water becomes quite invisible, its particles, he maintained, keep their distinctive form, quite different from that of the particles of air.

This was a notable intellectual advance, but some mechanism had to be found to keep the water-particles from falling out again, for it was entirely outside the orbit of seventeenth-century science to believe that moist air could be lighter than dry air at the same temperature and pressure, as in fact it is. Even Sir Isaac Newton's categorical

[1] For a more extended discussion see my *A History of the Theories of Rain* (London, 1965; New York, 1966), especially Chapters 1, 2, and 7.
[2] Aristotle, *Meteorologica*, Book I, Chap. 9.
[3] Descartes, *Les Météores* (Leyden, 1637), p. 239.

statement in 1717 that moist air is lighter[4] was simply ignored for decades, because it seemed clear to everyone that the moisture, having been *added* to the air, must make it heavier.

The real solution to the problem was found by Daniel Bernouilli, who in 1738 provided a kinetic theory of gases,[5] but this was more than a century ahead of its time. During most of the remainder of the eighteenth century, the fashionable opinion was that water *dissolves* in air, in exactly the same way that a solid dissolves in a liquid—a convenient analogy that ignored the fact that nobody, except the neglected Bernouilli, had any real explanation for the latter phenomenon. The first clear statement of this "solution theory of evaporation," which had been foreshadowed by several earlier writers, was made by Charles Le Roy of Montpellier in a paper[6] that we shall return to later.

The main competitor of the solution theory was the theory that water vapor is a chemical compound of water and "fire" or heat, the latter being almost everywhere considered in the eighteenth century to be a substance, extremely tenuous and almost or entirely without weight. Jean André Deluc was the most vocal partisan of this theory,[7] but Lavoisier supported it,[8] and by about 1800 the adherents to the solution theory were in the minority, except perhaps among the chemists. One important principle that had been established by 1790 was that aqueous vapors have the properties of gases, "and exercise them in complete independence of these fluids."[9] In other words, "the product of evaporation is always of the same nature, namely, an expansible fluid, which, either alone or mixed with air, affects the manometer by pressure and the hygrometer by moisture, without any difference arising from the presence or absence of air; at least without any perceived hitherto."[10]

At about this time, John Dalton had begun to study water vapor. It is unlikely that he had read Deluc's *Idées*, but he was aware of Joseph Priestley's conclusions about the interdiffusion of mixed gases.[11] Misinterpreting a demonstration of Newton's,[12] Dalton came to the

[4] In his *Opticks*, 3rd English ed. Reprinted 1952 by Dover Publications, Inc., New York; see p. 396.

[5] D. Bernouilli, *Hydrodynamica, sive de viribus & motibus fluidorum commentarii* (Strasbourg, 1738). The pertinent paragraphs will be found translated in *The World of Mathematics*, James R. Newman, ed. (4 vols., New York, 1956) II, 774–77.

[6] Le Roy, *Mém. Acad. Roy. Sci. Paris* (1751), pp. 481–518.

[7] *Recherches sur les modifications de l'atmosphère* (2 vols., 4°, Geneva, 1772; 2nd ed. 4 vols., 8°, Paris, 1784); *Idées sur la Météorologie* (2 vols., London, 1786; Paris, 1787).

[8] *Mém. Acad. Roy. Sci. Paris* (1777), pp. 420–32.

[9] Deluc, *Idées*, ¶5.

[10] Deluc, *Phil. Trans.*, Vol. 82 (1792), p. 424.

[11] Priestley, *Experiments and Observations relating to various Branches of Natural Philosophy* [etc.] (3 vols., London, 1779; Birmingham, 1781 and 1786). See for example Vol. III, p. 390.

[12] *Principia*, Book II, Prop. 23.

conclusion that the particles of each gas or vapor in a mixture are completely independent of those of all the others.[13]

By the time the meteorologists became aware of this, Dalton's reputation as a chemist seems to have overpowered the judgment of many of them, who were ready to accept even his application of his theory to the evaporation of water. If all the gases except water vapor were removed, he wrote,

> Little addition would be made to the aqueous atmosphere, because it already exists in every place, almost entirely up to what the temperature will admit; the evaporation of water would be essentially the same in that case as at present; only the full effect would take place in less time. In short this notion of pressure preventing the evaporation of liquids, which seems to have been taken as an axiom by modern philosophers, has been the cause of more error and perplexity perhaps than any other ungrounded opinion.[14]

It is quite surprising that Dalton did not see what he was asserting, namely that the whole atmosphere, from top to bottom, must be almost saturated "in every place." In the following year he went on to an even more astonishing assertion: "On my principle the density of the aqueous atmosphere at any height is totally independent of the density of the compound mass of air, and is to be ascertained by knowing the density of vapour at the earth's surface, and its specific gravity; in the same way as we would ascertain the density of the oxygenous or azotic atmospheres, or one of hidrogen, at any given height, having the like data."[15]

This so-called "Dalton's law" had a surprising longevity in spite of its fatal faults, especially in England but in other countries as well. Even in 1874 the great Austrian meteorologist J. von Hann considered it desirable to nail down its coffin.[16] It is interesting that nowadays the term "Dalton's law" refers merely to the additivity of gas pressures and vapor pressures in the same space.

Horace Benedict de Saussure, whom we shall meet again in the third section of this chapter, established by direct experiment that water vapor exercises a pressure of its own, its maximum value being dependent on the temperature.[17] Many workers measured the maximum vapor pressure as a function of temperature; the first were G. G. Schmidt[18] and John Dalton.[19] These determinations gradually increased in accuracy and completeness up to the time when Henri Victor Regnault published his impressive researches.[20] Interpolation formulas were devised by the mathematicians, beginning with J. von

[13] Dalton, *Mem. Manchester Lit. & Phil. Soc.*, Vol. 5 (1802), pp. 536–37.
[14] *Ibid.*, p. 545.
[15] Dalton, *Nicholson's J.*, Vol. 6 (1803), pp. 119–20.
[16] *Zeits. österr. Ges. Meteorol.*, Vol. 9 (1874), pp. 193–200.
[17] De Saussure, *Essais sur l'hygrométrie* (Neuchâtel, 1783), pp. 96–108.
[18] Schmidt, *Gren's neue J. Phys.*, Vol. 4 (1797), pp. 251–319.
[19] Dalton, *Mem. Manchester Lit. & Phil. Soc.*, Vol. 5, (1802) pp. 535–602.
[20] Regnault, *Compt. Rend.*, Vol 20 (1845), pp. 1120–66; 1220–37.

Soldner,[21] who used Dalton's results. In 1833 P. N. C. Egen was able to review twenty-nine such formulas, adding four of his own.[22]

The specification of the vapor pressure at a given time and place is one of the ways of stating the amount of water vapor in the atmosphere. Another way, adopted by De Saussure in his studies of hygrometry, is to state the weight of water vapor in unit volume of air. De Saussure used grains per cubic foot, and produced a double-entry table for various readings of his hair hygrometer[23] and temperatures from $-10°$ to $+30°$R.[24] Although he found by extensive and difficult experiments that the relative weights of vapor at various temperatures were about the same for all readings of his hygrometer,[25] he missed the simplifying concept of *relative humidity*, which is the ratio of the actual vapor pressure to the maximum vapor pressure at the existing temperature, usually multiplied by 100 and expressed as a percentage. J. A. Deluc came close to the idea of relative humidity in 1792, when he wrote:[26] "If . . . it is required to know the density of . . . steam, or its quantity in a given space; the observation of the temperature is required: as this, by previous experiments in order to determine the quantities of steam correspondent to its different maxima according to the temperature, will afford a coefficient for the *ratio* observed."

By 1800 Ludwig Achim von Arnim was using the terms "relative amount of water" (*relative Menge Wasser*) and "degree of saturation" synonymously,[27] and in 1803 L. W. Gilbert, the editor of the *Annalen der Physik*, put the matter quite clearly: "The degree of humidity depends on the ratio of the vapor actually present to that which is possible."[28]

2. Classification of instruments for measuring atmospheric humidity. The instruments that form the subject of this chapter are of several kinds, differing so greatly in principle that they will have to be discussed separately. In the order of their first invention, they are as follows: (1) those depending on the hygroscopic properties of various substances (hygroscopic hygrometers); (2) those depending on the formation of dew on a surface that can be artificially cooled (dew-point hygrometers); (3) those depending on the reduction of temperature produced by the evaporation from a moist surface (psychrometers); (4) those in which the moisture is taken from a known volume of air

[21] Soldner, *Ann. Phys.*, Vol. 17 (1804), pp. 44–81.
[22] Egen, *Ann. Phys.*, Vol. 27 (1833), pp. 9–40.
[23] See below, p. 103.
[24] De Saussure, *Essais*, p. 181.
[25] *Ibid.*, p. 119.
[26] Deluc, *Phil. Trans.*, Vol. 82 (1792), pp. 413–14. At this period Deluc habitually referred to water vapor as "steam."
[27] Von Arnim, *Ann. Phys.*, Vol. 4 (1800), p. 322.
[28] Gilbert, *Ann. Phys.*, Vol. 15 (1803), p. 167.

and weighed or otherwise measured (absorption hygrometers); and (5) miscellaneous instruments depending on other physical properties.

3. Hygroscopic hygrometers. The hygroscopic hygrometers can be divided broadly into two classes: those that make use of the change in weight as water vapor is absorbed by a substance, and those in which some change in shape or dimensions is measured. The second class can be further subdivided: there are hygrometers in which a change in length is the indicator, hygrometers using a change in twist, and hygrometers making use of other changes in form, as for instance the swelling of wood across the grain. The only hygroscopic hygrometers still used for serious meteorological purposes belong to the first of these three subclasses.

The hygrometer—or perhaps we should say hygroscope—in which a hygroscopic substance is weighed was described very nearly simultaneously by two writers in the middle of the fifteenth century. One of these was the immensely versatile Leone Battista Alberti, painter, poet, philosopher, musician, and architect, whose greatest work, *De re aedificatoria*, was finished in 1452.[29] In this book, which has gone through many editions, we may read that "We have found that a sponge is moistened by the humidity of the air, and so we make a balance with which we measure the heaviness and the dryness of the air and the winds."[30]

Nicolas Cryfts or Khrypffs was born at Cues on the Moselle in 1401, and became a Cardinal and Bishop of Brixen. Generally known to historians as Nicholas Cusanus, he was a highly original thinker, many of whose ideas were far in advance of his time. In the present context we are interested in his book *De staticis experimentis*, the autograph manuscript of which bears the date of its completion, August 23, 1450.[31] Nicolas described a hygroscope in the following passage: "If someone should hang a good deal of dry wool, tied together, on one end of a large pair of scales, and should balance it with stones at the other end in a place where the air is temperate, it would be found that the weight of the wool would increase when the air became more humid, and decrease when the air tended to dryness. Whence, by such a difference, he might weigh the air, and make probable conjectures about a change in the weather."[32] Apart from the optimism about forecasting, it ought to be noted that the Latin is entirely in the subjunctive mood, so that we may wonder if the Cardinal had actually made such a device. As to whether he and Alberti had had the idea

[29] G. Mancini, *Vita di Leon Battista Alberti*, 2nd ed. (Florence, 1911), p. 351.

[30] Alberti, *De re aedificatoria libri decem* (Argentorati [Strasbourg], 1541), fol. 148^{v} (Lib. X, cap. III).

[31] Henry Bett, *Nicholas of Cusa* (London, 1932), p. 90.

[32] *D. Nicolai de Cusa . . . Opera* (Basle, 1565), p. 176.

independently, we can only speculate, but one suggested a sponge, the other wool.

The invention of the hygrometer is commonly ascribed in the textbooks to Leonardo da Vinci, who was born two years after Nicolas of Cues had finished his book. Nevertheless Leonardo made two drawings of hygrometers of this kind which are preserved in the *Codex atlanticus*.[33] One, showing a sponge (*spugna*) counterbalanced by a weight, is accompanied by a note that may be translated "means of weighing the air and finding out when the weather will break."[34] Duhem found a good deal of evidence that Leonardo had read and taken notes on much of the work of Nicolas of Cues,[35] so that it is quite likely that he had noticed the reference to the hygroscope.

Before continuing the history of the balance hygrometer, let us follow that of hygrometry in general until about 1660. Santorio Santorre, whose contribution to thermometry I have noticed in Chapter 2,[36] described no less than three distinct types of hygrometer. The earliest dates from 1612:

> . . . we have found a very certain way of diagnosing the humidity of the air, that is to say how much of it there may be each day. This is to take salt of tartar, commonly called alum of the lees;[37] it is exposed to the air, but first it is weighed very exactly. Then in the morning it is weighed again. Now it always weighs more after exposure to the air, but considering the different weights we say that the greater the weight the greater the humidity, and the less the weight the less the humidity that reigns in the air. With this, therefore, we can perceive very exactly the ultimate degrees of its active and passive qualities.[38]

The reader will note that this is a different sort of hygrometer from those of Nicolas and Alberti. Instead of trying to determine the humidity at a given moment, Santorio integrated it, so to speak, over the night. Two years later he adds that the salt had first been dried in the sun,[39] and mentions, with no details, two other kinds of hygrometer: "the greater or lesser warping of very thin boards, especially of pear-wood," and "the contraction of lyre strings, or hempen cords."[40] Eleven years later he referred to this passage and described

[33] See *Il Codice atlantico di Leonardo da Vinci . . . ridotto e pubblicato dalla Regia Accademia dei Lincei* [etc.], 6 vols. (Milan, 1894–1904), Vol. 1, Plate XXXIIIB, and Vol. 2, p. 177; also Vol. 3, Plate DCCXXXIV.

[34] *Ibid.*, Vol. 2, p. 177.

[35] Pierre Duhem, *Compt. Rend.*, Vol. 148 (1909), pp. 685–87.

[36] Page 45 above.

[37] *Sumimus tartarum combustum, quod a vulgo dicitur alumen faecis*. This was interpreted by F. Burkhardt, *Die Erfindung des Thermometers* [etc.] (Basle, 1867), p. 9, as "alum," but I think that it was the tartar that gathers in wine barrels.

[38] Santorio, *Commentaria in artem medicinalem Galeni*. Pars III (Venice, 1612), Cap. LXXXVI, Particula III, col. 1209–10.

[39] Santorio, *Medicina statica* (Venice, 1614), Sect. II, Aphorism 4.

[40] *Ibid.*

two sorts of hygrometer using cords,[41] and as these are the very first such instruments to be described fully, I shall quote him at some length:

We have also two ways of measuring the departure of moisture and dryness from the natural state. These are the ones we mentioned in the fourth aphorism of the second section of our *Statica*.

The first way is explained by the third figure [Fig. 3.1], in which a cord is stretched out, or a lyre string if you prefer, provided that it is a thick one. The string is fastened to a wall, or in some other place. In the middle a lead ball is fixed, and a scale may be drawn nearby. When the air gets moister the cord contracts, but when the north wind dries it up, it loosens. Sometimes air from the south wets and shortens the cord so much that the ball rises to the letter *A*, but while the north winds blow they dry it until the ball reaches *B*. In this way, even if no wind is blowing, the degree of moisture or dryness that there may be in the air can be observed every day.[42]

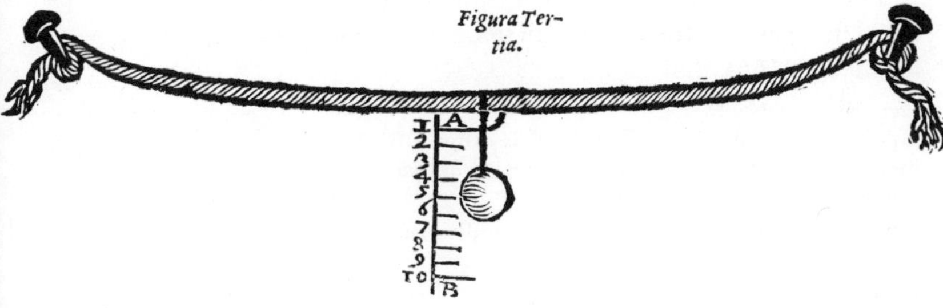

Fig. 3.1 The first string-hygrometer of Santorio.

His second hygrometer is less easy to believe in: "The second way is explained by the fourth figure [Fig. 3.2], which is patterned on a clock. I obtain a flaxen cord, very thick and long, because the thicker and longer it is, the better it will serve the purpose. The cord is shown at *C*. It is attached to a peg in the part behind, and therefore in humid air it turns the peg to the degrees that are marked, but when it is drying out in dry air, the cord loosens, and the peg moves to other degrees."[43] This one has had no offspring, in contradistinction to the first, which was the parent of a large family of hygrometers.

Santorio refers later in the book to a third hygrometer, a mysterious instrument about which he says only that it is "like a deep measuring cup, out of which comes a thread, from which is hung a leaden ball."[44]

[41] *Sanctorii Sanctorii . . . commentaria in primam fen primi libri canonis Avicennae* [etc.] (Venice, 1625), coll. 123–24.
[42] *Ibid.*, col. 23.
[43] *Ibid.*, cols. 23–24.
[44] *Ibid.*, col. 219.

Fig. 3.2 The second string-hygrometer of Santorio.

The first of Santorio's cord hygrometers had a good deal of development in Italy in the 1660's, first by Francesco Folli of Poppi, who learned in 1664 that

> His Highness the Grand Duke was looking into the way of making an instrument for distinguishing the degrees of moisture and dryness,[45] just as the thermometer had been invented a few years before. I certainly would not have thought of it myself, but on the following night I looked into the matter, and the next day I made one, and presented it [to Paolo Poltri]. This was in the year 1664. And when I came to live in Florence, which was in the year 1665, I presented one to his Highness my master, who was pleased to accept it and had a few made, which he immediately sent to several of the princes of Europe.[46]

This hygrometer had a strip of paper for the sensitive element; the middle point was pulled up by a weight acting on a cord that passed over a pulley. On the shaft of the pulley was a pointer moving over a scale. In the Museo di Storia della Scienza at Florence they have such an instrument (Fig. 3.3), believed to be the one that Folli gave the Grand Duke.[47] It is 72 cm long. An interesting feature is the use of a smaller idler pulley near the main one to increase the angle of contact. There are two other hygrometers of this general sort at Florence; one with a neat brass frame 28 cm long[48] may well be one of those that the Duke had made for his fellow princes.

Folli recognized that such an instrument would have a nonuniform scale, and so did Viviani.[49] In the remarkable collection of hygrometers at Florence, there are several ascribed to Viviani, some of

[45] This cannot have been the Grand Duke's condensation hygrometer (see below, p. 110), which had been invented nearly a decade earlier.
[46] Folli, *Stadera medica* (Florence, 1680), p. 114.
[47] Inv. no. 2434. The paper strip and the weight are missing.
[48] Inv. no. 2435.
[49] This is documented by Maria Luisa Bonelli, *Actes VIII Congrès Int. Hist. Sci.* (Florence, 1956), pp. 415–18.

Fig. 3.3 Folli's hygrometer (courtesy of Dr. M. L. Righini Bonelli).

Fig. 3.4 Viviani's hygrometer with a wooden frame (courtesy of Dr. M. L. Righini Bonelli).

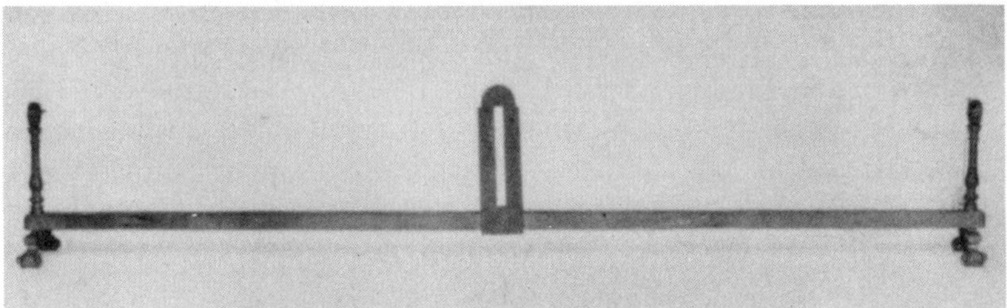

Fig. 3.5 The brass frame of another of Viviani's hygrometers (courtesy of Dr. M. L. Righini Bonelli).

which have uniform scales such as the large wooden one shown in Figure 3.4,[50] 2.60 meters long, and others nonuniform scales, as in the much smaller brass one (Fig. 3.5),[51] which has clamps at the ends to hold the paper strip. It will be seen that Viviani went back to the

[50] Inv. no. 799.
[51] Inv. no. 2437. There are two more, precisely similar, nos. 2438 and 2439.

simpler arrangement used by Santorio; Dr. Bonelli has found sketches of Viviani's showing such a construction.[52]

Another sort of hygrometer grew out of the observation that the awns or "beards" of various seeds twist or untwist as they become dry or moist. The common wild oat, *Avena fatua*, is an example, and children have no doubt amused themselves with this seed from time immemorial. Francis Bacon has a paragraph about a plaything of this kind,[53] in which he observes that such a seed can exert a considerable force. Emmanuel Maignan, or Magnanus, of Toulouse, whose account of Berti's experiments we have already heard about,[54] was the first, as far as I know, to describe and illustrate a hygroscope made with an oat beard;[55] he seems to have heard about it from Cardinal Giovanni Carlo de' Medici, who also figures in the early history of the barometer.[56] Maignan shows an elaborately decorated little instrument with its index under a glass top.[57] But that indefatigable globe-trotter Balthasar de Monconys records that two years before this, on November 6, 1646, he visited Torricelli in Florence, who among other things gave him "some wild oats with which to observe dry or humid weather."[58] Thus the common ascription of the instrument to Robert Hooke is incorrect, but Hooke improved it and tried to investigate its mode of action.[59] One of the difficulties with this instrument is that the free end of the awn makes more than one complete turn, a drawback circumvented in Hooke's instrument (Fig. 3.6) by the provision of a starwheel, advanced one tooth for each revolution of the pointer by a tiny pin fastened to the underside of the latter. Hooke also found other seeds, some even better than the wild oat, as for example "the beard of the seed of musk-grass, or *Geranium moschatum*."[60] He knew about the use of gut strings, but did not like them. In 1666, on June 27, Hooke "produced a new substance for a hygroscope, much stronger and better than the beard of a wild oat. It was the cod [pod] of a vetch, which was tried before the Society, and answered expectation."[61]

Hooke himself attributed the oat-beard hygrometer to Maignan, in his "*Method for making a History of the Weather*," published in 1667 by Thomas Sprat in his defense of The Royal Society.[62] It is also likely that someone mentioned Maignan's instrument at the meeting of

[52] Bonelli, *Actes VIIIe Congrès*, p. 415.
[53] Bacon, *Sylva Sylvarum* [etc.] (London, 1627), ¶494, p. 126.
[54] See p. 5 above.
[55] Maignan, *Perspectiva horaria* [etc.] (Rome, 1648), pp. 89–90.
[56] See p. 16.
[57] See also F. Sherwood Taylor, *Ann. Sci.*, Vol. 6 (1949), pp. 181–86.
[58] *Journal des Voyages de M. de Monconys* [etc.] (3 vols., Lyon, 1665 and 1666), I, 130.
[59] Hooke, *Micrographia* [etc.] (London, 1665), pp. 147–151.
[60] *Ibid.*, p. 151.
[61] Thomas Birch, *The History of The Royal Society* (4 vols., London, 1756–57), II, 100.
[62] Sprat, *The History of The Royal Society of London, for the improving of Natural Knowledge* (London, 1667), p. 173.

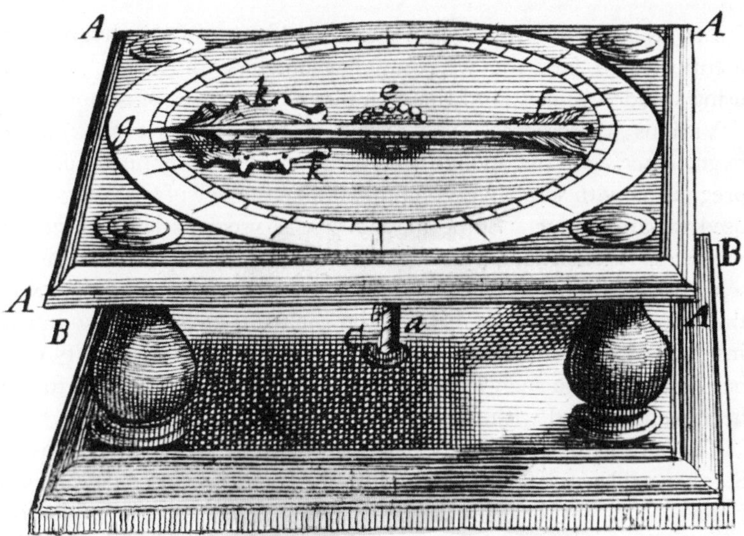

Fig. 3.6 Hooke's oat-beard hygrometer, from the Micrographia.

The Royal Society on October 7, 1663, for we read that "Mr. Hooke [who was the "curator of experiments" at that time] was ordered to bring in at the next meeting an hygroscope made of the beard of a wild oat, with an index."[63] Usually when an instrument devised by Hooke himself was under discussion, the phrase in Birch (and in the *Journal Books* from which it is compiled) would be "*his* [instrument]" rather than "*an* [instrument]."

On the same date "Dr [Jonathan] Goddard proposed for an hygroscope the contrivance of a lutestring with pullies and a cylinder; and was desired to try it."[64] This is a device that was reinvented several times, and the casualness of the passage suggests that hygroscopes may have been more common at that date than is generally believed. This impression is strengthened when we find the Accademia del Cimento in Florence, or rather Lorenzo Magalotti, its Secretary, introducing the Academy's hygrometer[65] with the remark that "although those that have been imagined at other times by divers clever people may be many and various," they are adding only one more.[66] It is known that, at least by 1656, Cardinal Giovanni Carlo de' Medici had a hygrometer at Rome in which a cord 8 or 10 *braccia* (4 or 5 meters) long, fastened at one end, passed horizontally to a pulley and was stretched by a weight. On the shaft of the pulley there was a pointer.[67]

[63] Birch, *History*, I, 311. [64] *Ibid.*

[65] See p. 110 below.

[66] *Saggi di naturali esperienze fatte nell'Accademia del Cimento* [etc.], (ed. princ. 1667) (3rd Florentine edition, 1841), p. 17.

[67] *Trattato della sfera di Galileo Galilei, con alcune prattiche intorno a quella . . . di Buonardo Savi* [anagram for Urbano Daviso] (Rome, 1656), pp. 194–196. The dedication to the Cardinal is dated March 20, 1656.

Let us now discuss briefly the later history of the hygrometers mentioned so far—the weighing hygrometers, the cords, the ones using wooden panels, and those employing the awns of various seeds.

Various hygroscopic substances were proposed for the balance hygrometer, but the sponge was a favorite,[68–70] sometimes after impregnation with some hygroscopic salt, like the one of Cruquius,[69] who used sal-ammoniac. Some ingenuity was used in devising suitable balances. Pickering,[70] who had been reading Boyle, made a balance sensitive to half a grain, with its axis extended and carrying a pointer that moved over a scale behind the shelter that he built round the instrument (see End Papers). It is far from clear how stability could be preserved with such large deflections, but Pickering seems to have been well satisfied with the instrument.

The problem of stability was attacked in a different way by Stephen Hales and J. T. Desaguliers,[71] who used a fusee, as shown in Figure 3.7. Another solution was the chain balance, a device that goes back at least to the time of Pascal. It was used by William Arderon,[72] with the slight difference that his chain was made of a row of fine split shot on a silk thread, "more or less of these being raised off a table or shelf" (Fig. 3.8). A chain balance of the modern type, in which equilibrium is reestablished by raising or lowering one end of the chain, which hangs in a loop, was adopted in Russia forty years

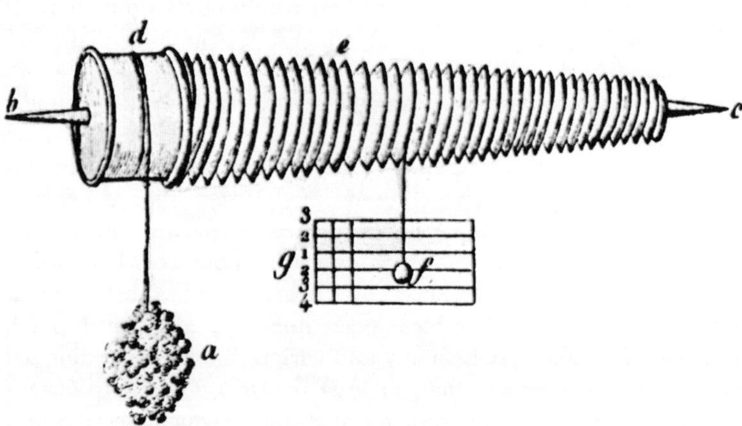

Fig. 3.7 The hygrometer of Desaguliers and Hales, 1744.

[68] Robert Boyle, *A Statical Hygroscope proposed to be farther tryed* [etc.] (London, 1673), pp. 1–7.
[69] Nicolaus Cruquius, *Phil. Trans.*, Vol. 33 (1724), pp. 4–7.
[70] Roger Pickering, *Phil. Trans.*, Vol. 43 (1744), No. 473, pp. 6–9. (No. 473 is erroneously numbered from page 1.)
[71] J. T. Desaguliers, *A Course of Experimental Philosophy* [etc.] (2 vols., London, 1734 and 1744), II, 299–300.
[72] Arderon, *Phil. Trans.*, Vol. 44 (1746), pp. 95–96.

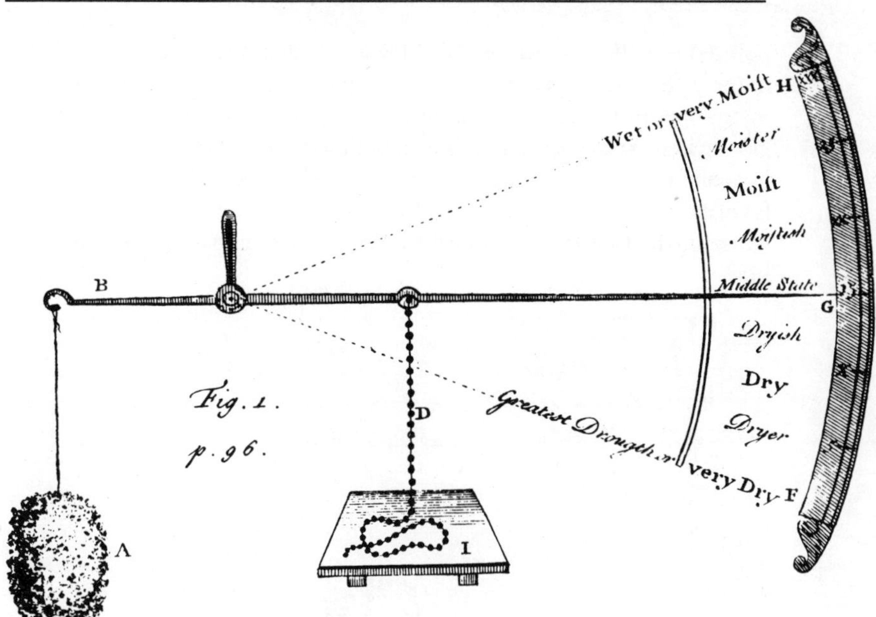

Fig. 3.8 Arderon's hygrometer, 1746.

later.[73] Instead of a sponge Inochodzow used a kind of stone, found on the left bank of the Volga, that absorbs and gives up moisture. The zero of the scale was given by the weight of the stone after it had been heated red hot, and the point of maximum humidity was determined by soaking it in water. Inochodzow gave credit for the chain balance to his fellow Academician G. M. Lowitz, who had died in 1774.

I shall leave these balance hygrometers with the note that in 1785 it was even proposed to use seaweed as an absorbing substance.[74]

The hygrometers or hygroscopes making use of the swelling of wood across the grain may be disposed of fairly briefly. In 1676 John Coniers or Conyers, an apothecary, described a huge instrument made of two pieces of deal each three feet square, fastened to a frame at their outer edges. The relative motion of their inner edges was transferred to a pointer by means that will be obvious from Figure 3.9.[75] An anonymous writer almost simultaneously suggested using a rack-and-pinion motion instead.[76]

Some very complex instruments of this class appeared, as for instance that of Teuber,[77] in which the swelling of four wide boards was accumulated by a complex system of levers. Pieter van Musschenbroek

[73] P. Inochodzow, *Acta Imp. Acad. Petrop.*, Vol. 2 (1778), pp. 193–204.
[74] De la Guerrand, *Lichtenberg's Mag.*, Vol. 3, part 2 (1785), pp. 159–60.
[75] Coniers, *Phil. Trans.*, Vol. 11 (1676), pp. 716–21. This instrument is also described and sketched in an entry dated April 15, 1675 in BM, Sloane ms. 958, fol. 119^{v}.
[76] Anon., *Phil. Trans.*, Vol. 11 (1676), pp. 647–53.
[77] Gottfried Teuber, *Acta Eruditorum* (1687), pp. 76–78.

pointed out the capital defect of these wooden hygrometers in 1739,[78] namely that the wood gradually becomes seasoned, and responds less to changes in humidity. So, he thought, do other substances, such as gut strings and cords, and he concluded that there were no good hygrometers. In spite of this pessimism the wooden hygrometer was favored by no less a man than Benjamin Franklin,[79] and the Science Museum in London has one[80] that is supposed to have belonged to him.

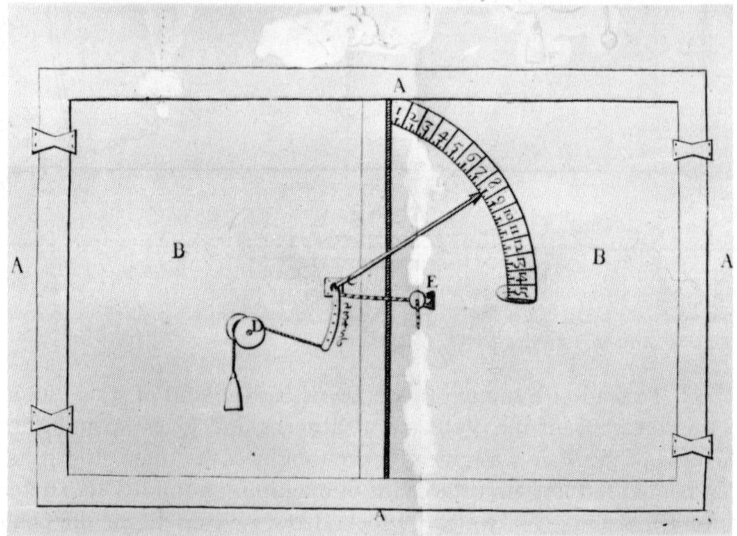

Fig. 3.9 Coniers' hygrometer, 1675.

There is another extremely simple type of wooden hygrometer, the origin of which I have not been able to trace beyond about 1770. At some time in the 1770's or 1780's the famous American instrument-maker and patriot David Rittenhouse made a hygrometer from two strips of mahogany glued together, one cut along the grain and the other across it.[81] A change in humidity alters the curvature of such a compound strip. At about the same time such an instrument was mentioned in England by J. H. Magellan,[82] who ascribed it to John Whitehurst, an eminent clockmaker of Derby who had been elected to The Royal Society.

There are two ways in which gut strings and cords can be used as indicators of humidity—either their changes in length can be measured, or else their twisting and untwisting. Robert Hooke was aware

[78] Musschenbroek, *Essai de physique* (tr. Pierre Massuet) (2 vols., Leyden, 1739), II, 699.
[79] Letter to Edward Nairne, dated "Passy, near Paris, Nov. 13 1780." In A. H. Smyth, *Writings of Benjamin Franklin*, Vol. VIII (New York, 1906), pp. 168–74.
[80] Inv. no. 1902–84.
[81] Brooke Hindle, *David Rittenhouse* (Princeton, N. J., 1964), pp. 84–85.
[82] Magellan, *Obs. sur la phys.*, Vol. 19 (1782), p. 353.

of both possibilities when he wrote the *Micrographia*,[83] but liked neither. I have already referred to the arrangements adopted by Santorio, which used changes in length. Of the large number of other schemes I shall illustrate only that of William Gould,[84] who described it as "made of a viol-string running upon pullies" (Fig. 3.10). As we have seen, Goddard had had the same idea twenty years before. Several variations of this, using bands of sheepskin instead of gut strings, were illustrated in 1686 by S. Foucher,[85] a Canon of Dijon, who dated his invention 1672. Foucher also had another hygrometer that seems to have been unique, and might as well be referred to here. Remembering the behavior of drum heads, he stretched a piece of parchment over a box and measured the motion of the center by means of a lever of large magnification. A small weight kept the parchment taut.[86]

Coming now to the hygrometers depending on the change in the twist of a cord, we find two distinct types, one having a rather long cord with a weight hung from it, the other a short piece of thick gut string fastened at one end and carrying a pointer at its other extremity. An instrument of the first kind was described by William Molyneux in the year of his election to The Royal Society.[87] A weight of about one pound, suitably marked, was hung from a length of whipcord (Fig. 3.11). After some experience with it, Molyneux shortened the string so that the greatest excursion of the index should be no more than one turn.

At this period the upper classes, especially in France, amused themselves by dabbling in the new natural philosophy and by collecting instruments. Like other furniture, these had to be ornate, and the *ébéniste* allied himself closely with the instrumentmaker, who, in France, was thought of merely as a workman and had no scientific standing.[88] Foucher makes suggestions for the decoration of a hygrometer:

> So we shall see Neptune coming out of the sea when humidity reigns, and hiding beneath the waves when the weather becomes dry. This Neptune will ride in a car drawn by six sea-horses, on the first of which will be a Triton with a trumpet, which will emerge first, after which the horses will appear, and then the car, inclined so as to reveal itself little by little, until Neptune is entirely visible . . . Sometimes only the end of his trident will be seen, and sometimes only the Triton or even merely the end of his trumpet. At this time the sun will come from behind a cloud and appear in the open as the

[83] Hooke, *Micrographia*, p. 151.
[84] Gould, *Phil. Trans.*, Vol. 14 (1684), pp. 504–6.
[85] Foucher, *Traité des hygrometres ou machines pour measurer la secheresse et l'humidité de l'air* (Paris, 1686), pp. 18–34.
[86] *Ibid.*, pp. 10–12.
[87] Molyneux, *Phil. Trans.*, Vol. 15 (1686), pp. 1032–35.
[88] On this, see Maurice Daumas, *Les instruments scientifiques aux XVIIe et XVIIIe siècles* (Paris, 1955), pp. 138–39.

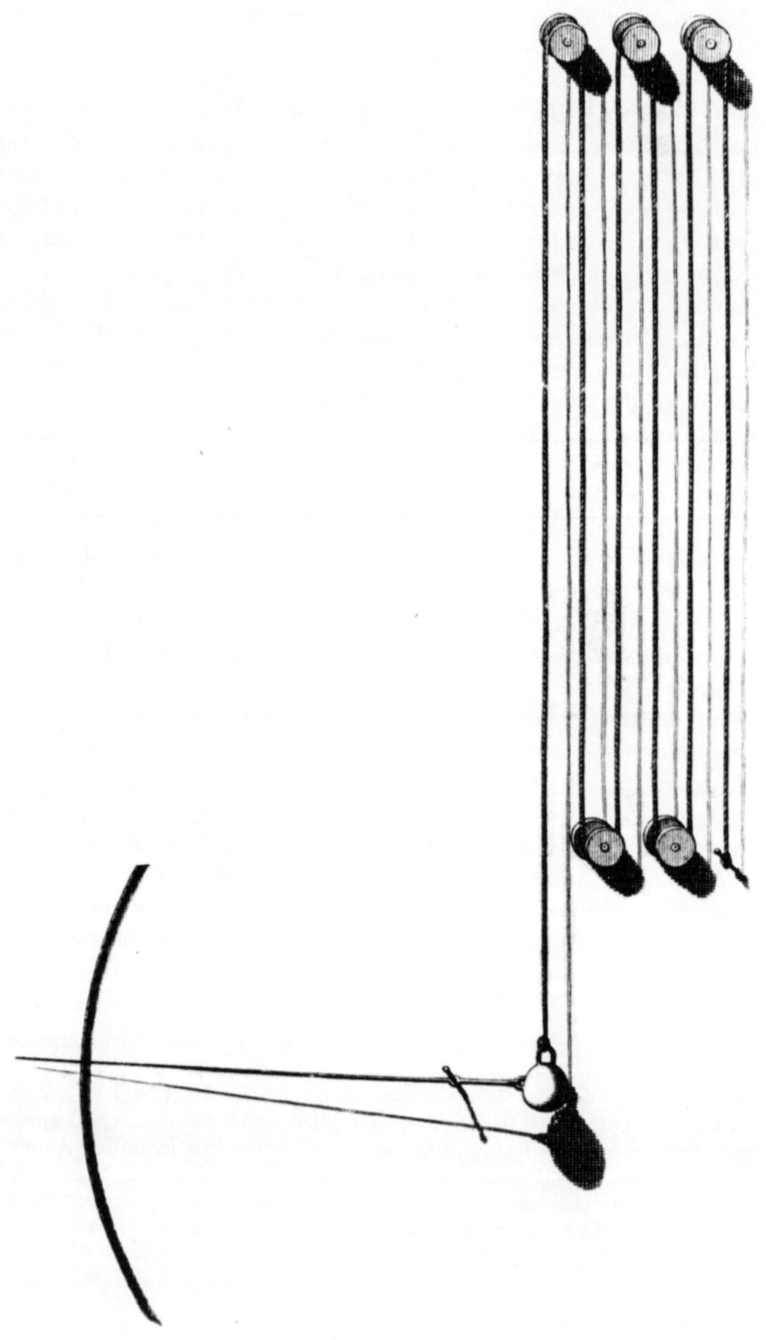

Fig. 3.10 Gould's viol-string hygrometer, 1684.

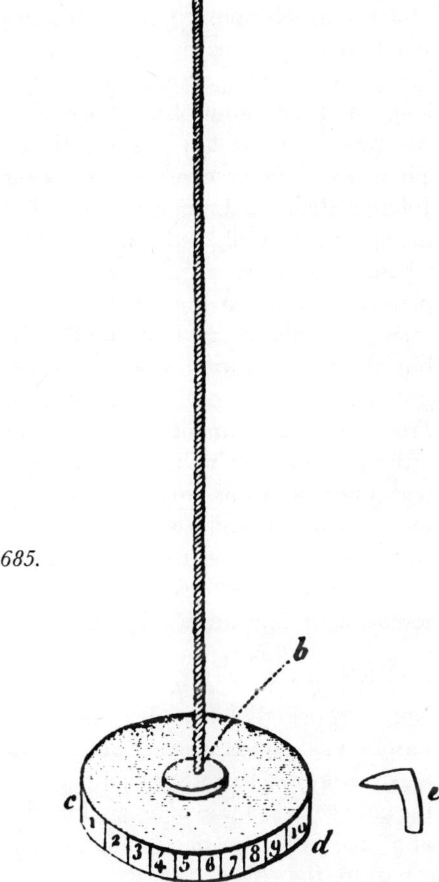

Fig. 3.11 Molyneux' hygrometer, 1685.

sign of an extreme dryness . . . it is well that Neptune should be represented in blue on a silver ground, and the sun in red against a gold background.[89]

The good Canon evidently knew his readers.

In the Hessisches Landesmuseum at Kassel there is an amusing hygrometer in which a model of a crocodile, nearly a meter long, is hung by a piece of rope from a simple wooden tripod about two meters high, and points its tail to a circular scale. This piece of garden furniture is assigned by the Director, Dr. Paul Kirchvogel, to the beginning of the eighteenth century.

[89] Foucher, *Traité*, p. 99.

The first description that I have found of the hygrometer with a short thick gut string was given by J. G. Leutmann,[90] who stated that he had used no other kind since 1708. The heavy viol string had previously been soaked for two or three days in a saturated solution of potassium carbonate (*cineres clavellati*), or in the liquid which comes out of this substance when it is left for some time in a damp cellar, or in a strong solution of Glauber's salts, and dried while supporting a small weight. This hygrometer was immensely popular, probably because of its compactness—the string need be only about one inch long—and thousands of them are to be found mounted on decorative barometers of the late eighteenth and early nineteenth centuries. They would deserve no further remark except that the celebrated Johann Heinrich Lambert took it into his head that they could be made to work well, and wrote a good deal about them.[91] Lambert, whose interests were mathematical rather than meteorological, explained how the twisted cord alters its angle of twist as it is made thicker by absorbing moisture.[92] He made extensive experiments, but did not seem to be worried by the appalling sluggishness of these instruments. He did not impregnate the strings with any salts. From such an eminent man the following idea seems strange: "In order that air may be humid, it is not enough that it should be loaded with many aqueous particles, but these particles must be coagulated into little drops, and these drops must attach themselves to the bodies that they touch. In this respect hygrometers indicate not so much the quantity of aqueous particles floating in the air as their disposition to coagulate and attach themselves to objects."[93] He persisted for a long time, bothering the important Augsburg instrumentmaker Georg Friedrich Brander in his attempt to have them made comparable.[94] Not surprisingly, Brander found that even when he followed Lambert's detailed instructions he was not successful.

An entirely different type of hygroscopic hygrometer, depending on the change in volume of a vessel made of some hygroscopic substance, appeared in 1687, when Guillaume Amontons showed such an instrument to the Académie Royale des Sciences.[95] This was almost certainly derived from the three-liquid barometer of Hooke,[96] and was made by Hubin, who was taken by Amontons to the meeting. As shown in the diagram (Fig. 3.12), a leather bag is tied to one end of a glass apparatus and filled with mercury. The changes in the volume

[90] Leutmann, *Instrumenta meteorognosiae inservientia* (Wittenberg, 1725), pp. 113–115.
[91] Lambert, *Hist. Acad. r. Sci. Berlin* (1769), pp. 68–127; *Nouv. Mém. Acad. r. Sci. Berlin* (1772), pp. 65–102. These essays were translated into German as *Hygrometrie, oder Abhandlung von den Hygrometern* (Augsburg, 1774).
[92] Lambert (1769), p. 97.
[93] Lambert (1772), p. 96.
[94] Their correspondence is in *Joh. Heinrich Lamberts deutscher gelehrter Briefwechsel* (ed. J. Bernouilli, 5 vols., Augsburg, 1782–84), *passim*.
[95] *Hist. et Mém. Acad. Sci. de 1666 à 1699* (11 vols., Paris, 1733), Vol. 2, p. 22.
[96] See p. 26 above.

of the bag with changes in humidity are indicated by the position of the interface between two immiscible liquids in a narrow tube. In the next decade Amontons modified this instrument by substituting for the leather bag a wooden chamber with one side made of thin horn.[97]

This sort of hygrometer was revived by J. A. Deluc in 1773, and somewhat simplified by the use of one liquid, mercury, instead of three.[98] In spite of his immense paper about it, Deluc quickly came to see the defects of such an instrument and abandoned it for an entirely different hygrometer that will be discussed later. The bulb

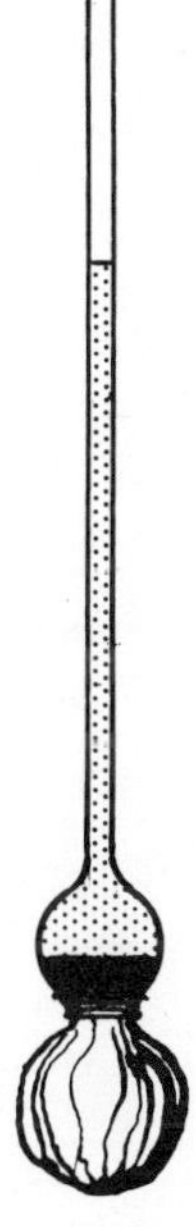
Fig. 3.12 Amontons' hygrometer, 1687.

of his instrument of 1773 was machined out of ivory; this was a tricky business, for it had to be very thin. A simpler substitute, the end of a goose quill, was used by Noël Retz, a physician of Arras,[99] who seems to have been far from dissatisfied with it.

In 1778 Jean Senebier, the city librarian of Geneva, said very truly that "we have had hygroscopes, and not hygrometers."[100] The truth of this statement and the eagerness with which a hygrometer was being sought are underlined by contemporary events at Mannheim. Encouraged by his chaplain Johann Jacob Hemmer, the Elector Palatine founded a meteorological society, the Societas Meteorologica Palatina, which in 1780 began to send sets of meteorological instruments to stations all over Europe, and even to a few places in America.[101] Hemmer, after considering Lambert's hygrometer, picked the one devised by Retz, and gave an elaborate description of its calibration.[102] Not unnaturally, there were arguments about this choice, and in 1781 this led the Academy to offer a prize for a hygrometer that had the possibility of calibration and a stable sensitivity. Eleven replies were received, and the prize was awarded jointly on October 20, 1784, to Giuseppe Toaldo of Padua and his nephew Vincenzio Chiminello.[103] And what was the winning instrument, described at enormous length by Chiminello?[104] Nothing but the hygrometer of

[97] Amontons, *Remarques et expériences phisiques sur la construction d'une nouvelle clepsidre, sur les baromètres, termomètres, & higromètres* (Paris, 1695), pp. 160 ff.
[98] Deluc, *Phil. Trans.*, Vol. 63 (1773), pp. 404–60.
[99] Retz, *Traité d'un nouvel hygromètre comparable, imité de celui de M. de Luc* [etc.] (Paris, 1779). Deluc (*Phil. Trans.*, Vol. 63 [1773], p. 450) had suggested that a quill might be used.
[100] Senebier, *Obs. sur la Phys.*, Vol. 11 (1778), p. 421.
[101] See Friedrich Traumüller, *Die Mannheimer meteorologische Gesellschaft (1780–1795). Ein Beitrag zur Geschichte der Meteorologie* (Leipzig, 1885).
[102] *Ephemerides Societatis Meteorologicae Palatinae. Historia et observationes anni 1781* (Mannheim, 1783), pp. 72–80.
[103] *Acta Acad. elect. sci. et eleg. litt. Theodoro-palatinae*, Vol. 6, pars physica (1785), pp. 8–9.
[104] Chiminello, *Giorn. enciclop. di Vicenza* (1785); Sept., pp. 97–114; Oct., pp. 81–114; Nov., pp. 81–109. It was also published as a pamphlet of 81 pages (Vicenza, 1785).

Retz, accompanied by elaborate and rather fantastic recipes for calibrating it and separating out the effects of temperature—for of course the instrument acted as a thermometer as well. One can only conclude that there had been a political struggle in the Academy, and that Hemmer had won.

We cannot help being surprised at the enthusiasm for hygrometers of this type shown even by such important people as Giovanni Battista Beccaria, who, in the intervals of his electrical researches, devised one for G. B. Lovizolo to construct and calibrate.[105] Beccaria's idea was to use a rye straw, attached to a capillary glass tube and filled with mercury. Lovizolo confesses disarmingly in a footnote, "I do not know whether an oat straw might be better." The enthusiasm was not shared in Geneva, where the hair hygrometer was being developed by H. B. de Saussure,[106] and we find Jean Senebier, the city's librarian, sending his observations to Mannheim with the note that "the Mannheim hygrometer deserves no confidence, if its variations are compared with those of Mr. De Saussure."[107] Hemmer, somewhat nettled, objected in a long footnote: "This author seems to assume De Saussure's hygrometer to be exact. But however many splendid advantages this instrument enjoys, we are a long way from believing it to be absolutely accurate in all its readings."[108]

In the succeeding half century there were several analogous hygrometers, including a revival of Deluc's ivory bulb by Sir John Leslie,[109] the most exotic being that of David Wilson of Dublin, who used the urinary bladder of a rat,[110] and later[111] provided the startling observation that London rats are "very subject to urinary calculi, which I do not find to be the case in other towns." Perhaps because of the rat-race?

In the years between 1780 and about 1830 an extraordinary number of hygrometers appeared, for most of which the reader may be referred to a dissertation by R. G. Bunsen.[112] Even the seed-awns had their day. Lieut. Henry Kater, an engineer serving with the British Army in India, was interested in the effect of humidity on terrestrial refraction, and in Mysore he found a grass, *Andropogon contortum* Linn., with a seed very suitable for use as a hygrometer.[113] He made and used a rough instrument. An elaborate hygrometer using this seed, with two dials and appropriate gearing, was made and sold by T. C.

105 Lovizolo, *Opuscoli scelti*, Milan, Vol. 4 (1781), pp. 253–69.

106 See page 101 below.

107 *Ephem. Soc. meteorol. Palatinae* for 1783 (Mannheim, 1785), p. 214.

108 *Ibid.*, p. 215, note.

109 Leslie, *Annals of Philos.*, Vol. 1 (1813), pp. 468–69.

110 Wilson, *Annals of Philos.*, Vol. 8 (1816), p. 154. There is one of these hygrometers at the University Museum, Utrecht (Inv. no. W.70).

111 *Ann. Philos.*, Vol. 9 (1816), p. 319*n*.

112 *Bunsen, Enumeratio ac descriptio hygrometrorum quae inde a Saussurii temporibus proposita sunt* (Göttingen, 1830).

113 Kater, *Asiatic Researches of the Asiatic Soc. of Bengal*, Vol. 9 (1809), pp. 24–31; 394–97.

Robinson of 38 Devonshire Street, Portland Place, London.[114] I reproduce Bunsen's figure of this (Fig. 3.13).

By this time there was no need of such peculiar hygrometers. Even while Hemmer was looking for a suitable instrument for his network of stations, Horace Benedict de Saussure was making his remarkable experiments with the human hair as a hygrometric element.[115] It is scarcely surprising that no one before De Saussure's time had thought of using such a fragile fiber as a hair. The idea seems to have been a pure intuition: "The invention of a comparable hygrometer had been one of my main occupations; I had tried various substances and methods, but nothing had satisfied me, when in 1775 I had the idea of using the hair in the construction of this instrument."[116] He wrote to Senebier about these researches, and Senebier sent the letter to the Paris journal *Observations sur la Physique*.[117] It was

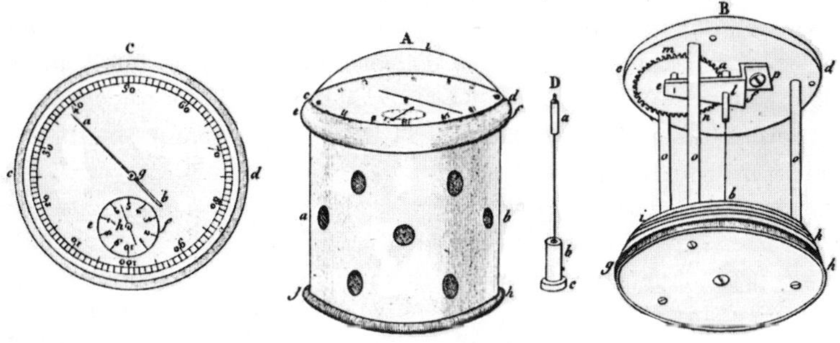

Fig. 3.13 Kater's hygrometer as made by Robinson about 1820.

a very modest and self-depreciatory letter, telling first how he degreased the hair by boiling it for a quarter of an hour in five or six ounces of water containing ten or twelve grains of caustic alkali and then how he obtained the point of extreme humidity by putting the hygrometer under a bell-jar of which the sides and bottom were wet. For the point of greatest dryness he took a piece of sheet iron, heated it red-hot, and sprinkled both sides with a mixture of powdered niter and cream of tartar; then, after letting it cool a little, he enclosed it in a dry jar with the hygrometer. He investigated the effect of changing the temperature at these extremes of humidity. In this way, he says, "By these different procedures I have obtained hygrometers which

[114] *Cf.* E. G. R. Taylor, *The Mathematical Practitioners of Hanoverian England, 1714–1840* (Cambridge, 1966), pp. 433–34. There are examples in the Science Museum, London (Inv. no. 1893–140) and in the University Museum, Utrecht (Inv. no. W82).

[115] De Saussure, *Essais sur l'hygrométrie* (Neuchâtel, 1783).

[116] *Ibid.*, preface, p. VI.

[117] Vol. 11 (1778), pp. 435–39.

move quite comparably, which are in no way thermometers, and of which the sensitiveness is as great as it is possible to desire."[118]

Unfortunately after a lapse of several months, or when the hair is exposed to a very dry atmosphere for a long time, it changes its calibration to some extent, and this, De Saussure admits, is a serious limitation to its use. He interprets its behavior as an indication that the hair holds two sorts of humidity, that which it picks up from the atmosphere, and that which is one of the constituent parts of the hair itself. In 1778 De Saussure certainly did not feel that he had solved the problem in a satisfactory manner. But in the winter and spring of 1781 he worked very hard at the hair hygrometer, and succeeded. ". . . I had the good fortune to discover the cause of the defect that had made me abandon [hair hygrometers], to find a remedy for that defect, and to determine with much precision the points of extreme humidity and dryness that I had seen the possibility of in 1776. Finally I gave these instruments a convenient and portable form."[119]

Apparently the source of the improvements was mainly a careful choice of healthy hair—that which is sold by the hospitals to the wigmakers was unsatisfactory—combined with care not to overdo the degreasing. He substituted sodium carbonate for sodium hydroxide, and found ways of ensuring that the hairs were equally exposed to the treatment along their whole length. He also found it necessary to avoid stretching the hair at any time, and to limit the tension under which it was kept in the instrument to three grains (about 200 mg).

De Saussure's two instruments are shown in Figure 3.14; the original one is on the right and the simpler and much superior portable one is on the left. They were made for him by the excellent instrumentmaker Paul of Geneva, and copied later by many other makers. There are a great number of the portable type in various museums, but the only one of the larger type that I can remember having seen is signed "Paul à Genève," and is in the Museo di Storia della Scienza at Florence,[120] but is incomplete.

De Saussure wanted to be sure that the purchaser of a hair hygrometer would not accidentally stretch the hair when taking the instrument out of its box, and there exists among his papers a corrected proof of a notice that was sent out with Paul's instruments, the illustration of which is clear enough (Fig. 3.15).[121]

The *Essais sur l'hygrométrie* contain much more than a description of his instruments, and their calibration. De Saussure made extensive experiments on the vapor pressure of water at various temperatures, not only in air but also in hydrogen and in carbon dioxide, and gave tables of water content in grains per cubic foot at a number of temperatures and for various readings of his hygrometer. Finally, in the

[118] *Ibid.*, p. 437.
[119] De Saussure, *Essais*, preface, p. VII.
[120] Inventory no. 5.
[121] Geneva, *Bibl. publ. et univ.*, Fonds De Saussure No. 56, envelope 4.

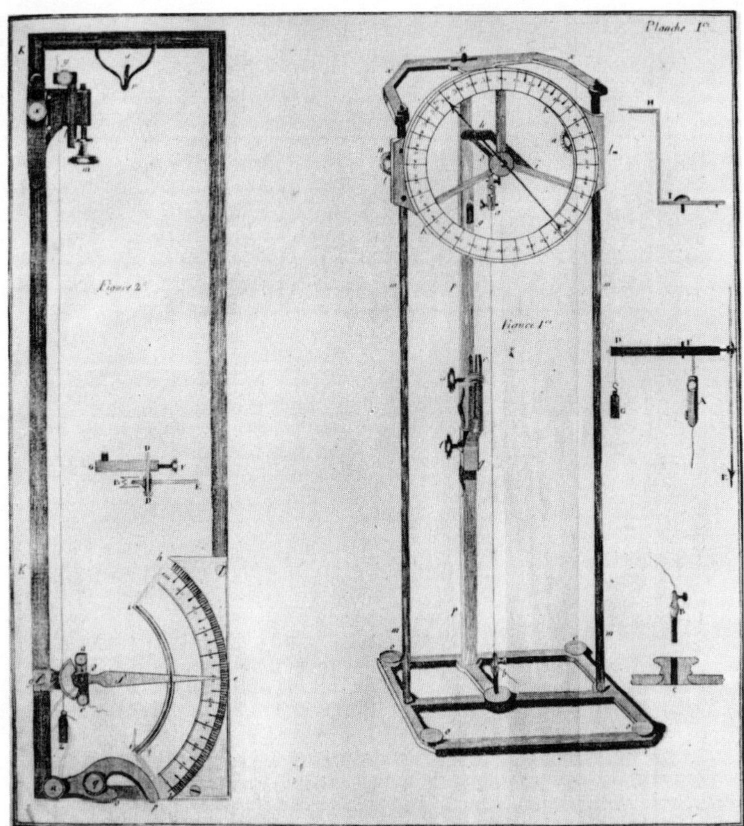

Fig. 3.14 De Saussure's hair hygrometers.

fourth and last "essay," he had a great deal to say about the role of water vapor in the atmosphere.[122]

Having thought of using the hair, De Saussure seems to have searched no further, a fact that was emphasized by the most persistent of his critics, Jean André Deluc, who had indeed tried many different substances for use in hygrometry, fastening at last on whalebone strips cut across the grain, or as he called them, "slips" of whalebone.[123] Apart from a much lower speed of response, this was probably a fairly useful choice, but Deluc made one decision that caused him and the science of meteorology a good deal of trouble. He got his extreme degree of moisture by immersing the whalebone in water, and categorically denied the validity of the process used by De

[122] *Cf.* Middleton, *A History of the Theories of Rain* (London, 1965; New York, 1966), Chapters 3 and 7.

[123] Deluc, *Idées sur la météorologie* (2 vols., London, 1786, and an almost identical edition, Paris, 1787), ¶¶31–98. (The paragraph numbers correspond in the two editions.)

AVIS

Qu'il convient de lire avant de ſortir l'Hygromètre de ſon étui.

Lorsqu'on veut mettre l'hygromètre en expérience, il faut commencer par le retirer de ſon étui, en prenant bien garde de ne toucher ni au cheveu, ni à l'aiguille, ni à la vis *C* qui eſt au haut de l'inſtrument.

Il faut enſuite le ſaiſir de la main gauche, en le tenant dans une ſituation verticale avec les doigts placés comme ils le ſont dans cette figure. Le doigt index doit s'appuyer légérement ſur l'aiguille & l'empêcher de ſe mouvoir pendant qu'on la dégage de la pince qui la tenoit aſſujettie, de peur qu'en retirant cette pince on ne donne à l'aiguille quelque grand mouvement qui puiſſe rompre ou tirailler le cheveu. Tandis que le doigt eſt dans cette poſition, il faut avec la main droite relâcher d'abord la vis *A* qui tient la pince ſerrée, puis la vis *B* qui aſſujetti cette même pince ; & retirer enſuite tout doucement le corps entier de la pince, de manière qu'en tournant ſur la vis *B* comme ſur un axe, elle ſe trouve au bas de l'inſtrument dans la poſition marquée par des points : on l'aſſujettit dans cette poſition en ſerrant la vis *B*, & l'aiguille étant alors parfaitement libre, on peut faire uſage de l'hygromètre.

De même lorſque l'on veut le remettre dans l'étui pour le tranſporter, il faut auſſi le tenir dans la ſituation que repréſente cette figure, abaiſſer l'aiguille avec le doigt juſques à ce qu'elle ſe trouve vis-à-vis du 40 ou 50e degré, afin que ſi le cheveu venoit à ſe deſſécher dans l'étui, il eût la liberté de ſe contracter. Tandis que le doigt tient ainſi l'aiguille aſſujettie, il faut tourner la pince juſques à ce que les deux becs ſupérieurs arrivent près de l'axe de l'aiguille, comme on les voit dans la figure, & embraſſent ainſi la double poulie autour de laquelle ſe roulent le cheveu & la corde du contre-poids. Il faut avoir attention que, dans le même tems, le contre-poids ſe loge entre les deux becs inférieurs de la pince. On ſerre alors la vis *A*, & enſuite la vis *B* juſqu'à ce que l'on ſente que l'aiguille, le contre-poids & la pince ſont ſolidement fixés.

Fig. 3.15 Instructions for the hair hygrometer (courtesy of the Bibliothèque publique et universitaire, Geneva).

Saussure.[124] Since Deluc marked his hygrometers (Fig. 3.16) 100 "degrees" at the point reached in water, whereas in saturated air they only attained about 80, it was not surprising that the two scales differed a good deal. Deluc, who really had worked hard on the

[124] *Ibid.*, ¶¶38–49.

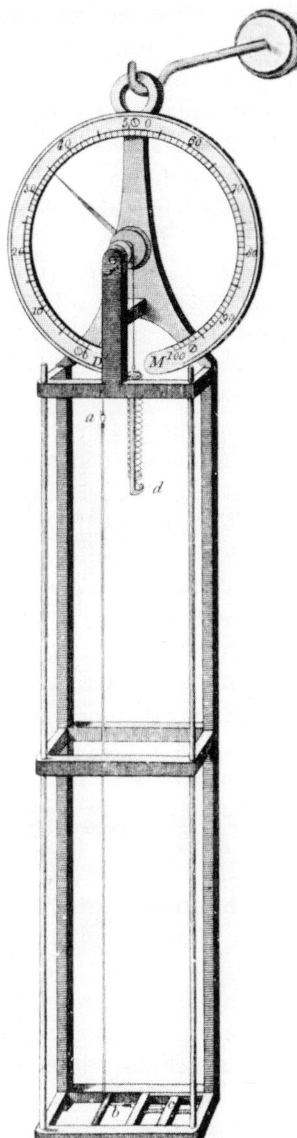

Fig. 3.16 Deluc's whalebone hygrometer.

subject, tended to pontificate, and his book cannot have been pleasant reading for De Saussure, who was equally sure of the superiority of his instrument and his methods. Deluc's criticism overcame the accustomed urbanity of his opponent, who lost no opportunity of publicizing his side of the argument. He wrote a defense of his instrument and sent it to the editor of the *Observations sur la Physique*, who printed it in its considerable entirety.[125] Not content with that,

[125] Vol. 32 (1788), pp. 24–45; 98–107.

he had it published separately in Geneva.[126] He sent a copy of the manuscript to Sir Joseph Banks with a covering letter[127] expressing his very strong feelings. The matter was even aired in the *Journal de Paris*.[128] According to Deluc, De Saussure went so far as to write "Before long I shall make it apparent how many of the objections of Mr. Deluc against this way of obtaining the extreme humidity are without foundation, and to what extent his new hygrometer is vicious and misleading."[129]

Deluc's feelings were indeed hurt, and he reproached De Saussure with "pronouncing against my hygrometer the laconic sentence that is now passing from mouth to mouth all over Europe";[130] surely an exaggeration. In his *Défense* De Saussure had made a further cutting remark, namely that the only three physicists who had attacked his *Essais* were inventors of hygrometers that differed from his.[131] As far as Deluc was concerned this was grossly unfair, for in his scientific work he was as disinterested as his fellow Genevan. He stuck to his guns, and although time has proved De Saussure to have been right, Deluc clearly had the better temper.

I shall not attempt to give the details of this controversy, which extended into the realms of meteorological theory,[132] and in which Deluc exhibited at enormous length his very considerable powers of self-deception. It is pleasant to report that he also went out of his way to praise the experimental skill of De Saussure, and his numerous meteorological observations. De Saussure was indeed an experimenter of great distinction, even in comparison with Lavoisier, Priestley, and Cavendish. His touchiness on this occasion may have been exacerbated by his bad health, which in December, 1785, led him to resign his professorship at Geneva.

Deluc's whalebone hygrometer was manufactured by many of the best instrumentmakers and is widely represented in European museums. Luke Howard selected it, largely because of its durability, for his famous observations in London, but he got his 100 "degree" point by exposing it "under a close glass to the vapour of water"[133]—De Saussure's method! John Farrar, the Hollis Professor of Mathematics and Natural Philosophy at Harvard, used it for six years (1800–06) in Cambridge, Massachusetts.[134] Alexander von Humboldt characteristically took several of each on his great expedi-

[126] *Défense de l'hygromètre à cheveu, pour servir de suite aux essais sur l'hygrométrie* (Geneva, 1788).
[127] BM, Add. MS. 8097, foll. 113ʳ–114ʳ.
[128] Deluc, *J. de Paris*, Dec. 12, 1787, p. 1489; De Saussure, *Ibid*, Jan 4, 1788, p. 18.
[129] *Vicieux et trompeur*. Quoted by Deluc, *Obs. sur la Phys.*, Vol. 31 (1787), p. 377. I have not found the original.
[130] *Obs. sur la Phys.*, Vol. 32 (1788), p. 139.
[131] *Défense*, p. 2.
[132] *Cf.* my *A History of the Theories of Rain*, Chap. 6.
[133] Howard, *The Climate of London* (2 vols., London, 1818–20), I, xix.
[134] Farrar, *Mem. Amer. Acad. Arts & Sci.*, Vol. 3, part 1 (1815), p. 385.

tion to South America, and found the hair hygrometers, made by Paul of Geneva, very stable in calibration for a period of three years.[135] He checked the 100-degree point of both the hair and the whalebone instruments by wetting them with liquid water. In spite of tradition, this makes little difference as far as the hair is concerned.[136] He complained about the sluggishness of the whalebone hygrometer in dry air, but seemed to think that it was very good near the saturation point, and praised it for the solidity of its construction and because it can be used in any position and is not troubled by wind.

Returning now to De Saussure's *Défense*; his two other critics were Chiminello, whom we have already met, and Giovanni Batista da San Martino, a Capuchin friar, who made a hygrometer out of the *tunica vellosa*, the innermost membrane of the intestines of an animal.[137] In view of this, Batista has been credited with the invention of the goldbeater's skin hygrometer,[138] but goldbeater's skin is prepared from the *outermost* membrane of the cecum of the ox. The first use of goldbeater's skin that I have found is in a publication of Marsiglio Landriani, which was posthumous and therefore not properly datable.[139] Landriani took a strip of goldbeater's skin, boiled it for a few minutes in a solution of soda, and then rolled it up into a cylinder about the size of a horsehair. He mounted it in an instrument much like that of De Saussure, but tried to make it indicate either minimum or maximum humidity by a mechanical variation.

De Saussure dismissed Giovanni Batista's hygrometer, which he referred to as being of goldbeater's skin (*baudruche*), with very little ceremony,[140] pointing out quite correctly that its inventor had not tested it for long enough, and treating the friar's rather silly objections to the hair hygrometer as they undoubtedly deserved. It is interesting that a century and a half later the speed of response of goldbeater's skin should have caused it to be preferred for use in meteorographs for the investigation of the upper air, until in its turn it gave way to the electrical hygrometer in such applications.[141] Meanwhile, it had been tried in hygrographs by the Paris firm, Richard Frères, as we are told by Colonel Sébert in his report to a French society.[142] "We may hope," he wrote, "that these instruments will keep their sensitivity for

[135] A. von Humboldt and A. Bonpland, *Voyage aux régions équinoxiales du nouveau continent* [etc.] (3 vols., Paris, 1814), I, 242–43.

[136] *Cf.* L. H. G. Dines, *Quart. J. Royal Meteorol. Soc.*, Vol. 68 (1942), pp. 260–61.

[137] Giovanni Batista [*sic*], "Saggio sopra un igrometro a tunica vellosa," *Opuscoli scelti sulle scienze e sulle arti* (Milan), Vol. 8 (1785), pp. 281–88.

[138] As by the editor (J. H. Voigt) of the *Magazin für die Neueste aus der Physik*, Vol. 6 (1789), Part 1, pp. 99–102, who interpreted *tunica vellosa* as *Goldschlägerhaut*.

[139] Landriani, *Giornale di Fisica* (Pavia), Decade 2, Vol. 3 (1820), pp. 111–16. Posthumous papers by Landriani appear in that journal from 1816 onward.

[140] *Défense*, pp. 55–60.

[141] See Chapter 10.

[142] Richard Frères, *Notice sur les instruments enregistreurs construits par Richard Frères, comprenant le rapport de M. le Colonel Sébert à la Société d'Encouragement pour l'Industrie Nationale, et l'exposé des perfectionnements et applications nouvelles* (Paris, 1889), p. 11.

a sufficient time; for a hygrometer of this sort made by the Richards in 1878, and which moves a pointer 35 cm long, always gives the same results and the same sensitiveness, although it has been taken at various times from extreme humidity to absolute dryness."

This report, so proudly reprinted, had been written in 1882, and the firm was using goldbeater's skin in their standard hygrographs in 1884,[143] but by 1889 this seems to have been abandoned in favor of an instrument (Fig. 3.17) employing a slip of cow's horn "cut in a certain direction and a twentieth of a millimeter thick,"[144] reminding us of Deluc and his slip of whalebone. They admitted that it was "a little lazy," but did not think this important where mean values were desired, "as in ordinary meteorology." They also advertised their hygrograph using a bundle of hairs, with correcting cams to make the scale linear in relative humidity—the first hair hygrograph.

Naturally the hair hygrometer suffered many other modifications, one of the first being due to the Paris instrumentmaker Jean François Richer or Riché, "brév
eté Ingénieur du Roi." In 1789 he made a hygrometer (Fig. 3.18) using eight hairs in such a way that, if the

Fig. 3.17 Richard Frères' hygrograph, 1889, using a slip of cow's horn.

levers did not stick, each would be under the same tension.[145] There is one of these instruments in the Science Museum, London (Inv. no. 1876–794), signed by Richer but having the system of levers at the top.

[143] *Zeits. Instrum.*, Vol. 4 (1884), p. 64.
[144] *Notice* (1889), p. 92. This had been patented (French Pat. 156,047) in 1883.
[145] Sage, *Obs. sur le Phys.*, Vol. 34 (1789), pp. 58–59.

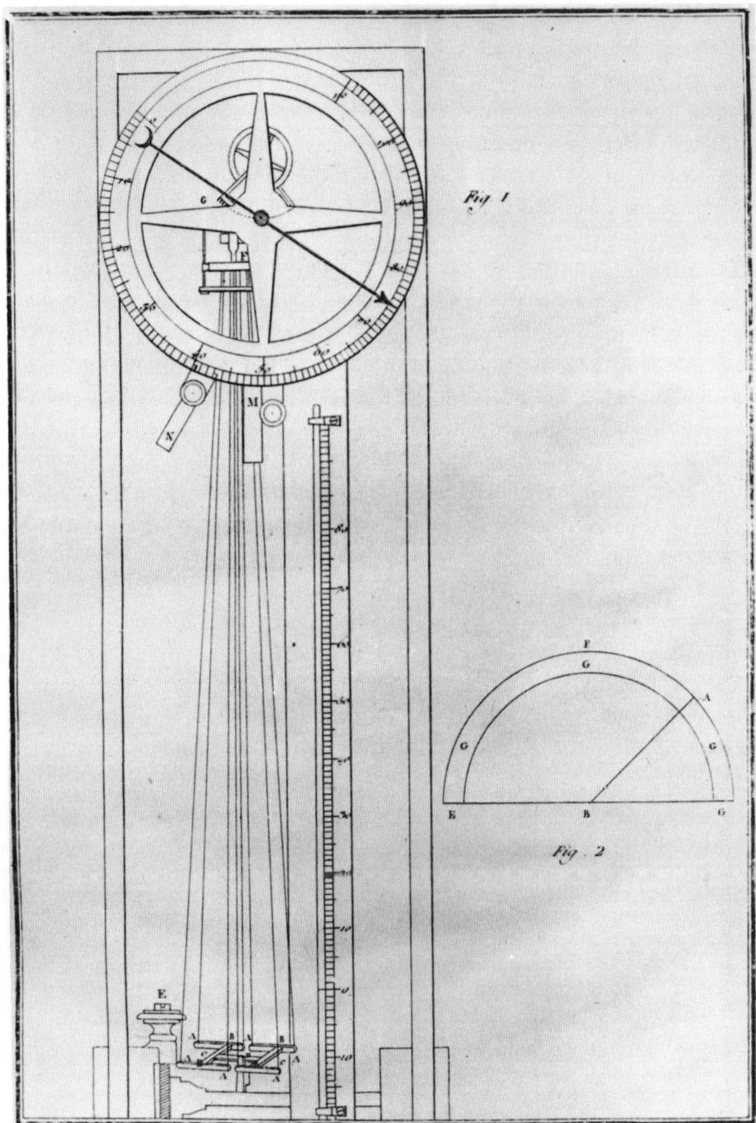

Fig. 3.18 Richer's hygrometer with eight hairs, 1789.

Of numerous later forms I shall mention only the one presented to the Paris Academy of Sciences in 1824 by Jacques Babinet.[146] It has three hairs hanging side by side, and each supporting a separate small weight bearing a fiducial mark (Fig. 3.19). The upper ends of the hairs are fastened to a bar that is movable by means of a micrometer screw, so that any of the three marks can be made to coincide

[146] See *Ann. Chim. et Phys.*, Vol. 26 (1824), pp. 367–70.

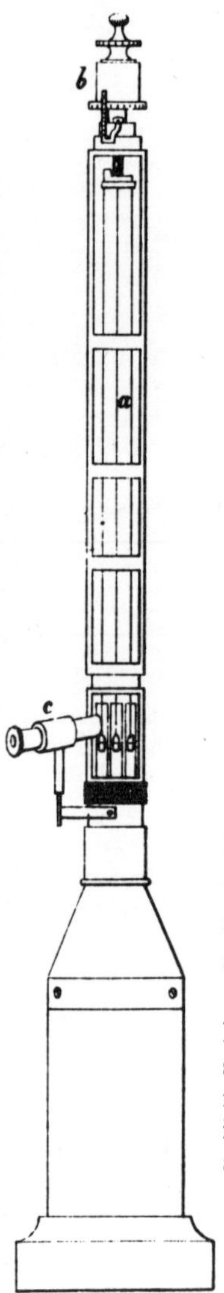

with the cross-hairs in the microscope. A glass cover is provided, which can be used to obtain a dry or a saturated atmosphere for calibration, for in the base of the instrument there is a vessel to hold a drying agent, or else some water. The Academy gave its approval, but such a design is misconceived, because the indications of the hair hygrometer are not precise or repeatable enough to justify such an elaborate construction.

As to what these indications actually mean, a satisfactory answer was provided in 1815 or 1816 by Gay-Lussac, who compared them with direct determinations of atmospheric water vapor by chemical methods,[147] giving the results in percentages of saturation. The instrument is more nearly logarithmic than linear. This nonlinearity was believed in some quarters to be a great disadvantage,[148] but nevertheless after about 1820 the hair hygrometer came to be the only hygroscopic hygrometer that was widely used for serious meteorological work. For the busy observer its best point was that it gave a *reading* at once, while the other types of hygrometer, which I shall now discuss, demanded an *experiment*.

4. Condensation hygrometers. Some time in 1655 Ferdinand II, Grand Duke of Tuscany, who was no mean natural philosopher,[149] noticed the condensed moisture on the outside of his drinking glass. The second of the contemporary "notes" of his experiments tells us the sequel:

N°. II. It was observed that when something iced was put in a glass, the surrounding air seemed to change itself into water; whence he thought of filling a covered vessel with ice, and when this was done the air was converted into water. This accomplished, he cast about for an easier way of doing it, and this was to make the instrument *D* [fig. 3.20], which was a hollow glass

Fig. 3.19 Babinet's hygrometer, 1824.

vase coming to a fine point at one end; it was filled with crushed ice, and held in a wooden tripod stand, with a beaker *E* below. The air that entirely surrounded it began to change into water, which dropped into the beaker from the pointed part.[150]

[147] J. B. Biot, *Traité de physique expérimentale et mathématique* (4 vols., Paris, 1816), II, 199–200. The results were made available to Biot during the printing of this work.

[148] *E.g.* by Auguste de la Rive, *Bibl. Univ.*, Vol. 28 (1825), p. 286.

[149] See also p. 51 above.

[150] G. Targioni-Tozzetti, *Notizie degli aggrandamenti delle scienza fisiche accaduti in Toscana nel corso di anni LX del secolo XVII* (3 vols. in 4, Florence, 1780), II, 163. See also Florence, Bibl. Naz., ms Gal. 259, fol. 25r.

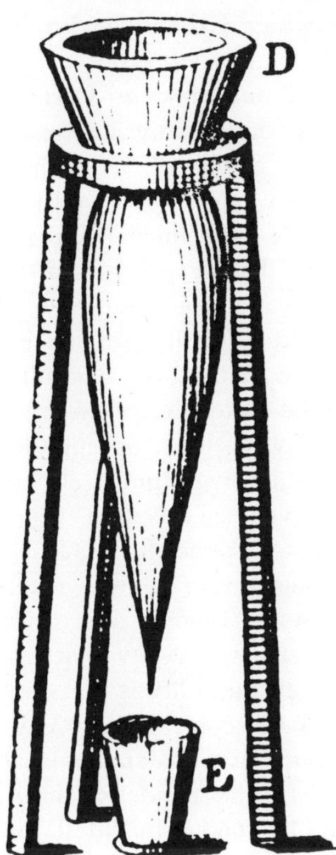

Fig. 3.20 The earliest condensation hygrometer, 1655.

On August 27, 1655:

N°. XII. Two vases filled with ice at the 18th hour. One, put in the cellar below His Serene Highness' apartments at the 18th hour, made 9 or 10 drops per minute. The other, left at the 18th hour in the great hall, made 11, 12, or 13 drops per minute.[151]

But on September 1, 1655, a vase in the shade, after producing a little water, became dry on the outside, although kept well supplied with ice. On this occasion "a rather strong south-west wind was blowing."[152]

It should be mentioned that there is another document in the Galileo manuscripts at Florence that assigns the date of September 11, 1657, to the Grand Duke's invention.[153] This is in a description of Ferdinand's experiments "collected to satisfy his own curiosity by Paolo Minucci" and is in Viviani's handwriting. However, the earlier date is confirmed by entirely independent evidence from

[151] *Ibid.*, II, 166; also Florence, Bibl. Naz., ms Gal. 259, fol. 14ᵛ.
[152] *Ibid.*
[153] Florence, Bibl. Naz., ms Gal. 259, fol. 13ᵛ.

Rome, where Urbano Daviso,[154] after describing the cord hygrometer already mentioned, reported that even while he was writing about this, Cardinal Giovanni Carlo de' Medici had been sent one of his brother's condensation hygrometers. The Cardinal, who shared the scientific curiosity of his fellow princes, wanted to be sure that the water collected by the cone came from the air, and not from inside the vessel. He used vinegar and snow, and then alcohol and snow, but the water that dropped from the cone tasted just the same as when water was used with the snow. Daviso then tells us that in a strong south wind (*scirocco*) a notable amount of water was collected, "which, when tried, had a taste like sulfur, or niter." Could this be the taste of distilled water that seems peculiar to those unaccustomed to it?

We may, I think, accept the date 1655 for the invention of the condensation hygrometer, although it has been thought by some that it may have been invented even as early as 1645, and that it should be ascribed to Torricelli.[155] Confusion may have arisen because of a "bread-and-butter" letter written on July 18, 1648 to Prince Leopold by Emanuel Maignan after a visit to Florence, in which he refers to "the other rare novelty that I saw . . . a very beautiful instrument with which humidity is measured."[156] In the absence of strong evidence to the contrary, we may assume that this was the oatbeard hygrometer that Maignan described in his *Perspectiva horaria*.[157]

News of the condensation hygrometer seems to have reached England by 1663 at the latest, for on June 12 of that year Balthasar de Monconys visited Sir Christopher Wren, and found a similar instrument with a characteristic addition. Monconys writes:

> He told me how to find out the dryness and humidity; for dryness being only a loss or diminution of humidity, it is enough to observe the quantity of the latter. Thus he puts a very big and wide glass funnel, with a very narrow spout, into a cellar, or a very humid and shady place. It is out of the way of the wind, but near a window, and hung from the ceiling, so that there is not the width of two fingers between the funnel and the roof or vault. For then, in proportion to the humidity that is in the air, this will condense to water, and distil drop by drop from the spout of the funnel into a very ingenious balance, that indicates very exactly the weight of anything.[158]

Monconys seldom understood completely what he was shown, and his illustration and description of this balance are quite incomprehensible. It is probable, of course, that the "funnel" was really a glass cone, and that it was filled with ice, or cold water. Wren, whose attitude to experimentation was far in advance of his time, would naturally look for a substitute for the counting of drops.

[154] *Trattato della sfera* (Rome, 1656), pp. 196–97.
[155] G. Boffito, *Gli strumenti della scienza e la scienza degli strumenti* [etc.] (Florence, 1929), pp. 67–68.
[156] Florence, Bibl. Naz., ms Gal. 275, fol. 16r.
[157] See p. 90 above.
[158] *Journal des voyages*, II, 54.

I have not been able to discover where Wren got this idea, if indeed he did not think of it independently. If it came from Florence, he was probably told about the Duke's hygrometer by Robert Southwell, a young man who had returned from the Grand Tour in 1661, and had shown Boyle a Florentine thermometer.[159]

By the time the Accademia del Cimento began to use it, the condensation hygrometer had become the elegant instrument shown in Figure 3.21, the top part being made of cork, sealed internally with pitch. Reproductions are to be found in large museums. It is commonly believed that the one in the Museo di Storia della Scienza is the original, but the catalogue[160] informs us that it is a modern replica.

Although numerical data are furnished by such an instrument, its results would undoubtedly be very sensitive to the motion of the air past it. An attempt to get round this difficulty was made about a century later by Felice Fontana, the curator of the "cabinet" of the then Grand Duke of Tuscany (this collection became the Museo di storia della Scienza). Fontana first cooled a piece of glass, or a closed bottle of known surface area, to the ice point, placed it on a balance, and noted the rate at which vapor was condensed on it. He then tried to get absolute values of humidity by enclosing a known volume of air in a glass box that could be quickly closed. This box contained a balance, onto which the cold body, wiped dry, was put just before the door was shut.[161] It seems to have been assumed that all the humidity in the air within the box was attracted to the cold body.

But before this time, a much better idea had occurred to a professor of medicine at Montpellier, Charles Le Roy,[162] who had been wondering how water, so much heavier than air, could remain suspended in the form of invisible vapor. He concluded that the process was similar to that by which solids may be held in transparent solutions by liquids. In a word, he believed that water dissolves in air; and besides the transparency of the resulting "solution," he adduced in favor of this hypothesis the experimental fact that more water can be held in this invisible state by warm air than by cold. To demonstrate this he simply sealed up a bottle of damp air and put it in a place where the temperature was falling. At some temperature, always the same as long as the bottle remained sealed, dew would appear in the bottle, and disappear when the temperature rose again. The temperature at which the dew appeared was the lowest temperature at which the air in the bottle could hold in solution, as he thought, all the moisture that happened to be in it when the bottle was sealed. So he generalized that air heated to a given degree of heat can hold in solution only

[159] See my *A history of the thermometer*, pp. 38–39.
[160] *Catalogo degli strumenti del Museo di Storia della Scienza* (Florence, 1954), p. 34.
[161] *Obs. sur le Phys.*, Vol. 9 (1777), pp. 196–98. This volume contains five articles describing the Grand-Duke's "cabinet."
[162] Le Roy, *Mém. Acad. R. Sci. Paris*, (1751), pp. 481–518.

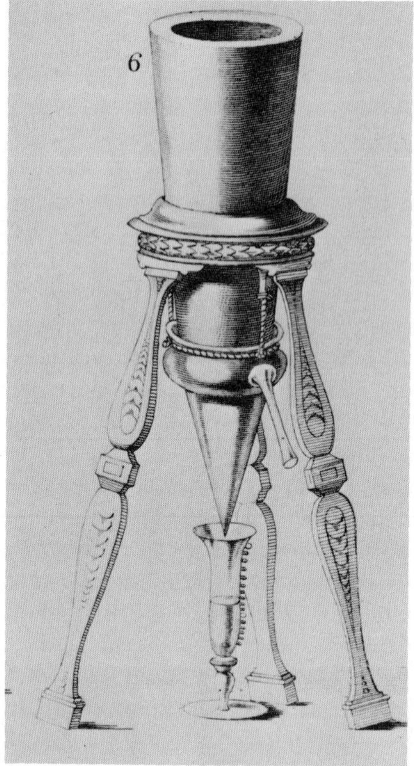

Fig. 3.21 Condensation hygrometer, from the Saggi.

a definite quantity of water.[163] This gave him a way of specifying unambiguously the amount of invisible moisture—that is to say all moisture except fog, mist, and cloud—in the air: "There is at all times a certain degree of cold at which the air is ready to release part of the water that it holds in solution. I call this temperature the *degré de saturation* of the air."[164] We now call it the dew point.

To measure it, he took water that was cold enough to produce condensation on the surface of the glass vessel that contained it. He then poured it into a second glass, letting it get a little warmer and continuing to measure its temperature, then back into the first glass, which had been wiped dry, and so on until it arrived at a temperature that would just not produce dew. This simple but not very accurate method was reinvented by John Dalton fifty years later;[165] he seems not to have known of Le Roy's work at the time.

It is surprising that this method was not sooner improved. De

[163] *Ibid.*, p. 486.
[164] *Ibid.*, p. 490.
[165] Dalton, *Mem. Manchester Lit. & Phil. Soc.*, Vol. 5 (1801), pp. 535–602.

Saussure[166] was well aware of it, but his enthusiasm for the hair hygrometer led him to undervalue other methods.

Two improvements were suggested in 1809. Soldner[167] described a thermometer with its bulb in the form of a thin plate; one side was to be moistened with ether to cool the bulb, and the formation of dew was to be observed on the other. This paper was published in Gilbert's *Annalen*, a circumstance that might lead one to suppose that it would be widely known, and yet at least four similar schemes were described in the third decade of the century.[168-171]

Much earlier, the chemist Berzelius had suggested a thermometer with a bulb preferably of steel, provided with a case of oiled silk to keep it dry, and the whole plunged into very cold water.[172] The silk case being removed after the thermometer had been cooled sufficiently, the temperature at which the dew disappeared was noted as the instrument warmed up. It is probable that the error in this procedure would depend on how far below the dew point the thermometer was originally cooled.

About ten years later Daniell[173] described his hygrometer derived from Wollaston's cryophorus, familiar now to generations of students of physics (Fig. 3.22). Two bulbs of thin glass, in one of which is the bulb of a sensitive thermometer, are connected by a wide tube. There is enough ether in the system to fill one of the bulbs half full; the remainder of the space is devoid of air. The bulb containing the thermometer is gilt and polished; the other one is covered with a thin cloth. If now all the liquid ether is in the bulb with the thermometer, and the other bulb is cooled (for instance by dropping some ether on it), a rapid evaporation of the ether inside the instrument will occur, lowering the temperature of the gilt bulb. The temperature indicated by the thermometer when dew appears is noted, and also the temperature when, while the cooling is interrupted, the dew again vanishes. The mean of these two temperatures is taken as the dew point.

Daniell[174] attempted to improve the instrument by making it of brass, except for the tube enclosing the thermometer scale. The greater conductivity of the brass was an obvious advantage, but it seems doubtful whether the necessary freedom from permanent gases could have been long maintained. Realizing this, Muncke[175] constructed an instrument with platinum bulbs and glass tubing.

Daniell, never one to hide his light under a bushel, was immensely

166 H. B. de Saussure, *Essais*, p. 56.
167 J. Soldner, *Ann. Phys.*, Vol. 32 (1809), p. 219.
168 D. F. Körner, *Ann. Phys.*, Vol. 70 (1822), pp. 139–40.
169 T. Jones, *Phil. Trans.*, Vol. 116 (1826), pp. 53–54.
170 "N.," *Edinburgh J. Sci.*, Vol. 4 (1826), p. 127.
171 J. Foggo, *Edinburgh J. Sci.*, Vol. 7 (1827), pp. 36–44.
172 J. Berzelius, *Phil. Mag.*, Vol. 33 (1809), pp. 39–42.
173 J. F. Daniell, *Quart. J. Sci.*, Vol. 8 (1820), pp. 298–336.
174 *Quart. J. Sci.*, Vol. 9 (1820), pp. 128–37.
175 G. W. Muncke, *Ann. Phys.*, Vol. 68 (1821), pp. 60–75.

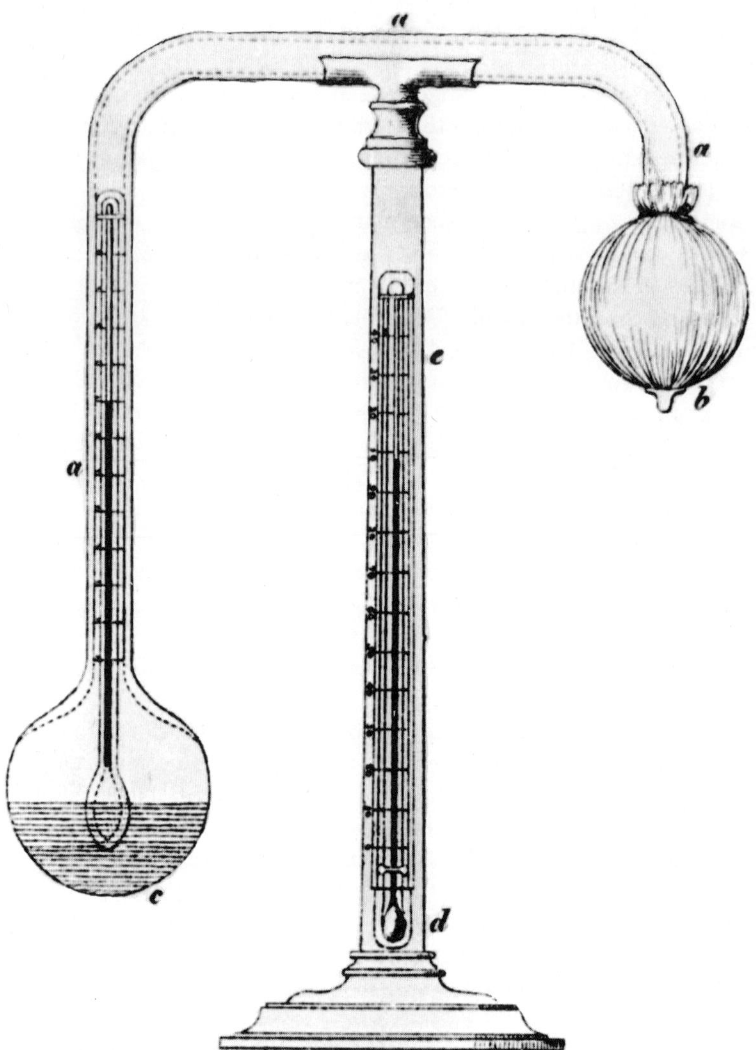

Fig. 3.22 Daniell's dew-point hygrometer.

proud of the new instrument. In defending it he seemed to feel it necessary to deprecate both the hair hygrometer and the wet-and-dry-bulb psychrometer,[176] the latter just becoming generally known.[177] Actually it probably gave much better results than the psychrometer, as the latter was used at the time; but it was not as good as he thought it was. It was pointed out by Regnault much later[178] that ether is seldom free from water, and that the process of cooling the instrument therefore adds water vapor to the air in its vicinity. This objection

[176] See p. 124 below.
[177] Daniell, *Meteorological Essays*, London (1823), pp. 186–201.
[178] Regnault, *Ann. Chim. et Phys.*, Vol. 15 (1845), pp. 194–95.

obviously applies with even more force to instruments such as those of Soldner and Jones. The temperature gradient in the ether between the thermometer and the surface on which the dew is deposited was another source of error, since this was different when the ether was cooling down and when it was warming up. This difference would produce a systematic error that might be important.

With delightful optimism, Daniell suggested that the hygrometer and the barometer together could solve the forecaster's problem. "By combining the observations of both instruments, we learn to modify their results, and by so doing *can hardly be deceived in the weather* for many hours in advance."[179]

Daniell's method of cooling the surface was very indirect, and indeed proved inadequate in hot and dry weather. The simpler solutions of the problem, referred to above, were defective. Another solution occurred to J. W. Dobereiner[180] in 1822; this was to bubble air through the ether, causing evaporation and cooling, the vapor of the ether being discharged at a safe distance. Dobereiner's work seems to have been forgotten, however, and in 1845 Regnault[181] described his hygrometer on this principle, of which Figure 3.23 will serve to replace a verbal account. There is no doubt that Dobereiner has the priority, but Regnault's is a name to conjure with. Besides giving the instrument a very useful form, he did a great deal of hard thinking about all the problems of hygrometry.

An excellent modification of Regnault's instrument was made by Alluard[182] in 1877 (Fig. 3.24). In this the ether is evaporated in a square gilded and polished brass tube. One side of the tube is framed by a band of brass, similarly finished and not quite touching the tube, so that the observer has two adjacent surfaces for comparison, on one of which dew does not form. This may have been the best dew-point hygrometer until very recent times, but it should not be forgotten that Francis Ronalds described an imposing hygrometer to the British Association in 1851,[183] in which part of the polished tube was arranged so that it stayed clear. This had an elaborate aspirator and was set up at Kew for routine observations.

A different sort of condensation hygrometer was that of Guiseppe Belli (Fig. 3.25).[184] In this instrument a hexagonal iron bar is fastened, with its axis vertical, in the cover of an insulated vessel filled with ice. The bar has an axial hole bored almost to its lower end, and this hole is filled with mercury. An enlargement of the lower end of the bar is

[179] *Quart. J. Sci.*, Vol. 8 (1820), p. 303 (my italics).
[180] *Ann. Phys.*, Vol. 70 (1822), pp. 135–38.
[181] *Ann. de Chim. et de Phys.*, Vol. 15 (1845), pp. 196–201.
[182] *J. Phys.*, Vol. 7 (1878), pp. 328–30. See also *Meteorol. Mag.*, Vol. 13 (1878), pp. 52–56.
[183] *B.A.A.S.*, Ipswich, 1851, *Report*, pp. 342–45.
[184] Belli, *Ann. Chim. et Phys.*, Vol. 15 (1845), 506–8. This is from his *Corso elementare di fisica sperimentale* (Milan, 1831).

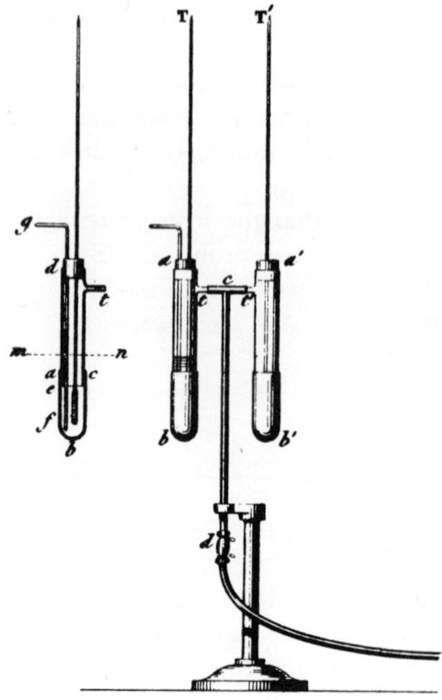

Fig. 3.23 Regnault's dew-point hygrometer.

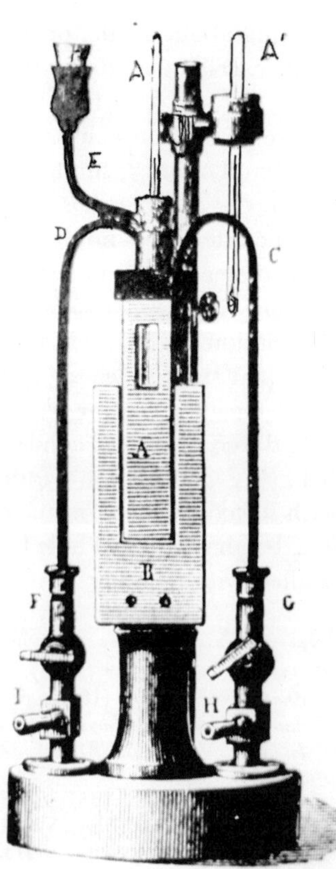

Fig. 3.24 Alluard's dew-point hygrometer.

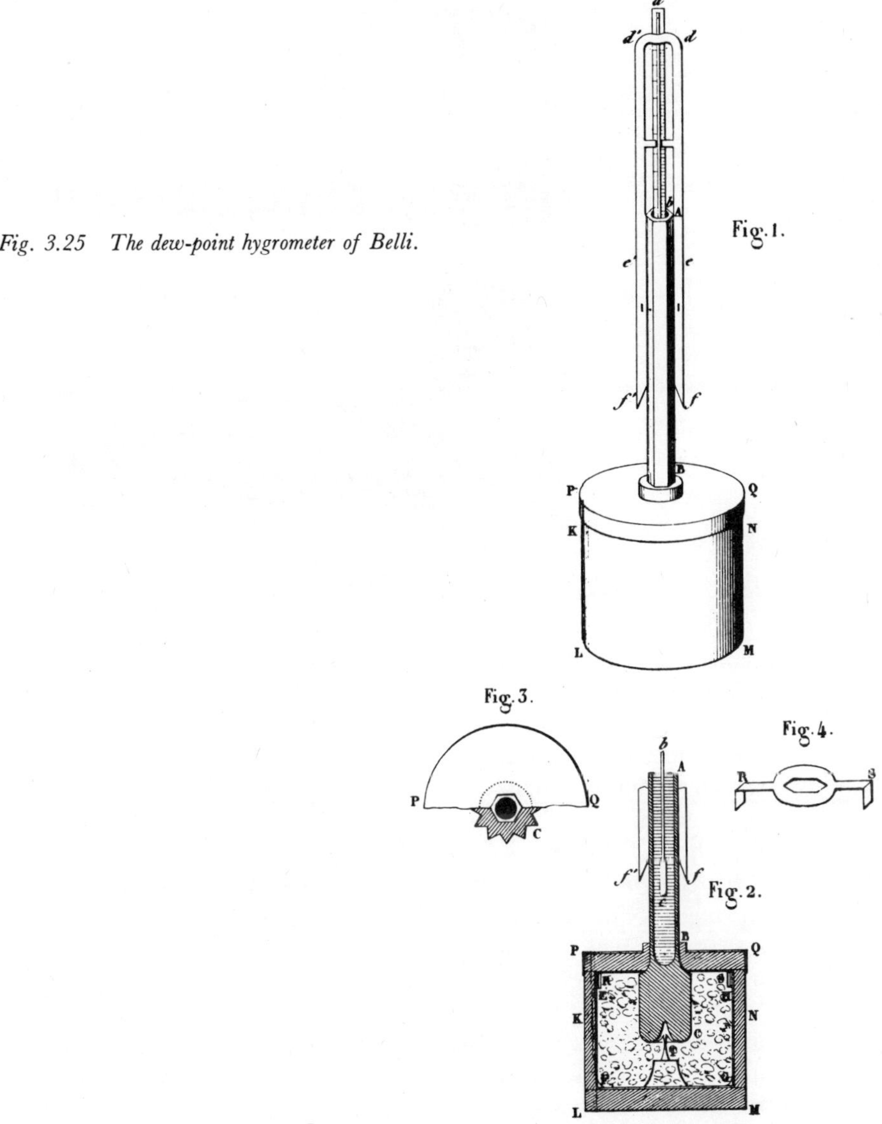

Fig. 3.25 The dew-point hygrometer of Belli.

grooved as shown at *C* in Belli's Figure 3, in order to give a larger area of contact with the ice. There will be a gradient of temperature along the bar, and the lower portion will become covered with dew as far as a line at the dew point. The temperature at this point is read by means of a thermometer plunged into the mercury, and attached to an external index to facilitate setting its bulb at the level of the dew-line. This was a very ingenious idea, and capable of fair accuracy, one would suppose, under steady conditions, with good ventilation.

Several other dew-point hygrometers were designed in the half

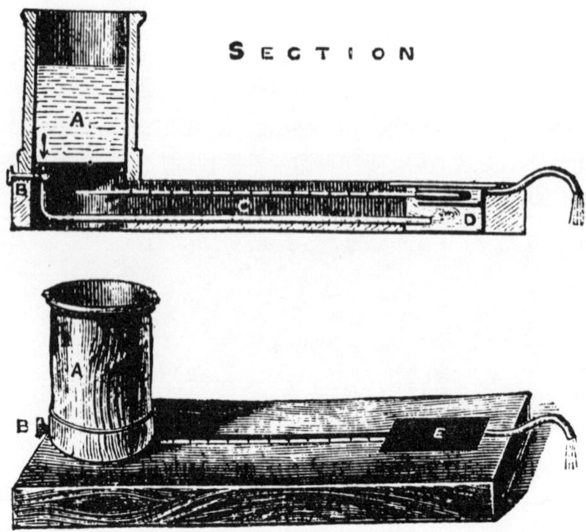

Fig. 3.26 The dew-point hygrometer of G. Dines, 1871.

century after 1840, the most original being that of Savary,[185] which had a bimetallic thermometer (gold and platinum) to measure the dew-point temperature. It cannot have been very accurate. The simple instrument of George Dines[186] is in complete contrast to this, and is shown in Figure 3.26, in which *A* is a small tank filled with cold water. By opening a tap *B* the water can be caused to flow beneath a very thin sheet of black glass *E*, just below which is the bulb of a thermometer. The dew can be made to appear and disappear by regulating the flow. Nine years later Dines described a slightly more elaborate instrument, which could also be cooled by evaporating ether and so used when the dew point is below 0°C.[187]

The instrument of Regnault, and Alluard's modification of it, were used at some large observatories for many years. Further improvement of the dew-point hygrometer had to await the development of the photoelectric cell, and is beyond the scope of this book.

5. The hygrometer, depending on evaporation from a moist surface, called the psychrometer. Cooling by evaporation must have been observed on the first occasion when a naked man swam a river and emerged on the other side. The phenomenon has long been used in tropical countries for cooling liquids, either by the use of a slightly porous vessel, or by wrapping a wet cloth around a jar and exposing it to the wind. At

[185] Described by Arago, *Ann. Chim. et Phys.*, Vol. 2 (1841), pp. 531–32.
[186] First described and illustrated in *Meteorol. Mag.*, Vol. 6 (1871), pp. 145–48.
[187] Dines, *Quart. J. Meteorol. Soc.*, Vol. 6 (1880), pp. 39–42.

some time before 1692 Robert Boyle was told about this practice by "an intelligent traveller."[188]

The scientific investigation of the phenomenon demands the use of the thermometer, and in the seventeenth and eighteenth centuries a number of writers noted that such an instrument indicates a lower temperature while its bulb is drying after being extracted from water. The reason for this was at first a mystery. The earliest account that I have found dates from 1681 and was written by the Jesuit Francesco Eschinardi, who reported that he

> found a most singular thing with the thermometer. When this was immersed in a glass of water in the month of April, 1681, in a closed room, it showed a greater cold than in the air of the room, when it was put in the liquid; but when it was taken out, it then showed a greater cold still. [He repeated the experiment several times.] . . . Up to now I have found no other reason for this than that which I published in Rome in 1660,[189] namely that the very thin film of water, with which the thermometer becomes covered on emerging from the glass, may be denser than ordinary water, and so may have more power to cool the thermometer than the water in the glass itself. In confirmation of this I found that when, immediately after taking the thermometer from the glass, I dried it with a woolen cloth far from the heat of the hand, the effect that I have described did not occur.[190]

He thought that an argument for "positive cold" might be derived from such phenomena.

In 1699 Guillaume Amontons noted in his paper on a heat engine[191] that when the bulb of his air thermometer was taken out of water the mercury index descended, "and afterwards rose again little by little as the bulb dried."[192] He concludes that "water that is ready to evaporate diminishes the spring of the air [in the thermometer] more than when it is in a large enough quantity to surround [the air] completely."[193] It will be seen that this brings evaporation into the picture.

In 1738 Georg Wolfgang Krafft found that a clean very thin cloth, wetted and hung up in the air, froze stiff when a thermometer suspended nearby was just above the freezing point.[194]

In a passage almost certainly written earlier than this, Musschenbroek takes up the problem: "In fact, if the thermometer is plunged into water, and if after taking it out we blow with the bellows against the moist bulb, it will be made much colder . . . even though the water into which the bulb was plunged was warmer than the liquid in

[188] Boyle, *The General History of the Air* [etc.] (London, 1692), p. 164.
[189] I have been unable to identify this publication.
[190] Eschinardi, *Lettera al Sig. Francesco Redi, nella quale si contengono alcuni discorsi fisicomatematici* (Rome, 1681), p. 40.
[191] Amontons, *Mém. Acad. R. Sci. Paris* (1699), pp. 112–26.
[192] *Ibid.*, p. 116.
[193] *Ibid.*
[194] Krafft, *Experimentorum physicorum praecipuorum brevis descriptio* [etc.] (St. Petersburg, 1738), p. 204.

the thermometer . . . Does not this result from the fact that water forcibly attracts heat (*le feu*), and, first incorporating itself with the heat of the thermometer, carries this heat away as soon as we begin to blow on it?"[195] The idea of the great affinity of water for fire or heat rose from the action of water in quenching fire, which was then universally considered to be a material substance.

Numerous and careful experiments were made in St. Petersburg by Georg Wilhelm Richmann, an able Swedish physicist whose death at the age of forty-two, while conducting an electrical experiment in a thunderstorm, excited tremendous interest. He concluded that the water on the bulb must be responsible for the cooling, and explained it by postulating the existence of "salts" in the air. which dissolve in the film of water and produce the cold.[196] He knew that the descent of the mercury was sometimes greater and sometimes less, but probably thought that this depended on the amount of "salts" in the atmosphere.

William Cullen, professor of chemistry at Edinburgh, had a student who noticed the cooling of a wet thermometer bulb.[197] The professor conducted extensive experiments, in which he showed that while many liquids cool the bulb, some, such as oils, have little or no effect, and that strong acids warm it, an effect that he correctly ascribed to absorption of water from the air. The cooling he ascribed with equal correctness to evaporation, and tried experiments *in vacuo*, also noting (simply as an empirical fact) the cooling of air when it suddenly expanded. But he did not suggest the use of the wet bulb as a hygrometer.

While Cullen was certainly the first to publish the correct explanation, René Antoine Ferchault de Réaumur very probably hit upon it independently, as it is unlikely that the publication of the Society in Edinburgh would have come to his notice before he died in 1757. In the Archives of the Académie des Sciences at Paris there is a bundle of papers that belonged to him, dealing with meteorological topics. Among these is a short draft entitled "Sur une singularité de la marche du thermometre."[198] "The cooling of the thin layer of water" (wrote Réaumur) "that moistens the bulb of the thermometer is not . . . produced by any salts . . . but [the water] continually disappears through evaporation. So this shows us that the vapors that rise from the water are raised at the expense of the heat in the

[195] P. van Musschenbroek, *Essai de physique*, tr. Pierre Massuet (2 vols., Leyden, 1739), I, p. 476.
[196] Richmann, *Novi Comm. Petrop.*, Vol. 1 (1747–48), pp. 284–290 [published 1750]. *Re* the supposed salts in the air, see my *A History of the Theories of Rain* (London, 1965; New York, 1966), pp. 73–74, 96.
[197] Cullen, *Essays and Obs.*, *Phys. and Lit.*, *of a Soc. at Edinburgh*, Vol. 2 (1755), pp. 159–71. Reprinted in his *Experiments upon magnesia alba . . . to which is annexed, an essay on the cold produced by evaporating fluids* [etc.] (Edinburgh, 1777).
[198] Acad. Sci., Paris, dossier *Réaumur*, *Thermomètre et baromètre* (Unnumbered item, 5 pp.). I have to thank Mme P. Gauja for drawing my attention to this file.

particles of fire in the layer from which these vapors come, so that the layer of water from which the vapors are detached is continually cooled."[199] This paper is undated and probably undateable, but one would suppose that Réaumur had read Richmann's article. In any event, he was as clear about the mechanism as contemporary concepts of heat would allow, but the application to hygrometry did not suggest itself to him.

Neither did it to Michael Christoph Hanow of Danzig, who in 1756 published the most closely reasoned of these mid-eighteenth-century accounts of the phenomenon.[200] He dismissed Musschenbroek's explanation on the grounds that if water attracts heat it should strongly cool a thermometer immersed in it. Nor could he believe, with Richmann, that such a minute amount of water could absorb from the air such a great deal of "salts," the existence of which he doubted. After many sorts of experiment he came to the conclusion that the cooling is due to the evaporation of the water. Then: "The cooling takes place most noticeably if (1) the air is not moist but very dry; (2) either some other air, or fresh air, comes to replace the old; or (3) in the same kind of air the evaporation is maintained for a long time; and (4) the material to be cooled, together with the film of water itself, is far warmer than the air in which the evaporation is to take place." The last condition seems trivial; but he continues:

> But if the air itself is already moist enough, or full of vapor, even if it is renewed and exchanged by the wind or in some other way . . . very little cooling can result.
>
> From the experiments it seems clear enough that the air that takes up the vapor and carries it away has a definite capacity, just as all other finite powers have. If it is saturated, so to speak (*Wenn sie gleichsam gesattiget ist*), with as much watery vapor as it can contain, it can take up no new vapor before the other is taken away or dispersed . . . Now as the cooling acts entirely according to the evaporation, we see the cause of the small and unimportant cooling in moist air or rainy weather, etc.; and on the other hand the more important cooling in dry air, which contains only a little of the watery vapors. It is the same if the air around the glass is frequently renewed.[201]

Hanow then describes several practical applications of the cooling, but a hygrometer is not one of them.

In fact, the first man to use this phenomenon to measure humidity seems to have been the famous Scottish geologist James Hutton, who died in 1797. In support of this ascription we apparently have to rely on his friend and biographer John Playfair, who wrote: "To one who considers meteorology with attention, the want of an accurate hygrometer can never fail to be a subject of regret. The way of supplying

[199] *Ibid.*, p. [3].
[200] Hanow, *Versuch Abh. naturf. Ges. Danzig*, Vol. 3 (1756), pp. 226–58.
[201] *Ibid.*, p. 252.

this deficiency which Dr Hutton practised was by moistening the ball of a thermometer, and observing the degree of cold produced by the evaporation of the moisture. The degree of cold *caeteris paribus* will be proportional to the dryness of the air, and affords, of course, a measure of that dryness."[202] But what if other things are not equal, and what do we mean by "proportional"? What was needed was a theory.

Hutton's compatriot John Leslie made some attempt to provide it, and he did see the convenience of having two thermometers side by side, with the bulb of one moist and that of the other dry. In 1790 he was living with Josiah Wedgwood, the manufacturer of pottery, and got him to have made "a cup of unglazed biscuit ware which is quite bibulous, about the size and shape of a pigeon's egg."[203] This was filled with water and hung in the air, and the difference in temperature of air and water was observed. In 1795 he got the idea of using two similar thermometers, wetting the bulb of one. Unfortunately he decided a priori that "the effect . . . depends entirely on the disposition of the air, and is not modified at all by agitation, or the frequent renewal of surfaces. Such means can only accelerate the term of equilibrium, in the same manner as wind brings a thermometer more quickly to the standard than still air . . ."[204] He also found this an obvious application of the differential thermometer, which he imagined he had invented.[205] Worried about the differential absorption of light or heat, he made the dry bulb of colored glass, and covered the other with paper and then with a thin silk of a color to match that of the dry bulb when the silk was wet. His instrument was graduated on a milligrade scale, that is to say in units of 0.1°C., and is shown in Figure 3.27. It would not show the actual temperature of the air, but in the elementary theory that Leslie developed this would not be important, for he considered the specific heat of air to be $\frac{3}{8}$ at all temperatures, and the latent heat of evaporation to be constant at 6,000 on his millesimal scale. He wrote that in equilibrium the heat lost by unit mass of air in falling from the temperature of the dry bulb to that of the wet bulb must equal the latent heat of evaporation of the moisture taken up.[206]

Carl Wilhelm Böckmann of Karlsruhe made comparisons of the hygrometers of Leslie, De Saussure, and Deluc,[207] using goldbeater's skin to cover the wet bulb. He also used separate thermometers as a substitute for the differential thermometer.

On March 6, 1815, Louis Joseph Gay-Lussac presented a theory of

[202] Playfair, *Trans. Roy. Soc. Edin.*, Vol. 5, part III (1805), p. 67, note.
[203] Leslie, *J. Nat. Phil., Chem., & the Arts*, Vol. 3 (1799), p. 462.
[204] *Ibid.*
[205] One was described by Johann Christoph Sturm, *Collegium experimentale sive curiosum* [etc.] (2 vols., Nuremberg, 1676 and 1685), I, 54; but it may have been invented even earlier.
[206] Leslie, *A short account of experiments and instruments, depending on the relations of air, to heat, and moisture* (Edinburgh, 1813), pp. 72–74.
[207] Böckmann, *Ann. der Phys.*, Vol. 15 (1803), pp. 355–76.

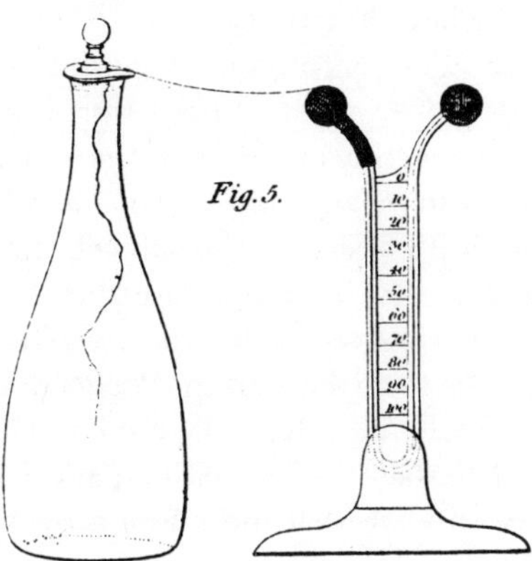

Fig. 3.27 Leslie's differential thermometer as a hygrometer, 1799.

the wet bulb to a meeting of the Academy of Sciences in Paris. This was published much later.[208] It was based on the same fundamental idea, but Gay-Lussac was more interested in the maximum cooling that could be produced by evaporation. As to hygrometry by this method, he pointed out that the cooling is a function of pressure, temperature, and humidity, so that "very extensive tables" would be needed. He thought the dew-point hygrometer simpler, and had not worked further on the subject of hygrometry.

The first theory of the wet bulb that had any real meteorological utility was published in August, 1822, by James Ivory.[209] He began by discussing what he calls Hutton's experiment, looking at it with a clear idea of the physical conditions. He assumed that there would be some ventilation, but not too much. He also recognized that an error would result if any heat was conducted down the stem of the wet bulb thermometer. Supposing this conduction to be negligible, he assumed that a steady state had been attained, in which all the air that came into contact with the wet bulb was saturated. In this condition the latent heat of vaporization of the water added to the air was equal to the heat given up by the air in falling from the temperature θ shown by the dry bulb to that of the wet bulb, θ', both expressed in centigrade degrees. Taking values of the specific heat of air and the latent heat of vaporization, from Biot's *Traité de physique*,[210] he arrived, after a good deal of algebra, at the approximate formula

$$e = e' - b(\theta - \theta')/1{,}200$$

in which e is the vapor pressure that is to be measured, e' the saturation

[208] Gay-Lussac, *Ann. Chim. et Phys.*, Vol. 21 (1822), pp. 82–92.
[209] Ivory, *Phil. Mag.*, Vol. 60 (1822), pp. 81–88.
[210] In 4 vols., Paris, 1816.

vapor pressure at the temperature of the wet bulb, and *b* the barometric pressure in inches of mercury.

Three years later Ernst Ferdinand August of Berlin attacked the same problem in the same way, and not surprisingly obtained similar results.[211] He also gave the instrument the unfortunate name *psychrometer*, which ought to mean "cold-meter," and by dint of writing a pamphlet[212] and several papers on the subject, managed to persuade his compatriots and almost everyone else that he had invented the instrument. But this was not his intention. When he submitted his 1825 paper to the *Annalen*, the editor, Poggendorf, told him about Ivory's work, as he freely admitted. It will be well to print a translation of the relevant passage: "The mathematical results that I have set out above agree so well with those found by Mr. Gay-Lussac for the cold produced by evaporation in dry air, which he tested by experiment, and with Mr. Ivory's formulas, about which I learned through the kindness of the Editor, that I think it worth the trouble to set them out in full, for as far as I know nothing of the sort has hitherto appeared in German."[213] August may clearly be freed from the charge of plagiarism.

If the Germans thought of the psychrometer as August's, the English got into the habit of calling the instrument "Mason's hygrometer," with even less justification. This was because of an absurdly pompous paper by a surgeon of Pentonville, John Abraham Mason,[214] who made no reference to Ivory or August, though he did mention Leslie. It may be of some interest to see the engraved illustration provided by Mason (Fig. 3.28).

There have been two sorts of attempts to refine the theory of the wet bulb: the purely theoretical and the partly empirical. There is no disagreement about the validity of a formula of Ivory's type:

$$e = e' - aP(\theta - \theta')$$

in which P is the barometric pressure in the same units as e and e', and a is a factor to be obtained either theoretically or empirically. Some meteorological services, for example the United States Weather Bureau, apply a small correction factor that takes into account the variability of the latent heat of evaporation with temperature; others do not.

Ivory's formula was tested by Kämtz[215] in 1834, Apjohn[216] in 1835, and Regnault in the 1840's. Regnault, who had himself determined new values of many physical constants and functions, showed by elaborate experiments that the wind speed had an influence on the

[211] August, *Ann. Phys.*, Vol. 5 (1825), pp. 69–88; 335–44.
[212] August, *Uber die Anwendung des Psychrometers zur Hygrometrie* (Berlin, 1828).
[213] *Ann. Phys.*, Vol. 5 (1825), p. 74.
[214] Mason, *Records of General Science*, Vol. 4 (1836), pp. 23–35; 96–111.
[215] L. F. Kamtz, *Ann. Phys.*, Vol. 30 (1834), pp. 43–71.
[216] James Apjohn, *Phil. Mag.*, Vol. 6 (1835), pp. 182–88; *ibid.*, Vol. 7 (1835), pp. 266–74; 470–73.

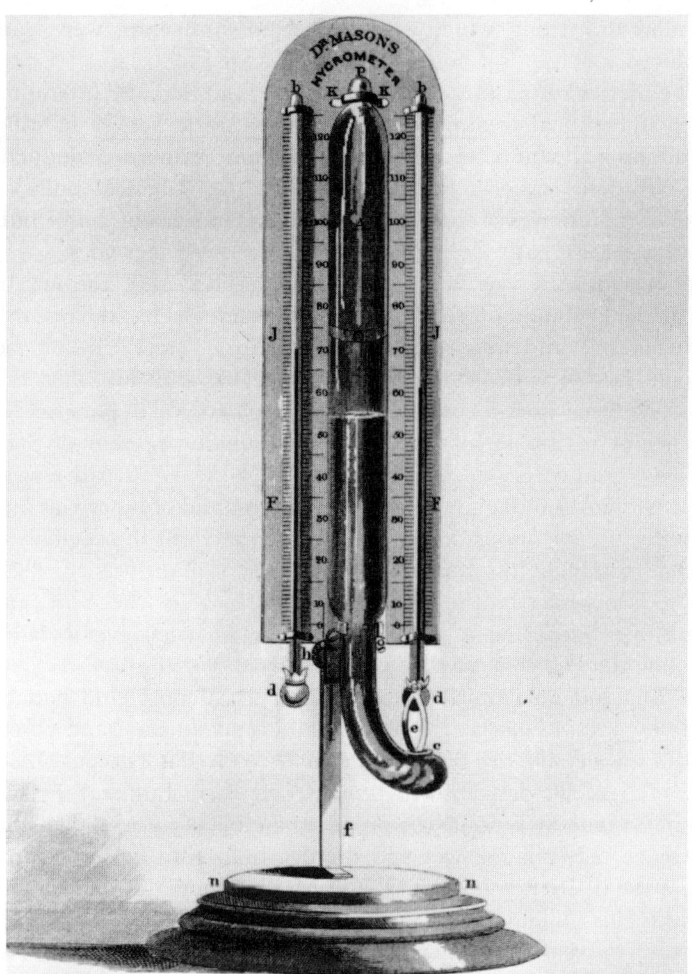

Fig. 3.28 "Mason's hygrometer," 1836.

formula.[217] He was at first inclined to produce a purely theoretical result, but in a later memoir[218] he decided that the ventilation and the surroundings have a great influence on the "constant" *a*, so that if the psychrometer is to be useful, it should be calibrated in the place where it is to be used. The great experimenter, unaware of the limitations imposed by the fluctuations in the atmosphere, was convinced at that time that "*The psychrometer should be considered as an empirical instrument*, analogous to De Saussure's hair hygrometer."[219] He believed that the only proper way to measure the humidity in the atmosphere was to absorb it and weigh it. However, his psychrometric

[217] V. Regnault, *Ann. Chim. Phys.*, Vol. 15 (1845), pp. 206–27.
[218] *Compt. Rend.*, Vol. 35 (1852), pp. 930–39.
[219] *Ibid.*, p. 938. The italics are from the original.

formulas for strong wind, moderate wind, and calm were widely employed.

The matter rested there until Maxwell[220] and Stefan[221] attempted a purely theoretical treatment, considering an atmosphere of still air surrounding the bulb, and taking into account radiation, conduction, and diffusion. One of the results is that (for spherical bulbs) the effect of radiation is proportional to the diameter of the bulb; but its importance is greatly decreased as the wind speed increases.

By this time it was fairly generally conceded that the wet-bulb thermometer should be ventilated for consistent results; the researches of Sworykin[222] did much to emphasize this. Ferrel[223] developed a formula for use with ventilated psychrometers, and published tables that have been used for many years throughout North America with very slight revisions; his work was an extension of that of Stefan, Maxwell, and Sworykin. It was found that the comparative success of the earlier formulae such as those of August and of Ivory was due to a completely fortuitous numerical equality between the coefficient of thermal conductivity of air and the coefficient of diffusivity of water vapor. This may be the greatest piece of luck (or the most serious misfortune, according to the point of view) that has ever befallen a scientific theory.

In England an empirical approach of a different kind had been made by James Glaisher.[224] By combining many thousand observations of the wet and dry bulb psychrometer with simultaneous observations of Daniell's dew-point hygrometer made in England, in India, and in Canada at Toronto, he prepared a table of *factors* by which the difference between the wet and the dry bulb is to be multiplied to obtain the difference between the dry-bulb temperature and the dew point. This "Glaisher factor" varies widely with the dry-bulb temperature, decreasing as the temperature rises. Tables of this sort were widely used in the British Empire for many years, but do not give good results with the ventilated psychrometer, for which they were indeed not intended. The method has, of course, no theoretical justification whatever.

Considering the psychrometer simply as an instrument, the possible variations are limited in so far as liquid-in-glass thermometers are used. They consist largely of means of providing adequate and perhaps standardized ventilation. A simple way of doing this, strongly advocated by the French astronomer Auguste Bravais,[225] was to whirl the thermometers around on a string or chain, and two

[220] J. C. Maxwell, *Encyclopaedia Britannica*, 9th ed., Vol. 7, s.v. "Diffusion."
[221] J. Stefan, *K. Akad. Wiss., Wien, Sitzungsber.*, Vol. 65 (1872), pp. 323–63.
[222] N. Sworykin, *Rep. Meteorol.*, Vol. 7, No. 8 (1881).
[223] W. Ferrel, "Report and tables for use with the whirled psychrometer," in *U.S. War Dept., Ann. Rep. Chief Signal Officer for 1886*, Washington 1887, p. 233.
[224] Glaisher, *Hygrometrical Tables* (London, 1847).
[225] Bravais, *Annuaire Soc. météorol. France*, Vol. 1 (1853), pp. 127–35.

thermometers mounted to facilitate this form a *psychromètre fronde*, or sling psychrometer. A much better instrument is the ventilated psychrometer introduced in 1892 by Richard Assmann,[226] in which a centrifugal fan, operated by clockwork, draws air past the bulbs of the thermometers and also between two polished concentric tubes surrounding them, thus almost eliminating the effect of radiation. The instrument is shown, partly in section, in Figure 3.29.

Psychrometers were also made to give a record, as in the celebrated Kew "photographic thermograph," actually a wet-and-dry-bulb instrument,[227] in which an air bubble in the mercury thread of each thermometer was imaged on a drum covered with photographic paper. A record could even be made with Bourdon tubes, with one surrounded by a wet cloth, as was shown by the Paris firm of Richard Frères (Fig. 3.30).[228] It cannot have been very accurate, and neither can that of Daniel Draper, director of the Central Park Observatory in New York, who used iron-zinc bimetallic strips twenty inches long.[229]

6. Miscellaneous hygrometers. There are a number of other possible ways of measuring humidity, but none of them seem to have led to useful meteorological instruments, if we except the electrical resistance hygrometers that will be referred to in Chapter 10. In the fever of electrical experimentation that marked the second half of the eighteenth century, it was to be expected that the application of electricity to hygrometry should have been thought of, especially in view of the difficulty of making experiments on static electricity in wet weather. Thus the Abbé Theodore Augustus Mann proposed the use of a standard electrical machine operated in a standard way, believing that the amount of electricity excited by it would be proportional only to the humidity.[230] At about the same period Alessandro Volta managed to write thirty pages of his elegant and persuasive Italian to suggest the use of an ordinary pith-ball or gold-leaf electrometer to measure humidity, noting the time of discharge between two fiducial marks, when the instrument was shunted by a strip of some hygroscopic substance.[231] He calibrated this by comparison with the hair hygrometer of De Saussure. If we can assume the absence of strong ionizing radiations, this was the first electrical resistance hygrometer.

[226] Assmann, *Abh. K. preuss Meteorol. Insts.*, Vol. I (1892), pp. 115–270.
[227] Royal Society of London, *Report of the Meteorological Committee for 1867* (London, 1868), pp. 27–54.
[228] *Notice sur les instruments enregistreurs construits par Richard Frères* [etc.] (Paris, 1889), pp. 118–20.
[229] Draper's remarkable recording instruments are described and illustrated in *Engineering*, Vol. 40 (1886), pp. 535–36.
[230] Mann, *Acta Acad. Theodoro-Palatinae*, Vol. 6, *pars physica* (1785), pp. 65–68 (published in 1790).
[231] Volta, *Mem. Math. Fis. Soc. Ital.*, Vol. 5 (1790), pp. 551–80.

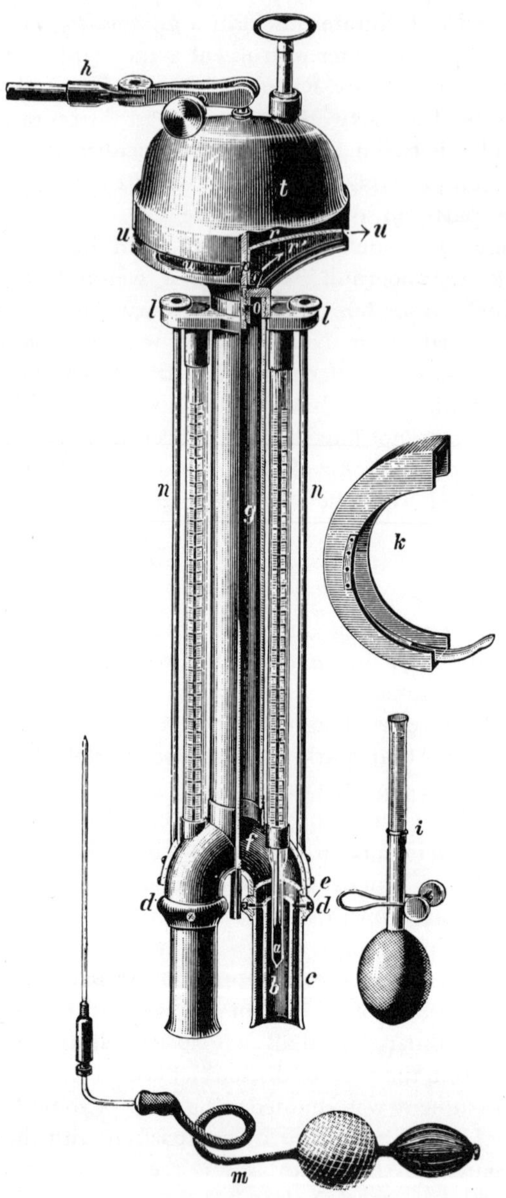

Fig. 3.29 Assmann's psychrometer, 1892.

Attempts were made from time to time to turn the "chemical" or "absorption" hygrometer into a useful meteorological instrument, especially after Regnault, in an extremely perfectionist mood, had

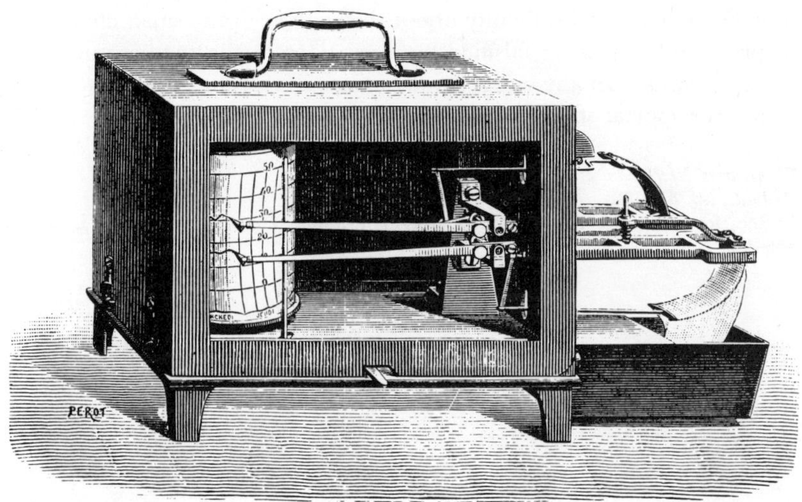

Fig. 3.30 Richard Frères' recording psychrometer, about 1885.

written a set of recommendations for the establishment of meteorological observatories,[232] in which, after reviewing the usual hygrometers, he wrote: "I think that we should give up the use of these instruments in observatories. It is preferable to determine directly, by weighing, the quantity of humidity that exists in the air during a given period of time."[233] However, he made the important point that the absorption tube, full of pumice wet with concentrated sulphuric acid, must be placed at the point where the humidity is to be measured, in order to avoid the errors due to a long entrance tube. The aspirator and the air meter can be indoors at any convenient place.

In the same year, 1871, E. H. von Baumhauer described a photographically recording instrument operating on the principle of a hydrometer floating in a bath of oil.[234] Air was led through this in known volume, and the moisture was absorbed by sulphuric acid on pumice stone inside the hygrometer body. The record would represent the time-integral of the moisture per unit volume. Three years later, M. Snellen modified this arrangement by hanging the drying tube on a balance, air being admitted and exhausted through oil-

[232] *Ann. Chim. et Phys.*, Vol. 24 (1871), pp. 225–58.
[233] *Ibid.*, p. 231.
[234] *Archives Néerlandaises*, Vol. 6 (1871), pp. 419–25.

sealed bells.[235] Von Baumhauer applauded this, but did not seem to think it better than his own.[236]

There is an alternative principle that may possibly be used in an absorption hygrometer, and that is to measure the fall in pressure resulting from the abstraction of the water vapor from a sealed volume.[237] The capital difficulty about this is to keep the small changes in pressure from being falsified by alterations in the temperature of the apparatus. In any event, none of these devices has come into use at meteorological stations.

[235] *Archives Néerlandaises*, Vol. 9 (1874), 477–79.
[236] *Ibid.*, pp. 479–80.
[237] See, for example, Max Thomas Edelmann, *Ann. Phys.*, Vol. 6 (1879), pp. 455–58.

IV

The Rain Gauge and the Atmometer

1. Introduction. A knowledge of the amount of water gained from the atmosphere by precipitation and lost by evaporation is of obvious utility in many fields, and the oldest measurements of rainfall that we know about were made in the interests of agriculture. In more recent times the calculations of hydroelectric and waterworks engineers have required increasingly detailed knowledge of the distribution of rain and snow.

The rain gauge, which has been called *hyetometer*, *ombrometer*, and *pluviometer* (the last is a word of mixed parentage) is essentially a very simple device. Like many other instruments, however, it has been forced to give a continuous record; the number of different recording gauges must run into hundreds.[1] Fortunately they are classifiable, and in this chapter I shall confine myself to the earliest representatives of each class, with the exception of one or two of special technical interest.

A great deal of the immense literature deals with the exposure of rain gauges, the comparison of gauges of various sizes and at various heights above ground,[2] and the shielding of gauges from the effects of wind.[3] I shall deal with this only insofar as it is an instrumental matter.

2. Early rain gauges. The rain gauge is one of those devices that were invented in the Orient and reinvented much later in western Europe. Early in the present century an Indian scholar, Jogindra Nath Sammadar, found a passage in a very ancient work in Sanskrit, the *Arthastra*, or Science of Politics, which shows that in the fourth cen-

[1] John C. Kurtyka, *Precipitation Measurements Study*, Illinois State Water Survey Division, Report of Investigation no. 20 (Urbana, Ill., 1953), lists about 160. This paper gives 1,079 references altogether, not all verified. It must be used with extreme caution.

[2] On this, see my *A History of the Theories of Rain*, pp. 97–99; 168–69.

[3] For a bibliography of this, see Charles F. Brooks, *Need for Universal Standards for Measuring Precipitation, Snowfall, and Snowcover*. International Association of Hydrology, Bulletin 23 (Riga, 1938).

tury B.C. rain was measured in several parts of India.[4]

The quantity of rain that falls in the country of Jangala is 16 dronas; half as much more in Anupanam countries; 13½ dronas in the country of Asmakas; 23 dronas in Avanti; an immense quantity in Apparantam, the borders of the Himalayas, and the countries where water-channels are made use of in agriculture.

When one-third of the requisite quantity of rain falls both during the commencement and closing months of the rainy season, and two-thirds in the middle, then the rainfall is considered very even.

A forecast of such rainfall can be made by observing the position, motion, and brightness of Jupiter, the rise and set of Venus, and the natural and unnatural aspect of the Sun.

From the Sun, the sprouting of the seeds can be inferred; from the position of Jupiter, the formation of grains can be inferred.

Three are the clouds that continuously rain for seven days; 80 are they that pour minute drops; and 60 are they that appear with the sunshine—this is termed rain-fall. Where rain falls, free from wind and unmingled with sunshine, so as to render three turns of a ploughing possible, there the reaping of good harvest is certain.

Hence, according as the rainfall is more or less, the Superintendent shall sow the seeds which require more or less water.

Sammadar notes that the *drona* is a measure of capacity, and so presumably some standard vessel was used to catch the rain. There is no indication that the Indians of this period had thought of expressing the rainfall as a depth of water, but this latter idea had arisen by about A.D. 100 in Palestine, as was shown by Hermann Vogelstein in a study of the agriculture of that period. He wrote: "The knowledge of the importance of rain for agriculture had already led to fairly accurate observations and measurements in the time of the Misnah [first two centuries A.D.] The depth of rainfall was measured with the help of a vessel. It seems to have amounted to one tefah (about 9 cm) in the first period of early rain; twice as much in the second, and three times as much in the third period."[5]

More than 1,000 years later the rain gauge, in a technically recognizable form, turned up in Korea. In 1910 Dr. Y. Wada, the Japanese director of the Korean Meteorological Observatory at Chemulpo, published an article in French about fifteenth-century Korean rain gauges.[6] In the annals of the period, he had found the following passage:

In the 24th year [1442] of the reign of King Sejo, the king caused a bronze instrument to be constructed, in order to measure the rain. This is a vase [30 cm] in depth and [14 cm] in diameter, standing on a pillar. The instrument has been installed at the Observatory, and each time that rain falls,

[4] Sammadar, *Quart. J. Roy. Meteorol. Soc.*, Vol. 38 (1912), pp. 65–66.

[5] H. Vogelstein, *Die Landwirtschaft in Palästina zur Zeit der Misnah, I. Teil: Der Getreidebau* (Berlin, 1894), p. 3.

[6] Wada, *Sci. Mem. Korean Meteorol. Observatory*, Vol. 1 (1910). Translated in *Quart. J. Roy. Meteorol. Soc.*, Vol. 37 (1911), pp. 83–86.

the officials of the Observatory measure the height with a scale, and make it known to the king. These instruments were distributed in the provinces and cantons, and the results of the observations were sent to the court.[7]

Apparently King Sejo was famous for his enlightened policies. Wada makes the point that there was a formal network of rainfall stations in the fifteenth century; this must have been by far the earliest meteorological network of any kind, antedating by two centuries the one established by Ferdinand II of Tuscany.

It seems very unlikely indeed that the three inventions of the rain gauge, in India, Palestine, and Korea, can be connected in any way. It is inconceivable that any of them was known to Benedetto Castelli, who in 1639 was interested not in agriculture, but in engineering. On June 18, 1639, Castelli wrote to Galileo from Rome,[8] telling him that he had been in Perugia, and that there had been a continuous rain that lasted for eight hours. He wanted to estimate how much the rainfall contributed to the outflow from Lake Trasimeno. He thought that one might suppose that the rain was uniform over the whole region, "and so, taking a glass vessel in the form of a cylinder, about a palm high and half a palm wide, and putting in a little water, enough to cover the bottom of the vessel, I carefully noted the level of the water, and then exposed it in the open air to receive the rainwater that might fall into it, and let it remain for the space of an hour."[9] He drew a short line on his letter to indicate how much the level in the vessel had risen.

It would be interesting to know whether any of the members of the Accademia del Cimento had heard about this. In any event, they did not follow it up, and the next rain gauge, in fact the next two, were made in England, both by that versatile genius Christopher Wren. It is characteristic of him that they were recording rain gauges, and that at least one of them was part of a meteorograph, an instrument that made a record of several meteorological elements. This will be described in Chapter 7, and I shall note here only that the rain gauge consisted of a mechanism for bringing a succession of containers under the discharge of a collecting funnel, one every hour; that it was certainly completed by June, 1663; and that by November of the same year Wren had seen the major defect of this scheme (namely that the rainwater in the containers might largely evaporate before the observer examined them) and had abandoned the idea.[10] Nevertheless it persisted, and in 1726 we find it in one of Jacob Leupold's immense books of machines, with an ingenious discontinuous motion provided by a one-toothed gear, as shown in Figure

[7] *Ibid.*, p. 84.
[8] Galileo, *Opere*, ed. naz. (20 vols., Florence, 1890–1909), XVIII, 62–66.
[9] *Ibid.*, p. 62.
[10] See pp. 246–49 below.

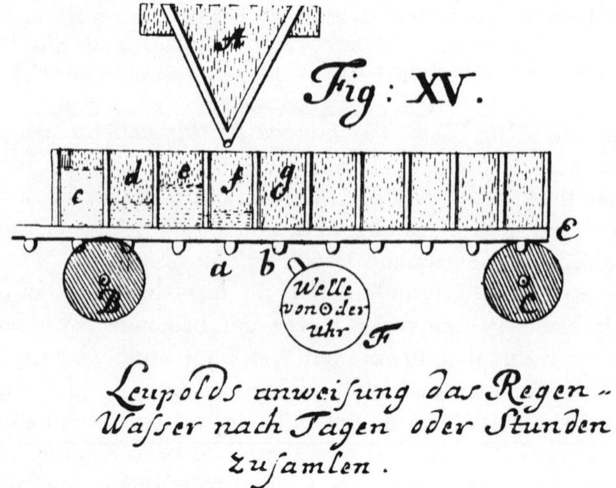

Fig. 4.1 One of Leupold's recording rain gauges, 1726.

4.1.[11] Leupold usually gave credit where it was due, so that we may assume that he had reinvented this himself. The idea reappears in the Erzgebirge in 1789 among an assemblage of recording instruments, but Christian Gotthold Hermann mounted twelve bottles in a circle on a horizontal disk, and brought them successively under the receiving funnel by clockwork, so that each remained for an hour.[12]

On January 22, 1661/2, Wren had shown The Royal Society "his Experiment of filling a vessel with water, which emptyed it self, fil'd at a certain hight."[13] This was the "tipping bucket," which is now familiar, and indeed was ancient in Wren's time, having been known to the Arabs in the dark ages. He could have seen it in a book by Isaac de Caus printed in 1644, where it forms part of the drive of a water-clock.[14] That Wren used it in a rain gauge is not noted in the above extract, but Thomas Sprat makes it clear: "[Wren] has contriv'd an instrument to measure the quantities of rain that falls: This as soon as it is full, will pour out it self, and at the years end discover how much rain has fallen on such a space of land, or other hard superficies, in order to the theory of vapours, rivers, seas, &c."[15] This presupposes some counting mechanism. We shall see later that Robert Hooke built the tipping bucket into a meteorograph, and it is

[11] Leupold, *Theatri statici universalis. Pars III, sive theatrum aerostaticum, oder: Schau-Platz der Machinen zu Abwiegung und Observierung aller vornehmsten Eigenschafften der Lufft* [etc.] (Leipzig, 1726), p. 298 and plate XVII.

[12] Hermann, *Mechanisch verbesserter Wind-, Regen- und Trockenheits-Beobachter* (Freiburg im Sachsen, 1789), pp. 68–75.

[13] Royal Society, *Journal Book*, Jan. 22, 1661/2.

[14] Isaac de Caus, *Nouvelle invention de lever l'eau plus hault que sa source* (Londre [*sic*], 1644), p. 19 and plate V.

[15] Thomas Sprat, *History of The Royal Society* (London, 1667), p. 313. This rain gauge was still in the "cabinet" of the Royal Society in 1681, for it is described by Nehemiah Grew, *Musaeum Regalis Societatis* [etc.] (London, 1681), p. 358.

always difficult to apportion credit for mechanical ideas between the two friends, Hooke and Wren, who undoubtedly stimulated each other's imaginations. Meanwhile it is remarkable that in the "Method of making a history of the weather," which Hooke wrote in 1664 for The Royal Society,[16] he does not ask for any measurement of rain.

3. Development of the simple rain gauge. Sprat's reference to "the theory of vapours, rivers, seas, &c." will underline the fact that the seventeenth-century measurements of rain were made not in the interests of agriculture or engineering, but to answer a purely academic question about the origin of rivers and springs, a question that had been debated since classical times. I cannot deal with the history of this problem here; it will be sufficient to state that there was doubt whether the rainfall and snowfall together are sufficient to account for the water carried by the rivers to the sea. The first European rainfall observations made in a systematic way are described in a book with a title that makes their purpose clear.[17] Though Perrault's rain gauge is not described, we learn that he measured the rainfall near the headwaters of the Seine for three years, obtaining an average of 19 [Paris] inches, 2⅓ lines (511 mm), and calculated from hydraulic data that this was much more than enough to maintain the river.

A few years later, Edme Mariotte had a similar experiment made by a friend who lived at Dijon, and with the same purpose, but he did give a brief description of the rain gauge.[18] A collector two feet square was exposed on a bracket six feet away from a window near the top of a house, and a pipe carried the rainwater into a cylindrical vessel from which, Mariotte notes, very little could evaporate. The same general construction was copied by Philippe de la Hire at the Paris Observatory, where the rain gauge was a tin vessel four square feet in area and six inches deep, with a bottom that sloped a little to one corner, where a short tube led the rainwater into a jug. After each rain the water was measured in a small cubical vessel.[19]

Meanwhile, observations were being made in England, first by Richard Townley at his house in Lancashire, whose rain gauge was "a round tunnel (*sic*) of 12 inches diameter," with a leaden pipe to discharge into a vessel in his room.[20] He measured the water in a cylindrical glass that had been graduated by weighing. Townley was the first to publish monthly totals, and he tells us that he had hoped that several other people might have instituted observations in other

[16] *Register Book II*, fol. 1ff.
[17] [Pierre Perrault], *De l'origine des fontaines* (Paris, 1674; 2nd ed. Paris, 1678).
[18] Mariotte, *Traité du mouvement des eaux et des autres corps fluides* (Paris, 1686), p. 30. Also in *Oeuvres* (2 vols., paged as one; Leyden 1717), p. 338.
[19] La Hire, *Mém. Acad. Roy. Sci. Paris* (1700), p. 6.
[20] Townley, *Phil. Trans.*, Vol. 18 (1694), p. 52. "Tunnel" is surely a misprint for "funnel."

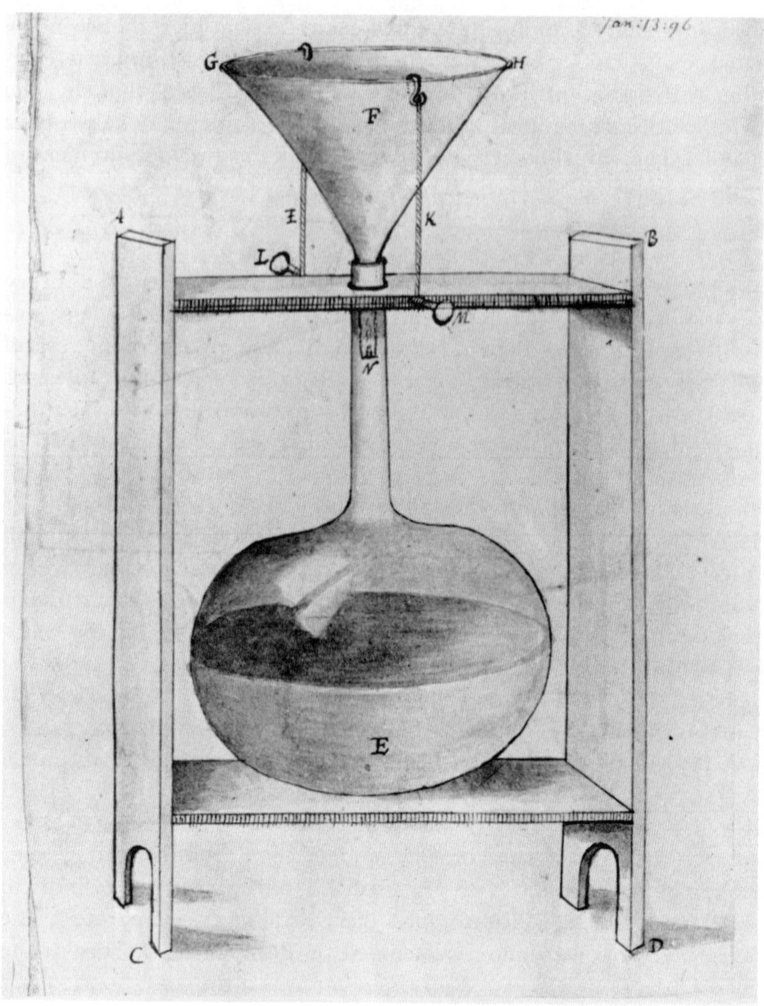

Fig. 4.2 The Gresham College rain gauge of 1695 (copyright of The Royal Society; reproduced by permission).

parts of England, but this does not seem to have happened at the time.

Nevertheless, Townley's paper may have been the inspiration for measurements at Gresham College in London, begun on August 12, 1695 and published anonymously in the *Philosophical Transactions* two years later,[21] with the earliest published figure of a simple rain gauge. On the manuscript of this paper[22] it is ascribed to W. Hunt,[23] and I

[21] *Phil. Trans.*, Vol. 19 (1697), p. 357.
[22] Royal Society, *Classified Papers* V, item 1. Copyright by the Royal Society; quoted by permission.
[23] But it must surely have been Harry Hunt, the "Operator" to The Royal Society from 1676 to 1696.

reproduce his original drawing in Figure 4.2. The drawing is dated "Jan. 13. 96," which almost certainly means 1696/7, and is accompanied by the following explanation, possibly the earliest detailed description of a rain gauge: "*A.B.C.D.* is a frame to support the glases *E.* is a large bolt head with a neck of 20 inches longe and Capable of holding 2 galons—*F.* is a funel, whose diameter being 11. inches and 4/10 from *G.* to *H.*—*I.K.* are two stays or pack threads, which are strain'd by two pins *L.M.* to hold the funel stedey against the high winds. *N.* the pipe of the funel at *N.* being not wider than 1/5 of an inch thorrow, which the evaporation can be but little." It is noteworthy that the rainfall is reported in pounds, ounces, and grains, with only the annual total converted to a depth of water.

Gauges varied enormously in area. The Rev. Mr. Horsley of Widdrington in Northumberland used a funnel thirty inches in diameter.[24] To avoid having to weigh the water he measured it in "a cylindrical measure and gage," the measure exactly three inches in diameter, the "gage" apparently a graduated stick; "or, instead of a gage, the inches and divisions may be mark'd on the side of the cylindrical measure." Here we have the two common ways of performing the operation at the present time, though the "measure" would now be made slightly larger in order to allow for the cross-sectional area of the stick. Horsley's way of doing it almost coincides with that recommended by the Secretary of The Royal Society in his famous general "invitation" to make weather observations,[25] except that the funnel should be about two or three feet wide. The problem of exposure was considered: "The funnel should be sited so that, whatever wind may blow, no part of the rain may be intercepted by a building or other obstruction."[26]

At the other extreme of size was the "ombrometer" of Roger Pickering (see End Papers), which had a funnel only one inch square, discharging directly into a graduated glass tube about half an inch in diameter. This was calibrated by means of a measure holding one cubic inch, and graduated in thirty-seconds of an inch of rain.[27] This use of a graduated glass tube as an integral part of the gauge had been made by Johann Georg Leutmann,[28] but his funnel was one foot square, and the tube was graduated in pounds and half-ounces. It was made like a burette, with a stopcock permitting small amounts of water to be drawn off into a narrower tube graduated in eighths of an ounce. Leutmann preferred to give his results in pounds of water per square foot. The winters were cold in his village near Wittenberg, and "so that in the winter time the funnel should not get blocked by

[24] Horsley, *Phil. Trans.*, Vol. 32 (1723), pp. 328–29.
[25] James Jurin, *Phil. Trans.*, Vol. 32 (1723), pp. 422–27.
[26] *Ibid.*, p. 426.
[27] Pickering, *Phil. Trans.*, Vol. 43 (1744), No. 473, pp. 1–17. (No. 473 is erroneously numbered from page 1.)
[28] Leutmann, *Instrumenta meteorognosiae inservientia* (Wittenberg, 1725), pp. 127–32.

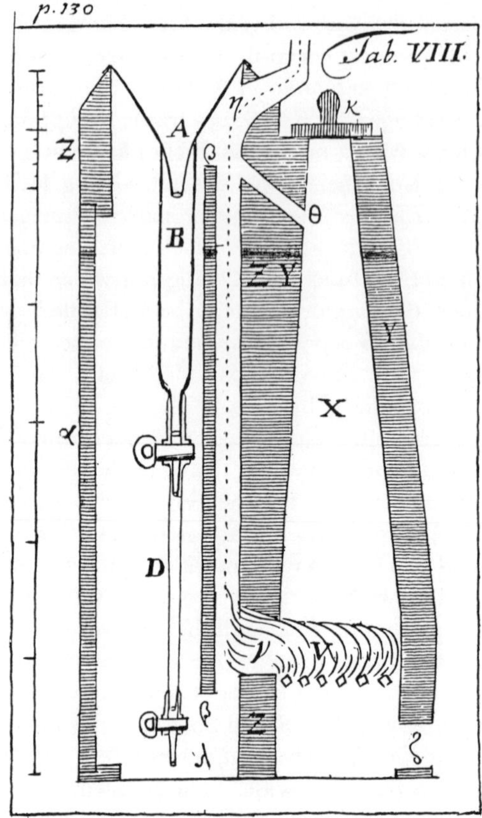

Fig. 4.3 Leutmann's heated gauge, 1725.

snow or frozen water, a sort of tower *Z* [fig. 4.3] may be built of stone, as high as the hyetometric instrument and wide enough so that the funnel may exactly fill the top. On one side may be built a furnace *Y* of the kind called 'lazy Harry' . . . It should be about a foot wide and high enough to hold a fire for twelve hours."[29] If this was built, it was probably the first heated rain gauge, but in any event it may cause us to reflect on the blessings of gas and electricity.

For the next thirty years or so most people seemed content to follow the lead of the Paris Academy[30] or of Jurin. Grischow,[31] who reports that rainfall began to be measured only in 1728 at Berlin, was aware of the difficulty of catching snow, "which may easily be expelled by the fury of the wind, through the aperture that has just caught it." So that more of the snow could be caught, Pieter van Musschenbroek,[32]

[29] *Ibid.*, p. 130.
[30] For example, J. F. Weidler, *Observationes meteorologicae atque astronomiae ann. MDCCXXIIX et MDCCXXIX* [etc.] (Wittenberg, 1729), pp. 19–21.
[31] Augustin Grischow, *Miscell. Berol.*, Vol. 4 (1734), pp. 349–51.
[32] Musschenbroek, *Novi comment. Acad. imp. sci. Petrop.*, Vol. 8 (1760–61), pp. 367–91.

writing in his old age a budget of good advice to observers, recommended a brass rain gauge, one Rhenish foot square, with sides eight inches high, and a pyramidal bottom. Although this was not generally recognized until much later, the high vertical sides were also important in preventing the loss of rain by splashing.

Describing the meteorological instruments used at The Royal Society's house in 1776, Henry Cavendish[33] showed his concern about splashing, but had a different solution:

> The vessel which receives the rain is a conical funnel, strengthened at the top by a brass ring, twelve inches in diameter. The sides of the funnel and inner lip of the brass ring are inclined to the horizon, in an angle of above 65°; and the outer lip in an angle of above 50°; which are such degrees of steepness, that there seems no probability either that any rain which falls within the funnel, or on the inner lip of the ring, should dash out, or that any which falls on the outer lip should dash into the funnel.

In the 1790's Thomas Garnett was thinking about the effect of splashing, and also about evaporation, rejecting on the latter account a drastic cure for splashing suggested by the famous blind philosopher John Gough of Kendall. This was to fit "a linen strainer of a conical figure" to the mouth of the gauge; "this flexible funnel should be stretched by a weight or string fixed to its apex within the vessel." Garnett preferred "a perpendicular rim, an inch or two high, fixed to the rim of the funnel."[34] He also favored square gauges, believing that the wind produces violent eddies in cylindrical ones.

Gauges with a cylindrical vessel above the funnel, that is, funnels with very deep rims, were tried by the versatile L. A. J. Quetelet in Belgium, beginning about 1840.[35] Designed to catch snow, they had a large wetted area, and the resulting evaporation reduced the catch in light rains. A fact established by Quetelet was that the inner surface of gauges should not be painted, for if they are, large drops adhere and eventually evaporate. In rain totalling not more than 10 mm, a painted funnel registered less than half as much as an unpainted one.[36]

The worry about splashing was not universal in Cavendish's time, and the rain gauges sent out to various stations from Mannheim by the Elector Palatine were essentially like the old Paris gauge.[37]

Luke Howard, through his long-continued observations of the weather, and especially through his book *The Climate of London*,[38] had great influence in England. His gauges, much like that of Hunt except that the copper funnel had a turned brass rim, were still being offered for sale in the last quarter of the nineteenth century (Fig. 4.4).

[33] Cavendish, *Phil. Trans.*, Vol. 66 (1776), pp. 384–85.

[34] Garnett, *Royal Irish Acad. Trans.*, Vol. 5 (1794), pp. 257–63. I have not found any publication of Gough's idea.

[35] Quetelet, *Sur le climat de la Belgique* (Brussels, 1849–57), Part 5 (1852), p. 2.

[36] *Ibid.*, p. 3, note.

[37] *Ephemerides Soc. meteorol. Palatinae anni 1781* (Mannheim, 1783), pp. 82–84.

[38] Two vols., London, 1818–20; 2nd ed., 3 vols. London, 1833.

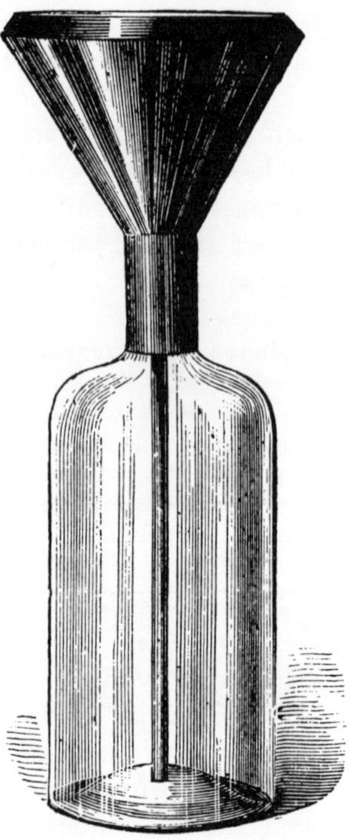

Fig. 4.4 Howard's rain gauge.

Howard also used an all-metal gauge, and after 1811 adopted the practice of sinking the instrument into the ground, "bringing the mouth of the funnel nearly to the level of the turf."[39]

This introduces the question of the proper exposure of a rain gauge, a problem with an immense literature. How high above ground should a gauge be? In the eighteenth century it was noticed that gauges farther from the ground caught less rain, and though it was soon suspected that this might be the effect of wind, it was 100 years before the explanation was everywhere accepted.[40] Most of the early gauges were on roofs, but Leutmann's was an exception. Howard's idea gave an opportunity for rain to splash into the gauge from its surroundings, but in 1842 Thomas Stevenson, the celebrated lighthouse engineer, who was the father of Robert Louis Stevenson, showed how this could be avoided.[41] He began by pointing out that the measurement of rainfall is of importance not only to "the interesting but uncertain science of meteorology," but to agriculture and civil engineering as well, which seems to take us back to the Koreans and to Castelli. After an excellent discussion of the errors of rain gauges and

[39] *Ibid.*, I, xxiii.
[40] *Cf.* my *A History of the Theories of Rain*, pp. 98–99; 168–70.
[41] Stevenson, *Edinb. New Philos. J.*, Vol. 33 (1842), pp. 12–21.

particularly of the effects of wind, he showed two forms of gauge with the mouth sunk level with the turf, one of which is illustrated in Figure 4.5. Splashing is prevented by an annular bristle brush *fg*.

The "pit gauge" did not catch on, partly because gauges are needed in large numbers, and its installation is expensive, and partly because it is not very suitable for snowy climates. Many years later G. J. Symons observed that "Mr Griffith reported that it was a very difficult gauge to work, and that its chief efficiency was in collecting insects."[42] However, arguments like those of Stevenson did result in the general adoption in the British Isles of a standard height of only one foot—above "short turf"—for the mouth of a rain gauge. This

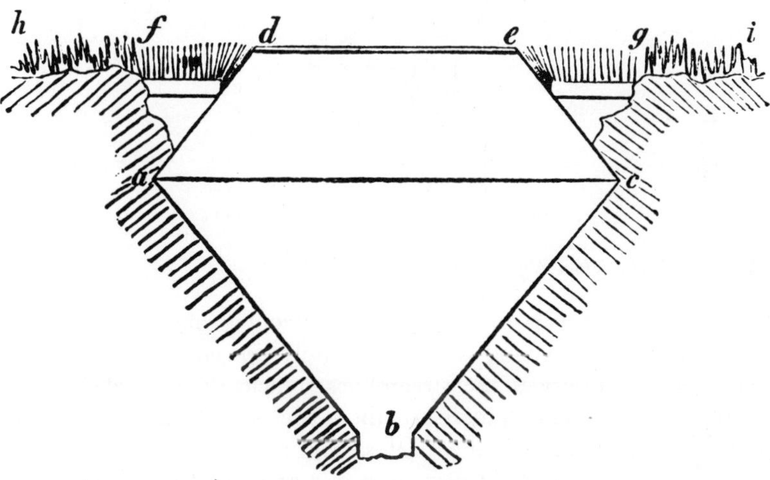

Fig. 4.5 A rain gauge of Thomas Stevenson, 1842.

had its effect on the design of recording gauges. In most of the snowier countries gauges came to be taller, from approximately 70 to 200 cm above the ground.

The measurement of snow was soon recognized to offer peculiar problems, which have not even yet been solved to everyone's satisfaction. In 1788 Franz Carl Achard provided a table four by ten feet in extent in the middle of the garden of the Berlin Academy's house, and a vertical brass scale in the center of the table, but he admitted that this was useless in windy weather.[43] A carpenter's rule has been the commonest of snow gauges until the present day. Catching snow in ordinary rain gauges, even if they have very deep collectors, is seldom satisfactory because of the effect of wind, for the turbulence produced by the gauge itself is of importance. In an effort to render the flow of

[42] *Quart. J. Roy. Meteorol. Soc.*, Vol. 17 (1891), p. 181.
[43] Achard, *Mém. Acad. R. Sci. Berlin* (1788), p. 100.

air across the gauge more nearly horizontal, the Director of the Smithsonian Institution soldered a "circular plate of tin" to the gauge "like the rim of an inverted hat, three or four inches below the orifice of the gauge."[44] This was about 1858. Twenty years later F. E. Nipher invented his shield, which has been the subject of endless experiment, discussion, and improvement ever since.[45] It is shown in section in Figure 4.6. The upper, wider part of the cone is lined with

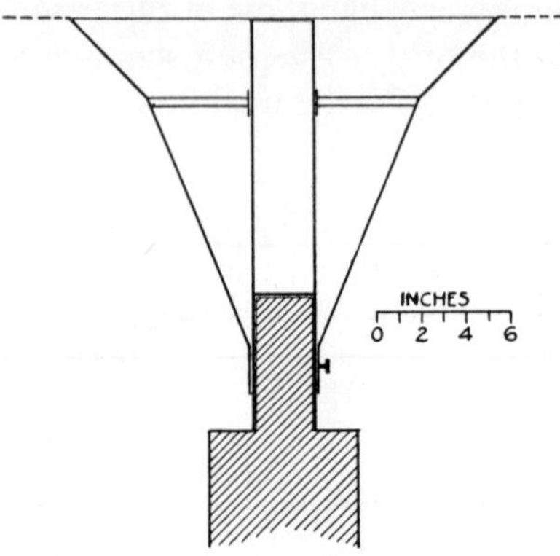

Fig. 4.6 The original Nipher shield, 1878.

wire mesh, and this is extended into a horizontal ring. By the use of the wire mesh, splashing is almost entirely prevented, and such an arrangement improves the streamlines sufficiently to make its use highly desirable, though there are still difficulties with snow, especially wet snow. I shall not deal with this subject further, but instead refer the reader to an excellent discussion and annotated bibliography by Charles F. Brooks.[46]

4. The "vectopluviometer." A vectopluviometer is an instrument designed to determine the direction from which the rain is coming, or to measure the distribution of rain between various directions, or both. It may simply measure the amount of rain from each of, say, eight directions, or it may attempt to indicate the average angle to the vertical made by the raindrops collected by it. The earliest vectopluviometer of which I am aware combined the functions of the last two classes, and was the invention of an instrumentmaker named Kerr.[47] It had two circular openings, one horizontal, the other in a vertical plane and faced into the wind by a vane. A pipe from each

[44] *Scientific writings of Joseph Henry* (2 vols., Washington, D.C., 1886) II, 262.
[45] Nipher, *Proc. A.A.A.S.* (St. Louis meeting, 1878), Vol. 27 (1878), pp. 103–8; or in German in *Zeits. österr. Ges. Meteorol.*, Vol. 14 (1879), pp. 250–54.
[46] Note 3, this chapter.
[47] *Annals of Philosophy*, Vol. 7 (1816), pp. 387–88.

collector was bent so that it discharged into two concentric sets of eight receivers. In 1837 a rather similar instrument was described by the Reverend Thomas Knox, but it had only the vertical collector and one set of eight compartments.[48]

In 1840 the geologist John Phillips thought of another sort of vectopluviometer, which had no moving parts.[49] This was in effect an assemblage of five rain gauges, one facing upward, and the other four facing north, east, south, and west, respectively. From the amount of rain collected by these, the average azimuth and inclination of the raindrops could be calculated. Vectopluviometers have not been much used, although it seems possible that they could interest city planners.

5. Classification of recording rain gauges. Like most of the other meteorological instruments, rain gauges have been made to provide a record on a chart, showing the incidence of precipitation as a function of time. The most useful classification of recording gauges is probably that of G. J. Symons,[50] who devoted his life to the organization of rainfall measurements in the British Isles. Symons divided all recording gauges into those in which a float, rising in a vessel as rainwater accumulates, moves the recording pencil, and those in which the weight of the collected water is used in some way to make a record.

These two main classes can be subdivided according to the following scheme.

1. Float Gauges
- (a) The record limited by the capacity of the float chamber, there being no automatic provision for emptying it, and none for a multiple traverse of the chart by the pencil.
- (b) The record still limited by the capacity of the chamber, but with mechanical arrangements for returning the pencil to zero one or more times.
- (c) The float chamber emptied automatically at a predetermined level, usually by a siphon, returning the pencil to zero.

2. Balance Gauges
- (a) The record limited by the capacity of the vessel in which the water is collected.
- (b) The collecting vessel emptied when full by a siphon, the pencil returning to zero.
- (c) The tipping-bucket.

This is fairly comprehensive, but tipping-bucket gauges ought to be further classified as follows: gauges in which the least count corre-

[48] Knox, *Phil. Mag.*, Vol. 11 (1837), p. 260.
[49] Phillips, B.A.A.S., Glasgow, 1840, *Sections*, pp. 45–47.
[50] Symons, *British Rainfall* (1879), pp. 33–41.

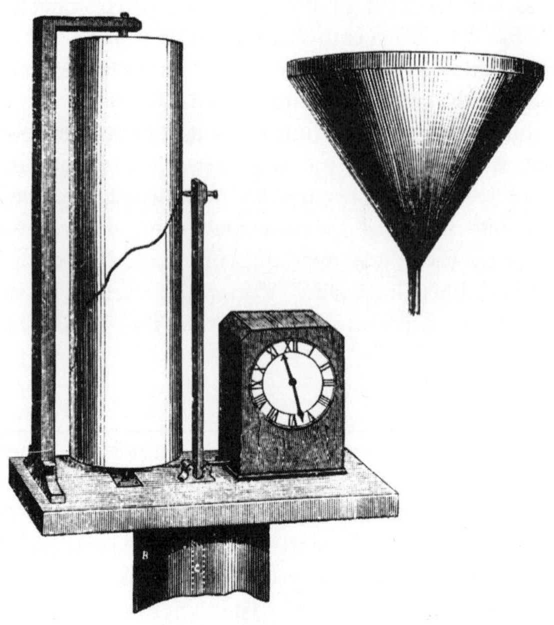

Fig. 4.7 Bevan's float rain gauge, 1827.

sponds to one filling of a bucket (c_1); and gauges in which the bucket is continuously weighed while it is being filled, the tipping being used only to return the pencil to zero (c_2). It will be seen that there is a close similarity between classes 2b and $2c_2$.

Because of the risk of damage by frost, float gauges are unpopular in countries with a severe winter climate, for example in Canada.

6. Recording rain gauges using a float. In 1782 John Hyacinthe Magellan, a Portuguese who had settled in London and had been elected to The Royal Society, described an extraordinary meteorograph,[51] though I have found no indication that it was ever built. It was to contain a recording rain gauge with a float chamber, and the recording pencil was to be attached directly to a rod rising from the float. My skepticism is reinforced by the circumstance that the funnel for catching the rain was required to be of the same diameter as the cylindrical float chamber so that the magnification of the record would be unity.

Apart from this, the earliest recording rain gauge using a float was probably that of Bevan.[52] The illustration (Fig. 4.7), from Symons'

[51] Magellan, *Obs. sur la Phys.*, Vol. 19 (1782), pp. 108–25; 194–212; 257–73; 341–56; especially pp. 354–55.
[52] Bevan, *Phil. Mag.*, Vol. 2 (1827), pp. 74–75.

paper,[53] appears to have been constructed from the description; at all events it should only be noted that the float chamber had to be emptied manually to such an extent that the recording stylus would start again at zero.

This simple type of recording rain gauge (Class 1a) is obviously not susceptible to much significant variation. Class 1b is only a little more versatile. Its first representative was part of a meteorograph[54] which will be referred to again in Chapter 7. According to Symons in the article already cited, water from a funnel one foot square was led into a float chamber. The float was surmounted by a rod carrying six inclined planes at suitable intervals, and as the float rose in the chamber, one end of a lever operating the pen arm slid sideways along one of the planes until it went off the edge and dropped on to the next one, returning the pen to zero.

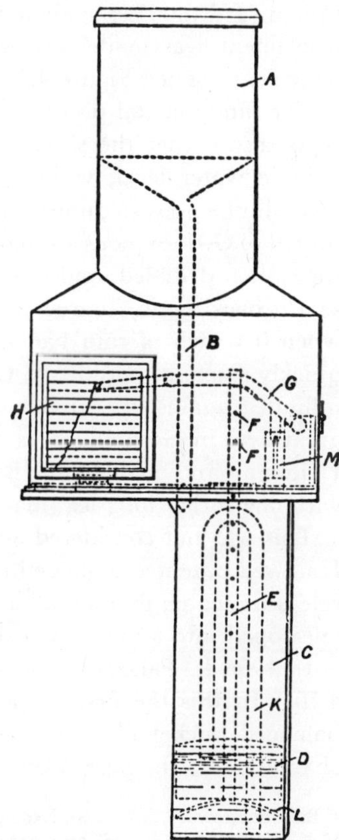

Fig. 4.8 Halliwell's "hyetograph," 1907.

The "hyetograph" of F. L. Halliwell[55] was a more sophisticated development of the same idea, and won favor, especially in the British Isles, by its simple construction (Fig. 4.8). The descent of the

[53] *British Rainfall* (1879), pp. 33–41. Symons quotes the date as 1817.
[54] George Dollond, B.A.A.S., Southampton, 1846, *Sections*, p. 17.
[55] British Patent 27,174 of 1908. See also H. R. Mill, *British Rainfall* (1908), pp. 22–27.

pen arm after each half-inch of rain is cushioned by an oil-filled dashpot.

The third subclass (1c) of float gauges, in which the float chamber is emptied periodically by a siphon, is a large one. The earliest seems to have been the invention of Robert Beckley, an assistant at Kew Observatory who had a remarkable flair for the design of instruments. Beckley's gauge, first described to the British Association in 1869 by the Superintendent at Kew,[56] is shown in Figure 4.9. It is strongly constructed with a frame of cast iron, and great care was taken to keep the weight-driven pendulum clock dry, even to the extent of using an oil seal at *S*. Water collected by the funnel *C* is led to the chamber *E*, which is supported by a thick iron cylinder *F*, floating in a bath of mercury and attached rigidly to the recording pencil *K*. As the chamber *E* fills it sinks down, and when it is full it is emptied by the siphon *MN* and rises again to its zero position. This well-thought-out instrument was made by the instrumentmaker James Hicks, from whose catalogue[57] Figure 4.9 is taken.

The fundamental problem in the design of a siphoning rain gauge is to ensure that the siphon empties the chamber quickly at some definite water level, without any dribbling; after 1875 there were several ingenious solutions. In 1886 the firm of Richard Frères submitted to G. J. Symons a siphoning rain gauge, but Symons found that the siphon dribbled, and returned it. The resourceful Jules Richard sent it back again, "cured." The cure was an electromagnet which, when 0.4 inch of rain had fallen, gave a strong push to the float, quickly starting the flow.[58] Of course the requirement for a battery militated against the success of this idea. A simpler scheme, used a great deal in Germany, was the recording gauge devised by Gustav Hellmann and made by the Berlin firm of Fuess,[59] which had a siphon with one rather long leg, made of a narrow glass tube.

This was not considered adequate in England, and in 1901 F. L. Halliwell patented a gauge in which the float, at the top of its travel, released a catch that let fall a siphon. The shorter limb of the siphon telescoped into a mercury seal, starting the flow at once.[60]

In 1906 L. Palazzo found the ingenious solution shown in Figure 4.10. In this the first drops from the siphon *FF* fall into a small balanced bucket *H* in a closed chamber *L*. The bucket soon discharges down the pipe *N* and the resulting suction starts a full flow.[61]

[56] Balfour Stewart, B.A.A.S., Exeter, 1869, *Sections*, p. 52. See also *Meteorol. Mag.*, Vol. 6 (1871), pp. 53–55, and *Rep. Meteorol. Comm. Roy. Soc.* for 1869, pp. 36–39.

[57] *James J. Hicks' Illustrated and Descriptive Wholesale Catalogue of Standard, Self-Recording, and other Meteorological Instruments* . . . London, James J. Hicks, 8, Hatton Garden, n.d. [about 1875]. I am obliged to Mr. A. L. Maidens for drawing my attention to this work.

[58] Symons, *British Rainfall* (1898), pp. 22–23.

[59] Hellmann, *Meteorol. Zeits.*, Vol. 14 (1897), pp. 41–44.

[60] F. L. Halliwell, British Patent 8214 of 1901 (Apr. 22).

[61] Palazzo, *Riv. Meteorol. Agrar.*, Rome, Vol. 26 (1906), pp. 837–42. See also *Zeits. Instrum.*, Vol. 27 (1907), pp. 202–3.

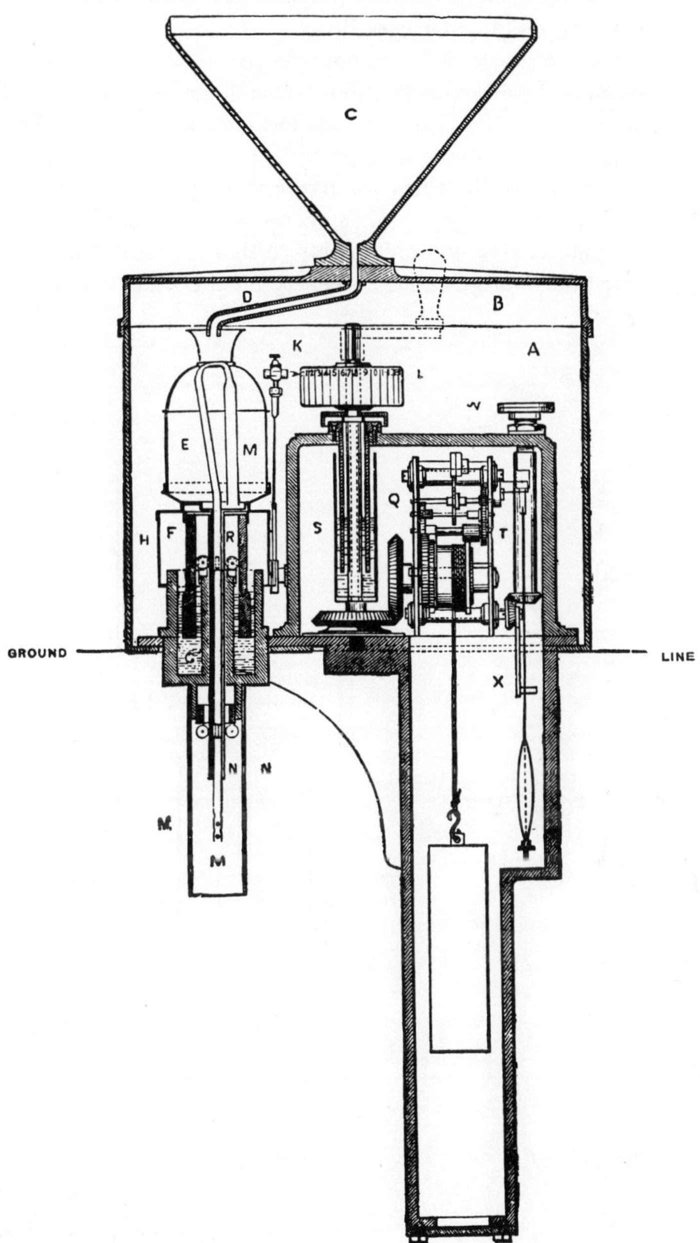

Fig. 4.9 Beckley's recording rain gauge, 1869.

The clock is cleverly arranged around the float rod, and the record is made in rectangular coordinates.

Of the many other siphon gauges, two will be mentioned. The simpler is the "natural-siphon" rain gauge of Negretti & Zambra, an invention of Halliwell,[62] which has a coaxial siphon. The upper end of the outer tube is formed by a polished concave glass cap. The upper end of the thick discharge tube, polished to a rounded edge, comes very close to the glass, forming a very narrow annular aperture with all its area compressed into a short vertical distance (Fig. 4.11). The area is large enough to carry off the water rapidly.

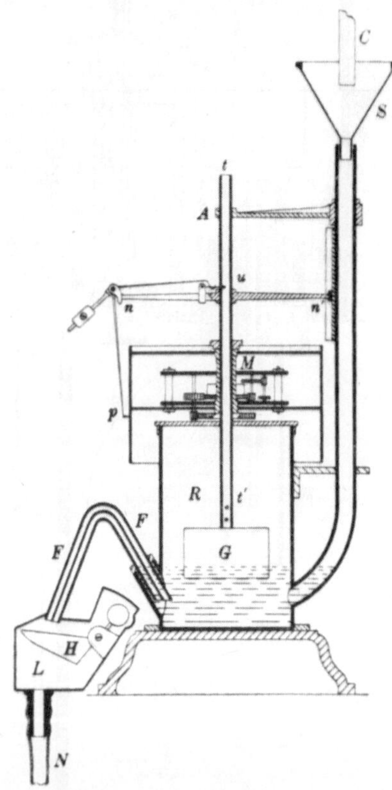

Fig. 4.10 Palazzo's recording rain gauge, 1906.

The "tilting-siphon" gauge of W. H. Dines[63] was intended to provide a positive siphoning action (Figure 4.12). In this the float-chamber *A*, the float *B*, and the siphon *D* are mounted on knife edges at *C*. When the float rises to the top of its travel, it strikes a lever attached to the trigger *E*, which releases the entire assembly and

[62] Halliwell, British Patent 167,016, May 11, 1920.
[63] Dines, *Meteorol. Mag.*, Vol. 55 (1920), pp. 112–13. *Collected Papers*, pp. 112–13. It was first used at Benson in 1915.

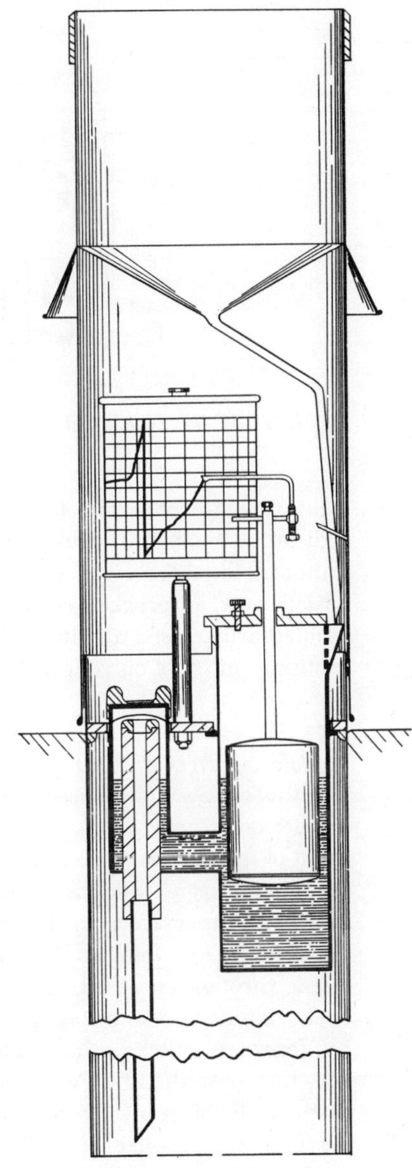

Fig. 4.11 The "natural-siphon" rain gauge (courtesy of Negretti & Zambra).

allows it to tip toward the right, emptying out just enough water to allow the float to return to zero. At that point the counterweight F returns the mechanism to its original position, where it is held by the trigger E until the float again releases it. One traverse of the pen corresponds to 0.20 inches or 5 mm of rain.

One other rain gauge using a float ought to be mentioned, as it

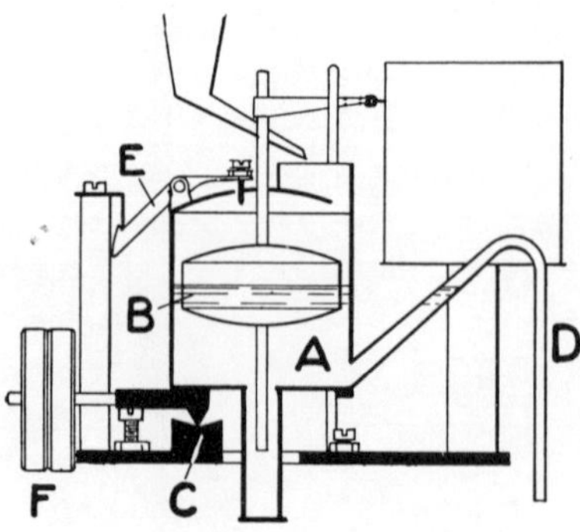

Fig. 4.12 W. H. Dines's "tilting-siphon" rain gauge.

seems to be the only one of its class designed for electrical recording at a distance.[64] In this instrument a float in the measuring vessel of a standard Signal Service rain gauge was attached to a chain that passed over a sprocket to a counterweight. Pins on the sprocket momentarily closed a switch once for each 0.05 inch of rain, and the resulting pulses of current operated a distant chronograph.

7. *Balance gauges other than tipping-bucket gauges.* As we saw in section 5 of this chapter, there are two sorts of balance gauges; of these the simpler one seems to have been invented more recently, probably as a result of the need for a recording gauge that would measure precipitation in any form. Such a gauge was developed by S. P. Fergusson at Blue Hill Observatory in the 1880's,[65] and greatly improved as time went on.[66] The final form is shown in Figure 4.13. Rain or snow falling through an aperture eight inches in diameter is collected in a bucket supported on a specially designed spring balance. The levers *L,L* form a parallel motion. There is a very ingenious linkage to the pen, thus described by Fergusson: "The end of the recording style *P* on the right of its pivot is heavier than the end carrying the pen, and the link *I* is heavier than I^2; consequently continuous downward motion of the frame *B* and its connections will cause the pen to move alternately backward and forward until it has crossed the record sheet four times."[67] The design of the mouth of the gauge and a dashpot containing glycerine reduce oscillations caused by the wind, and for use in the summer a funnel can be inserted in order to prevent evaporation.

[64] C. F. Marvin, *Science*, Vol. 11 (1888), pp. 97–98.
[65] Fergusson, *Amer. Meteorol. J.*, Vol. 5 (1888–89), p. 321.
[66] Fergusson, *Monthly Weather Rev.*, Vol. 49 (1921), pp. 379–86.
[67] *Ibid.*, p. 379.

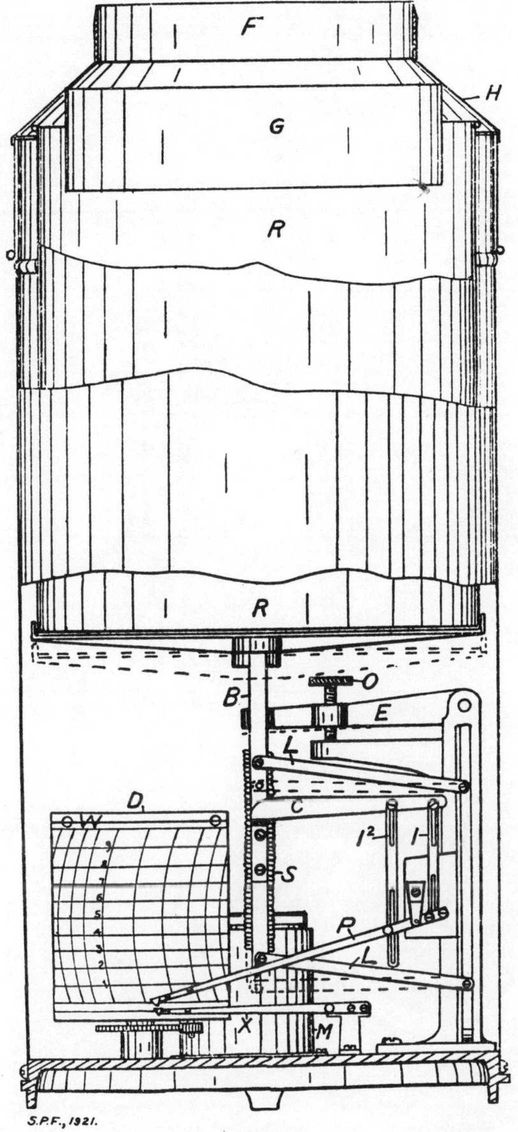

Fig. 2.

Fig. 4.13 Fergusson's recording rain and snow gauge.

About 1890 C. F. Marvin of the U.S. Weather Bureau, who had a passion for electrical instruments that record at a distance, invented a rain and snow gauge which was essentially a steelyard with the vessel for the precipitation on the shorter arm, and a weight that was moved along the longer arm by an electrically operated ratchet and pawl whenever the balance needed to be restored.[68] At the same time a pen

[68] U.S. Weather Bureau, *Instructions for using Marvin's weighing Rain and Snow Gauge* (Washington, D.C., 1893), pamphlet.

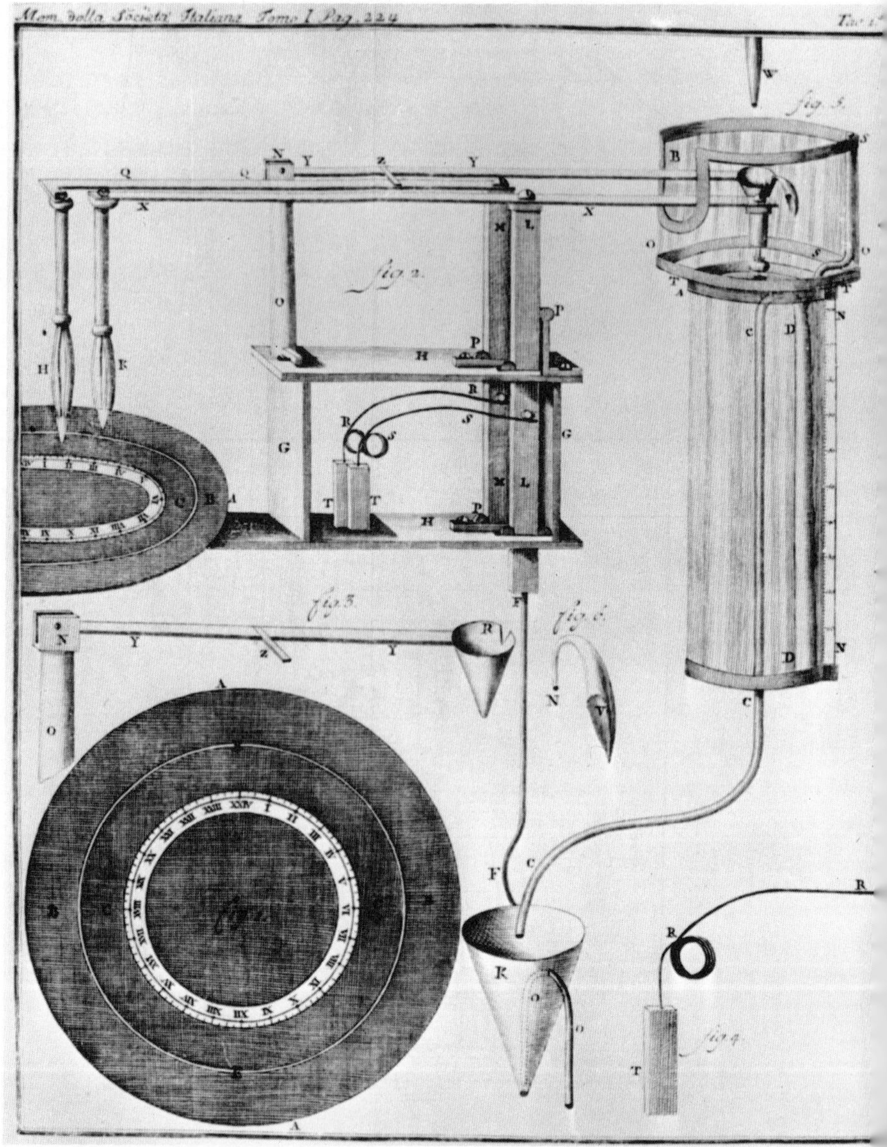

Fig. 4.14 Landriani's rain recorder, 1782.

was moved a corresponding amount in the distant recording mechanism.

The sort of gauge in which the vessel that is weighed is emptied by a siphon when it is full has a precursor in a peculiar instrument, described by Marsiglio Landriani in 1782,[69] which is in a class by itself. This instrument (Fig. 4.14) records the duration as well as the

[69] Landriani, *Mem. Mat. fis. Soc. Ital.*, Vol. 1 (1782), pp. 203–24.

amount of rain on a horizontal plate revolved once in 24 hours by a clock. Whenever water is running into the small cup *P*, the pencil *H* is depressed, marking the duration. The water—and there is plenty of it from the nine square feet of funnel—siphons out into a glass vessel holding the equivalent of one third of a line (1/36 inch) of rain. When this is full, the siphon *DC* empties it into the conical vessel *K*, and its weight depresses a pencil also marked *K*, as well as interrupting the flow of water into the glass vessel, until the conical vessel *K* is emptied by the siphon *OO*. The result was to be a dot for every 1/36 inch of rain, but it is hard to believe in the success of this improbable mechanism.

In the 1830's Follett Osler developed a combined anemograph and recording rain gauge. The anemograph will be dealt with in the next chapter. The rain gauge part of it was apparently modified some time before 1855, the later version having a collecting chamber, mounted on springs and discharged by a siphon after a quarter of an inch of rain had fallen.[70] The rapid discharge of the vessel was ensured by a small tipping bucket in a closed chamber, as in the instrument of Palazzo already described. The idea is shown very schematically in Figure 5.11, page 190.[71] Gauges of this type have not found wide application, though it would be entirely logical to put Beckley's gauge in this group, if it is considered as a hydrostatic balance with mercury as the liquid.

8. *Tipping-bucket gauges.* I have already noted that the first use of the tipping bucket in a rain gauge was made by Wren in about 1660, and also that there are two distinct forms of tipping-bucket rain gauge: a simpler kind that merely counts how many times the bucket discharges, and a more sophisticated construction in which the bucket is weighed while it is filling. Rather surprisingly, the second kind appeared as early as 1670. There seems to be no good reason to discuss them separately.

Robert Hooke devised at least two types of tipping-bucket rain gauge, as we learn from the posthumous publication of some of his miscellaneous papers.[72] Hooke had drawn a diagram (Fig. 4.15) with which he did about three pages of formal geometry to show how the center of gravity of the bucket moved as water was poured into it. Derham then tells us that in April, 1670, Hooke invented two different mechanisms for a recording rain gauge. In the first the bucket was counterpoised by "a string of leaden bullets" that all rested on a table when the bucket was empty, and were lifted off one by one as it

[70] The earlier version, illustrated in plate 37 of the 14th *Report* of the B.A.A.S. (York, 1844), had double tipping buckets on a Roman balance.

[71] It is fully described in John Drew, *Practical Meteorology*, 2nd ed. (London, 1860), pp. 274–76.

[72] W. Derham, ed., *Philosophical Experiments and Observations of the late eminent Dr. Robert Hooke* [etc.] (London, 1706), pp. 43–47.

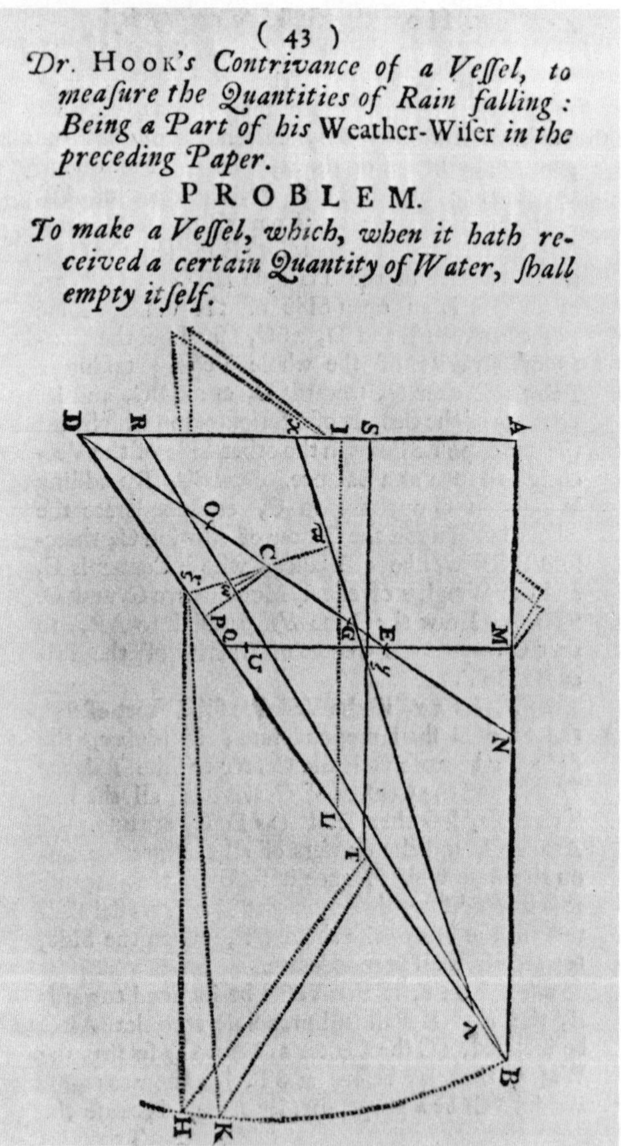

(43)

Dr. HOOK's *Contrivance of a Veſſel, to meaſure the Quantities of Rain falling: Being a Part of his* Weather-Wiſer *in the preceding Paper.*

PROBLEM.

To make a Veſſel, which, when it hath received a certain Quantity of Water, ſhall empty itſelf.

Fig. 4.15 Hooke's diagram of the tipping bucket.

filled. He rejected this because the motion proceeded by jumps, and it is interesting that nobody in the seventeenth century realized that a much smoother motion could be obtained by hanging a chain in a narrow loop. Hooke's second idea was to counterpoise the bucket by "a cylinder immersed into water, mercury, or any other fluid." As the bucket filled, the counterpoise was drawn out of the liquid, progressively reducing the upward force on the cylinder, and maintaining the equilibrium. When the bucket tipped and emptied, the counterpoise sank to its original level. This was a very sophisticated idea with the advantage that the motion was exactly proportional to the added weight. As may be deduced from Figure 4.15, these mechanisms were invented for use in Hooke's weather-wiser or weather-clock, which will be discussed further in Chapter 7.

The next tipping-bucket rain gauge that I have come across is in one of the celebrated books of machines compiled by Jacob Leupold.[73] This is a somewhat inferior design (Fig. 4.16), because the weight of the water in the bucket had to overcome not only the counterweight *L* but also the friction of a ratchet and pawl. Note the valve that retains the water in the funnel *A* while the mechanism is tipping.

These instruments had a single tipping bucket. In 1827 John Taylor described a rain gauge with three buckets mounted on a wheel, and a train of gears for counting,[74] but as the buckets were held by the friction of an arm against a pin, it cannot have been very satisfactory. The idea of putting two prism-shaped vessels back to back so that they oscillate between stops, alternately filling and emptying, entered the literature in 1738.[75] This was a huge affair, providing power to drive the bellows for an iron furnace. It was almost a century later that the oscillating buckets appeared in a rain gauge, apparently almost simultaneously in Switzerland and in England. In 1830 Johann Kaspar Horner, a professor in Zurich, described the gauge shown in Figure 4.17,[76] calling the double bucket "das Schiffchen." A ratchet-and-pawl mechanism operated a counting wheel, which advanced only at every second tip; it had 50 teeth, and was loose on a shaft. Behind it, fixed to the shaft, was a 51-tooth wheel, which fell one tooth behind for each revolution of the 50 tooth wheel.

The English version, attributed to Samuel Crosley, seems to have been invented in 1829, but I have not been able to find any account bearing that date, the earliest that has come to my notice dating from ten years later.[77] In any event the Science Museum in London possesses an example of this instrument,[78] somewhat similar to that of Horner except for the arrangement of the gearing and recording dials.

Three sorts of improvement came later in the century. One was to cause the tipping buckets to revolve a spiral cam that raised a pen arm, as in the simple apparatus of Negretti & Zambra, sufficiently described by reference to Figure 4.18.[79] Another was to make each motion of the buckets close an electric circuit for a moment, a counter or recording pen being operated at a distance by an electromagnet. An instrument of this sort was sent to the 1876 exhibition at South Kensington

[73] Leupold, *Theatri statici universalis. Pars III*, p. 298.
[74] Taylor, *Phil. Mag.*, Vol. 70 (1827), pp. 406–8.
[75] Martin Triewald, *Phil. Trans.*, Vol. 40 (1738), p. 231–38.
[76] Horner, *Schweigger's J.*, Vol. 59 (1830), pp. 36–43.
[77] Robert Carr Woods, *Trans. Meteorol. Soc.*, Vol. 1 (1839), p. 35.
[78] Inv. no. 1893–159.
[79] *Meteorol. Mag.*, Vol. 34 (1899), pp. 36–38.

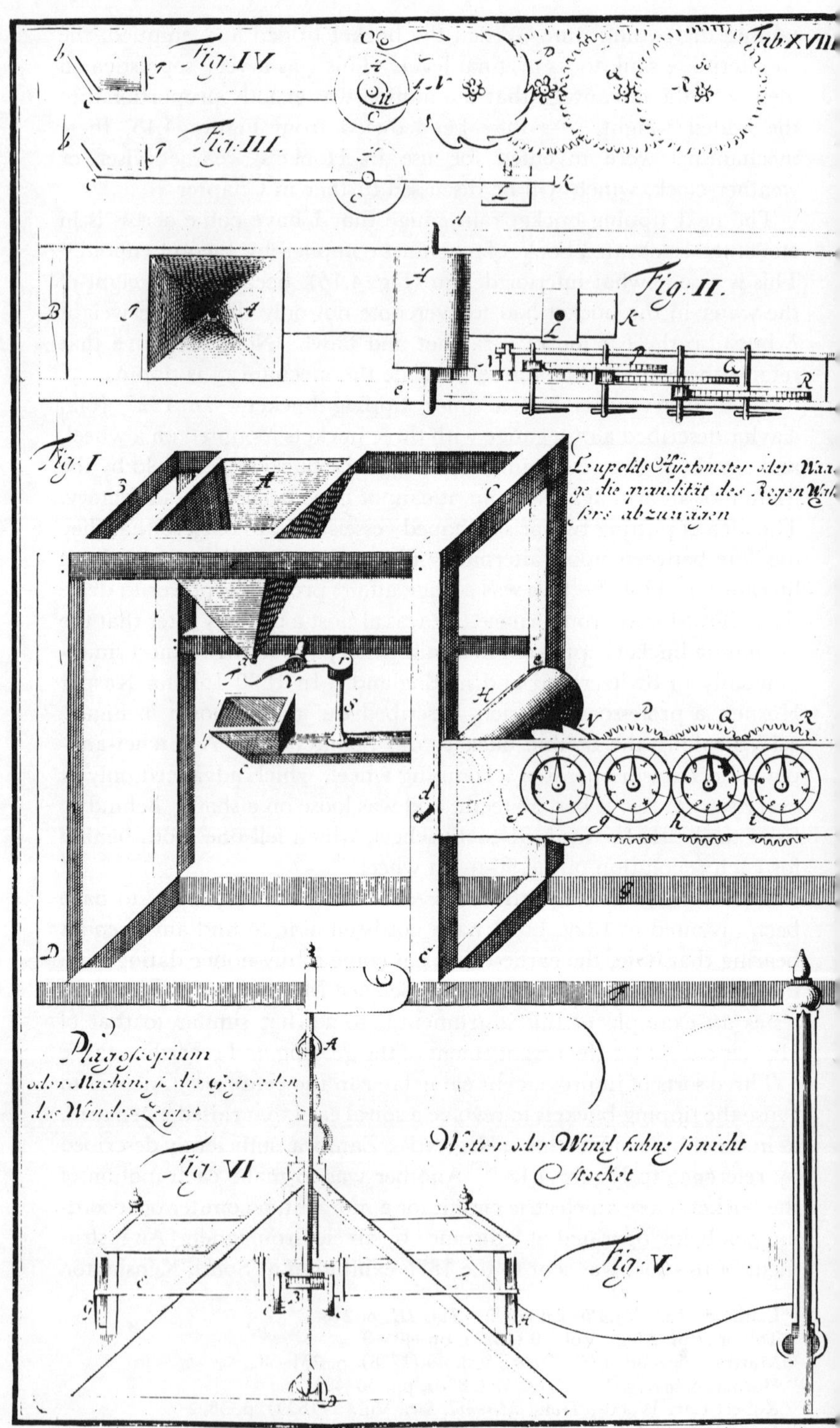

Fig. 4.16 Leupold's tipping-bucket rain gauge.

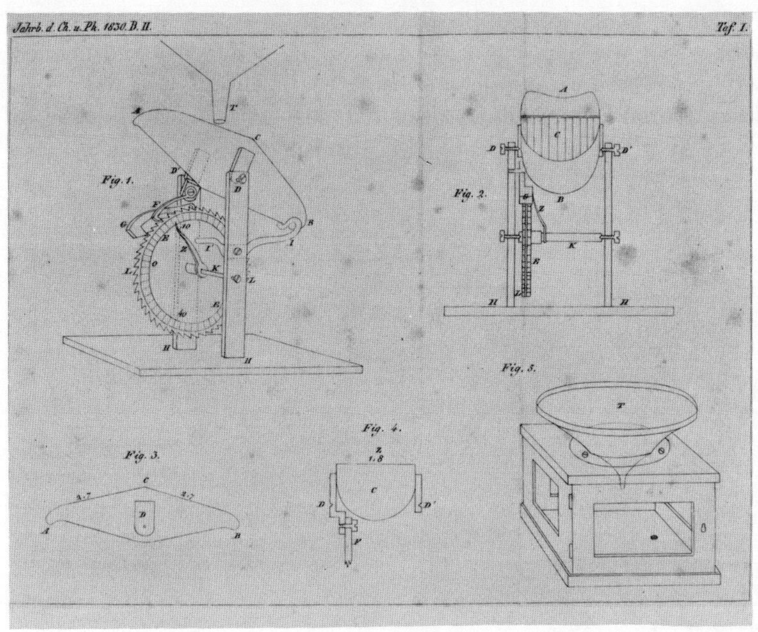

Fig. 4.17 Horner's rain gauge, 1830.

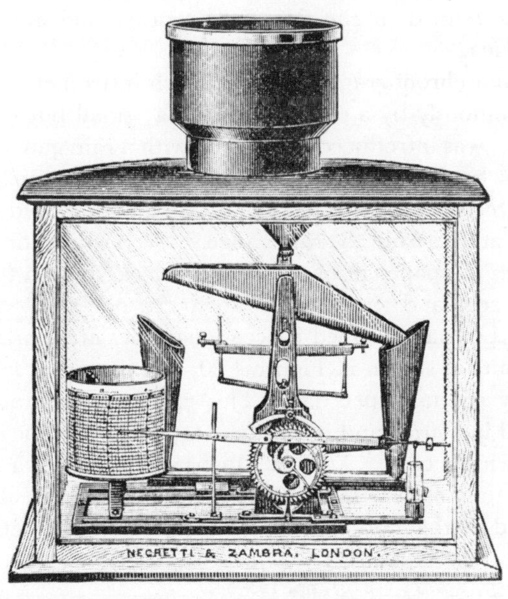

Fig. 4.18 Negretti & Zambra's recording tipping-bucket rain gauge.

by Yeates & Son of Dublin (Fig. 4.19).[80] In 1898 W. H. Dines,[81] noting that multiple contacts often occur, designed an instrument with a platinum wire projecting below each bucket. One or other of these normally touched an insulated pool of mercury, but for an

Fig. 4.19 Yeates & Son's electrical rain gauge.

instant when the bucket is tipping they both made contact, closing a circuit. The distant recorder had an electromagnet and a ratchet and pawl to revolve a spiral cam. Electrical rain gauges of this kind have been further refined in the twentieth century and are common in North America.

In the usual chronograph used with such instruments the paper is moved continuously by a clock. In 1889 a "small but not unimportant novelty" was introduced.[82] For use with a rain gauge, A. Sprung of the Deutsche Seewarte and the Berlin instrumentmaker R. Fuess arranged a recorder in which a band of paper is moved 2 mm each time a pulse arrives from the rain gauge.[83] At the same time the pen is moved across the paper in one hour by a clock, and on the hour it is returned to zero and the paper is advanced one step. In this way showers can be exactly timed if occasional lines are marked with the date and hour, as shown in Figure 4.20, and if no rain falls, only 4.8 cm of paper are used per day. This excellent apparatus was still working well in 1907,[84] and it is rather surprising that the idea has not been more widely used. At Kew Observatory there is a variation of this for the study of very heavy rain, in which the chart is on a drum that is turned by the motion of a float. Once a minute a time mark is

[80] *Meteorol. Mag.*, Vol. 11 (1876), pp. 153–54.
[81] *British Rainfall* (1898), pp. 17–18.
[82] A. Sprung and R. Fuess, *Zeits. Instrum.*, Vol. 9 (1889), p. 95.
[83] Sprung had first suggested this in 1882 in connection with an anemometer. See *Zeits. Instrum.*, Vol. 2 (1882), p. 207.
[84] Sprung, *Zeits. Instrum.*, Vol. 27 (1907), p. 341.

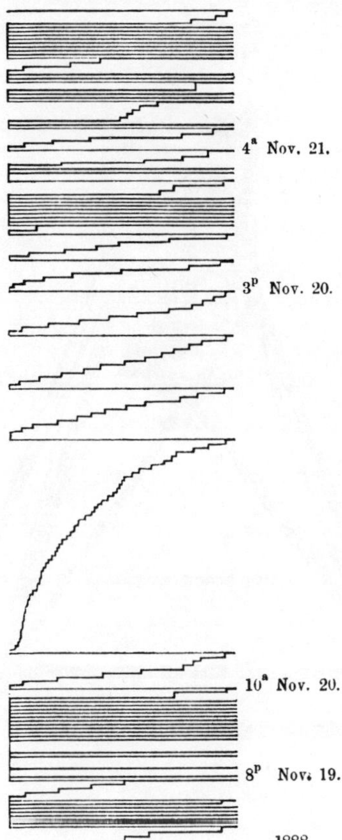

Fig. 4.20 Record made by Sprung's rain gauge.

made on the record, and the distance between the marks indicates how much rain fell during the minute. As the drum rotates it is also advanced axially.

The third sort of improvement was to mount the double tipping bucket on a balance, as Robert Hooke had done with the single bucket in the seventeenth century.[85] First done by Follett Osler before 1844, this improvement was executed with great elegance by Richard Frères in the 1880's.[86] Large tipping buckets are mounted on a Roman balance, and gradually descend until the buckets tip after 10 mm of rain have accumulated. The entire instrument is shown in Figure 4.21.

Meanwhile, the single bucket had its devotees. It was supported on a gravity balance by Karl Kreil in Prague[87] (Fig. 4.22) and on a

[85] See p. 156.
[86] Richard Frères, *Notice sur les instruments enregistreurs* [etc.] (Paris, 1886), pp. 24–26.
[87] Kreil, *Entwurf eines meteorologischen Beobachtungs-Systems für die österreichische Monarchie . . . nebst einem Anhange enthaltend die Beschreibung der an der K.K. Sternwarte zu Prag aufgestellten Autographen-Instrumente* [etc.] (Wien, 1850).

Fig. 4.21 Richard's recording rain gauge, 1886.

spring balance by Daniel Draper in New York[88] (Fig. 4.23). Connoisseurs of meteorological instruments will recognize the particular and indeed unique style of each author.

9. Rate-of-rainfall recorders. While in principle the instantaneous rate of rainfall can be derived from the slope of the record of a recording rain gauge, in practice the rate of fall varies so very widely that it is not feasible to do this for heavy rains and at the same time provide a legible record of light drizzle. For this reason, special instruments have been devised.

These are of two types: those that count drops of water falling from a specially constructed orifice, and those that use what amounts to a calibrated weir. Figure 4.24 shows a diagram of the drop counter of W. J. E. Binnie, described in 1892,[89] in which a piece of cambric is tied over the orifice *E* with the idea of producing uniform drops, which fall one by one onto a little plate *L* and make a momentary contact at *M*. The funnel was very small, about two inches in diam-

[88] *Amer. Meteorol. J.*, Vol. 1 (1884–85), pp. 487–88.
[89] Binnie, *Quart. J. Roy. Meteorol. Soc.*, Vol. 18 (1892), pp. 6–12.

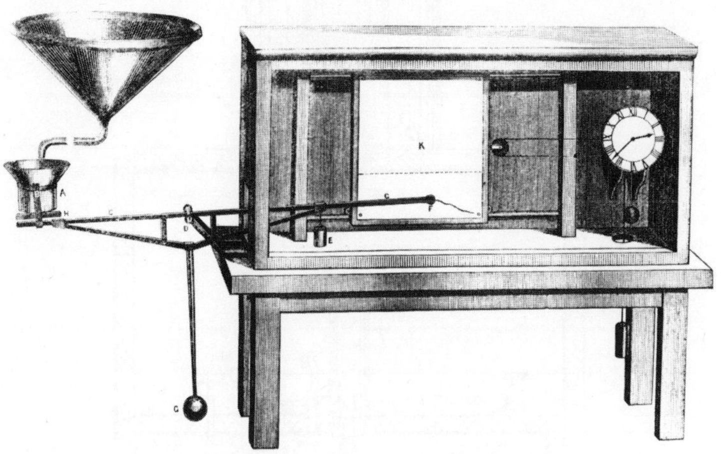

Fig. 4.22 Kreil's recording rain gauge.

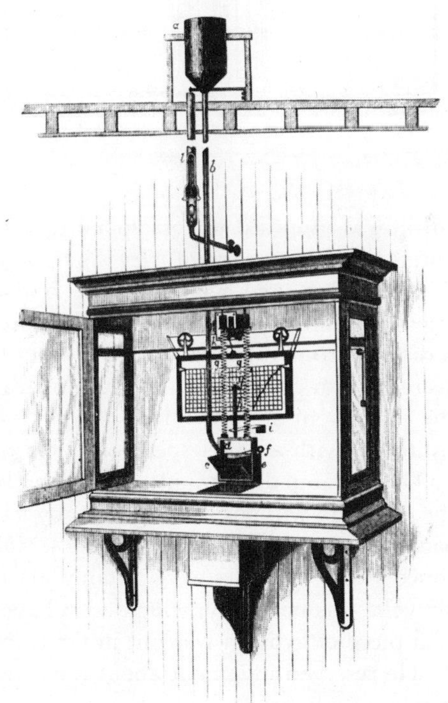

Fig. 4.23 Draper's recording rain gauge.

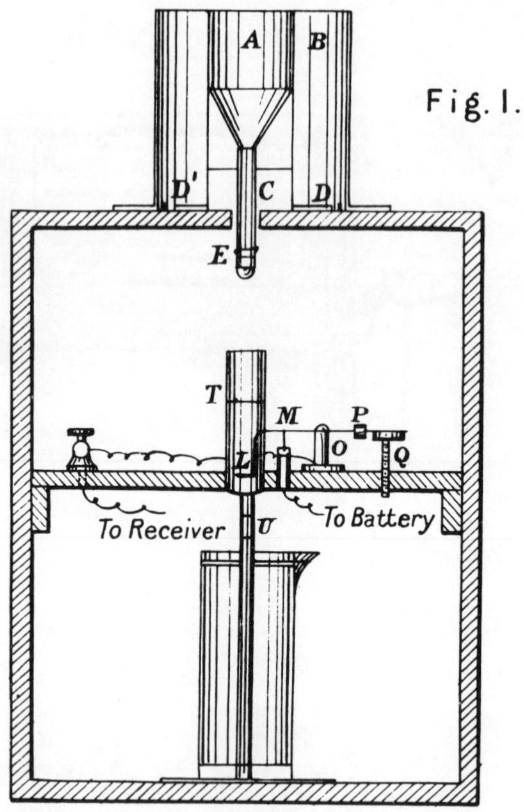

Fig. 4.24 Binnie's rate-of-rainfall recorder, 1892.

eter, and each drop was supposed to correspond to 0.01 inch of rain, so that the advantage of this instrument over the tipping bucket is not apparent. W. Gallenkamp used a larger funnel and more frequent drops.[90] The mechanism will be evident from Figure 4.25*a*. Dots were made on a distant drum, arranged to revolve four times an hour; the pen was lowered 2 mm per revolution. An apparatus of this sort is still part of the equipment at Kew Observatory. Later,[91] a Morse printer was substituted, with a paper speed of 22 cm per minute.

Sprung built a Gallenkamp drop counter, and discussed it in 1907.[92] He found that he could do without moving parts by letting the drops fall into the space between two electrodes (Fig. 4.25*b*); they would fall through at once if the electrodes were suitably spaced. A battery giving about 105 volts was necessary, which must have been dangerous, but if a small piece of gypsum was put in the S-shaped tube, 40 volts would do. He reserved judgment about the constancy of drop size at different rates of fall.

[90] Gallenkamp, *Meteorol. Zeits.*, Vol. 22 (1905), pp. 1–10.
[91] Gallenkamp, *Zeits. Instrum.*, Vol. 28 (1908), p. 34.
[92] Sprung, *ibid.*, Vol. 27 (1907), pp. 340–43.

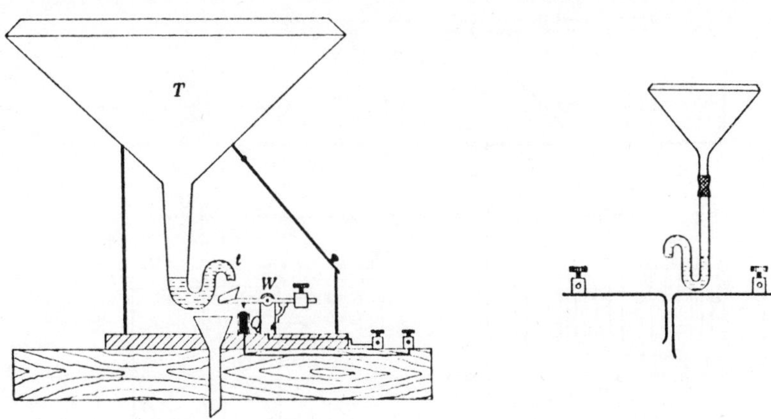

Fig. 4.25 (a) *Gallenkamp's drop counter, 1905;* (b) *Sprung's modification, 1907.*

Gallenkamp at once agreed that this was a difficulty.[93] Nor did he enjoy counting dots, and went over to the recorder shown in Figure 4.26, which is of the second type. After passing through a trap to remove air and dust, the rainwater enters a tube r, in which there is a long narrow slit. This tube is surrounded by another tube r_1, partly supported by the float H and carrying the pen F. This tube is prevented from revolving by a pin sliding in the slit. Because the water that enters r has to escape through the slit, the tube r_1 will rise until enough of the slit is exposed.

An adjustable orifice is also used in the better known recorder of Jardi,[94] which, while it is the same in principle, is a little less delicate. In this instrument the water from a large receiver falls through the tube A (Fig. 4.27) into the float chamber B. The float C has an appendix of diminishing cross-section extending through an aperture into a lower chamber D. The water can flow into D only through the annular space between the bottom of B and the tail of the float. The higher the float rises, the larger this space will be, and the float will continue to rise until as much water flows out of B as enters at A. The motion of the float is transmitted to a pen by a system of levers. By Torricelli's principle, the volume of water flowing out per second is proportional to the area of the opening and to the square root of the depth of the water in the chamber B. The tail of the float is given a shape that will produce a linear relationship between the rate of flow and the motion of the float.

I shall end this section with a reference to a gauge that was in-

[93] Gallenkamp, *ibid.*, Vol. 28 (1908), pp. 33–38.
[94] Ramon Jardi, *Servei meteorologic de Catalunya, Notes d'estudi*, Vol. I, no. 2 (Barcelona, 1927).

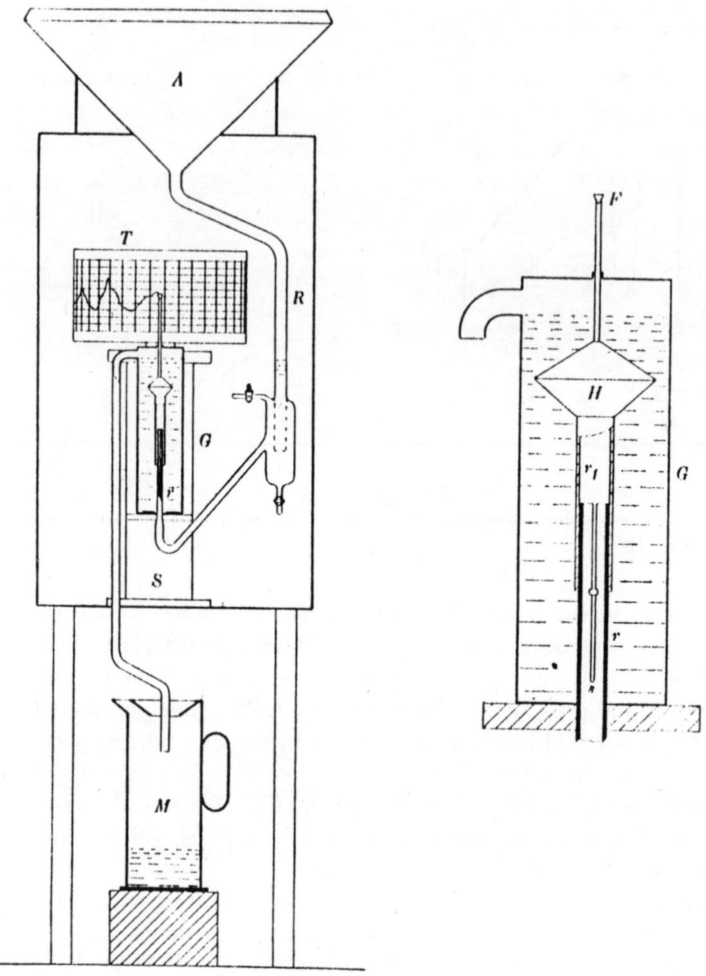

Fig. 4.26 Gallenkamp's later rate-of-rainfall recorder, 1908.

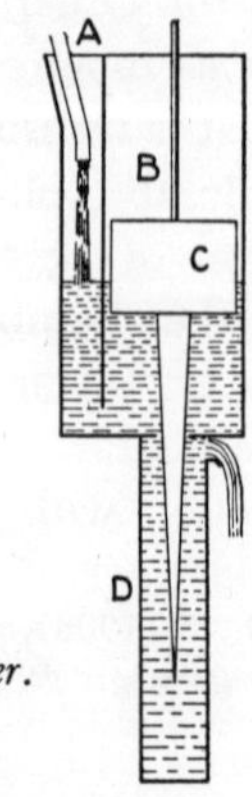

Fig. 4.27 The Jardi rate-of-rainfall recorder.

tended to determine the distribution of intensities of rain, and was described in 1853 by the remarkable amateur meteorologist P. H. Maille.[95] In this instrument (Fig. 4.28) the funnel is prolonged into a tube that bends round and terminates horizontally, producing a jet that will travel farther the more intense the rain, so that the water can be caught in a number of separate compartments.

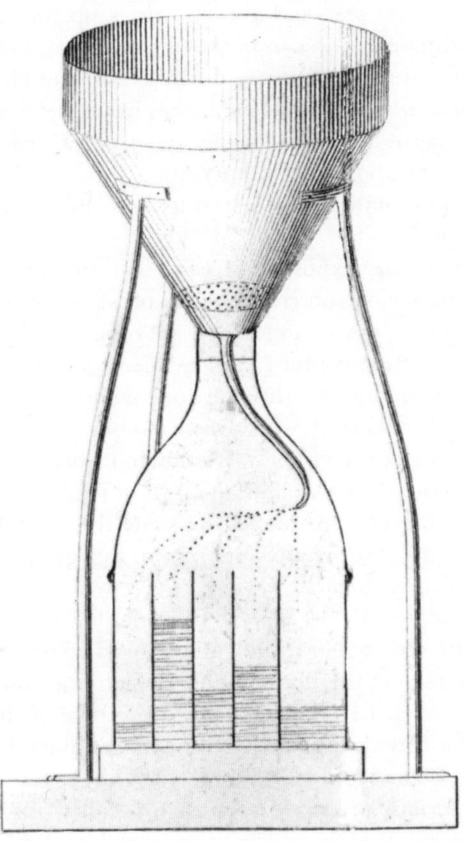

Fig. 4.28 Maille's rain-intensity gauge, 1853.

10. The Pluvioscope. About 1860 an instrument was built in France for the express purpose of timing the beginning and ending of light rain.[96] It received the name pluvioscope. A drum driven by a clock carried a specially prepared paper under a small rectangular aperture

[95] Maille, *Nouvelle théorie des hydrométéores, suivie d'un mémoire sur l'électricité atmosphérique, et d'un autre sur la pluviométrie* (Paris, 1853), p. 296. Concerning Maille, see W. E. K. Middleton, *Isis*, Vol. 56 (1965), pp. 320–26.
[96] Hervé-Mangon, *Annuaire Soc. Météorol. France*, Vol. 8 (1860), pp. 183–84.

in a horizontal plate. The paper was prepared by dipping it in a solution of iron sulphate, letting it dry, and then rubbing it with a fine powder made from gall nuts mixed with gum sandarac. Each drop of rain striking such a preparation left a circular black spot.

11. Evaporation gauges. It has been recognized since ancient times that of the water brought down by the rain, part runs into the rivers and at last into the sea, and the rest dries up and ascends again. Numerical estimates of the amount of water evaporated were first made in the seventeenth century. In 1687 Edmond Halley hung on a balance a pan of water about eight inches in diameter and four inches deep, with a thermometer in it. By means of a "pan of coals" the water was kept at about the temperature of the air in summer, and Halley tried to estimate from the results how much water evaporates from the oceans.[97]

By about 1730 the exposure of cans of water had become fairly common among serious observers, some of whom found in them opportunities for displays of ingenuity. As early as 1725 Leutmann described his "exatmoscopium" or "evaporatorium," in which a tin vessel six inches square and three or four inches deep was suspended from a "Pernault balance," which seems to have been two sectors of a large wheel on an axis having low friction in its bearings.[98] The vessel was counterbalanced by a weight, and it is difficult to understand how stability was achieved. At about the same time Stephen Hales was trying to estimate the evaporation from the ground by weighing glazed pans filled with earth.[99]

G. W. Richmann, in St. Petersburg, adapted the principle of the hydrometer to an evaporation gauge, producing the remarkable instrument shown in Figure 4.29, in which the variations in the weight of the vessel *AABB* are balanced by those of the buoyancy of the part of the three supports *gg*, *hh*, and *ii* immersed in the lower tank.[100] The magnification could be very large, but friction would have been a serious problem. Richmann called his instrument an *atmometer*, a word that is generally ascribed to Sir John Leslie,[101] although H. B. de Saussure used it,[102] and also the form *atmidometer*. The word *atmometer* is to be preferred to *evaporimeter* because of its unmixed etymology.

Richmann knew quite well that his elaborate instrument was merely an exercise in design, for in the very same number of the *Commentarii* he described how he had found more evaporation from deep

[97] Halley, *Phil. Trans.*, Vol. 16 (1687), pp. 366–70.
[98] Leutmann, *Instrumenta*, p. 144.
[99] Hales, *Vegetable Staticks* (London, 1727), p. 53.
[100] Richmann, *Novi Comm. Petrop.*, Vol. 2 (1749), pp. 121–27.
[101] Leslie, *A short Account of Experiments and Instruments depending on the Relations of Air to Heat and Moisture* (Edinburgh and London, 1813), p. 81.
[102] Saussure, *Essais sur l'hygrométrie* (Neuchâtel, 1783), p. 246.

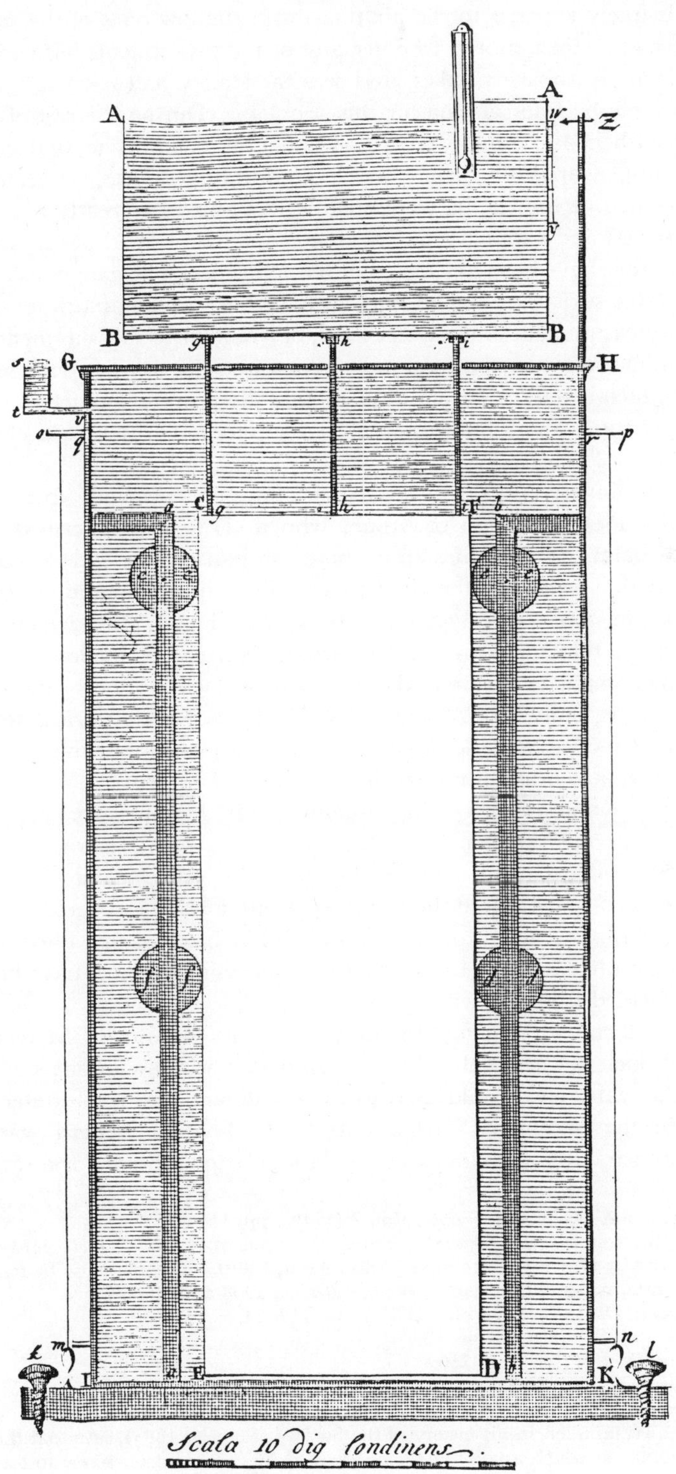

Fig. 4.29 Richmann's atmometer.

vessels freely exposed to the air than from shallow ones of the same material.[103] He doubted whether any of the experiments with small vessels apply to rivers, lakes, and seas, as Halley had tried to apply them. He had no solution for this problem. During the rest of the eighteenth century there were a number of attempts to find rules governing evaporation from vessels of differing depths, areas, and materials, but the results were chaotic.[104] This fact greatly reduces the interest and value of a description of the many small evaporation gauges that have been devised from time to time. The gauges using a free water surface could be read by weighing at suitable intervals, direct measurement of the water level, or volumetric measurement of the amount of water that had to be added in order to restore the level to a fiducial mark. In addition to this there are atmometers that measure the evaporation from a porous surface of earthenware, cloth, or paper.

An atmometer that operates by weighing can be made to give a record. Pietro Moscati of Milan, who in 1781 had described[105] a special balance supporting an evaporation pan on one end and provided with a weight on the other that could be moved in and out by a micrometer screw, later made an atmograph (Fig. 4.30) in which the pan was on the short arm of a steelyard.[106] A pencil on the longer arm recorded on a flat chart moved sideways by a clock. Almost a century later G. W. Hough of Albany, New York, used an electrical servo mechanism of his own design for a similar purpose,[107] but this can no more have given useful results than did that of Moscati.

Small evaporation vessels, usually made of brass or copper, were numerous. Pieter van Musschenbroek's "evaporatorium" or "exatmoscopium" was a brass box six inches square and eighteen inches high; he measured the water level with a brass scale held to one wall and pushed down to the bottom.[108] The box was first filled to a depth of only twelve inches, so that the water surface must have been well shielded from the wind.

The official atmometer distributed by the Mannheim Meteorological Society was a cubical brass box with a window in one side so that the water level could be read by a scale and vernier "similar to that in the barometer," fixed outside.[109] That of Achard was a twelve-inch cube, with an external glass tube up one side for a water

[103] Richmann, *Novi Comm. Petrop.*, Vol. 2 (1749), pp. 134–44.

[104] The interested reader should consult the excellent *Annotated Bibliography of Evaporation* by Grace J. Livingston (Washington, 1909). This may also be found in five instalments in the *Monthly Weather Review*, 1908 and 1909.

[105] Moscati, *Opuscoli scelti*, Vol. 4 (1781), pp. 117–23.

[106] *Ibid.*, Vol. 5 (1782), pp. 46–53.

[107] *Nature*, Vol. 9 (1874), p. 250.

[108] Musschenbroek, *Novi comm. Petrop.*, Vol. 8 (1760–61), pp. 367–91.

[109] Hemmer, *Ephemerides Soc. meteorol. Palatinae anni 1781* (Mannheim, 1783), pp. 84–85. Traümuller, in his history of the Society (Leipzig, 1885), says that there was another in which a point could be lowered by a micrometer screw to touch the water.

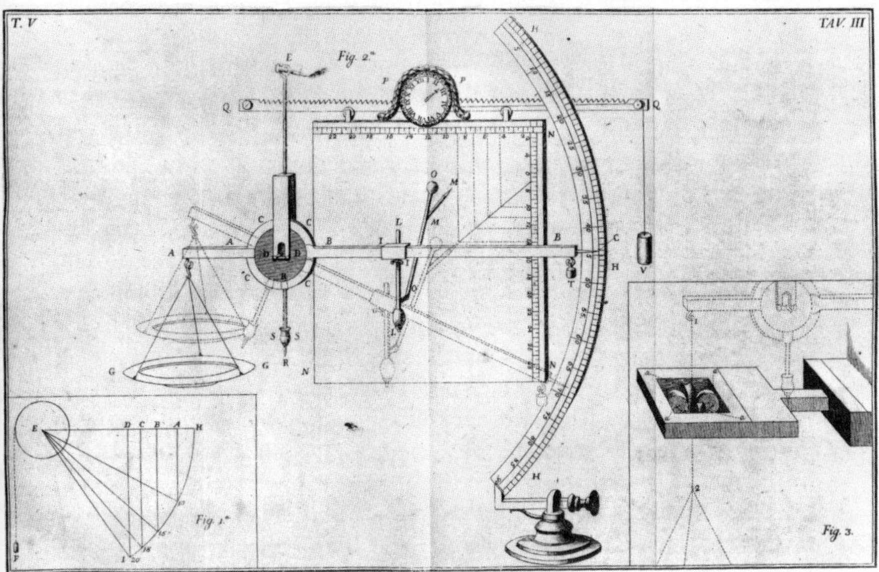

Fig. 4.30 Moscati's recording atmometer.

gauge, having a brass scale in inches and lines and a vernier measuring to 0.1 line.[110] Achard says that the tube is large enough not to change the level by capillary action. Luke Howard used a cylinder 5 inches in diameter and 1½ inches high, in which an etched glass diagonal scale was supported vertically.[111]

In an attempt to keep the temperature of the water from changing too rapidly, Richmann built a leaden vessel 20 inches in diameter and 5 inches deep, with a 5-inch-square compartment in the middle.[112] All but the inner chamber was covered and nearly airtight, and the water was kept at approximately the same depth in both parts of the vessel. With similar intentions, Herculanus Schwaiger at Hohenpeissenberg planted a brass box in the middle of a large box of earth.[113] Nearly a century earlier, Sédileau[114] had had two vessels, each 3 square feet in area, sunk into the ground on the terrace of the Paris Observatory, one being used as a rain gauge, the other as an evaporation tank.[115] This was reported in 1692. In time it became clear to a few people that such tanks, sunk in the ground and the larger the better, would give a much more precise idea of the actual evaporation from

[110] Achard, *Mém. Acad. R. Sci. Berlin* (1788), pp. 101–2.
[111] Howard, *The Climate of London* (2 vols., London, 1818–20), I, xxiv–xxv.
[112] Richmann, *Comm. Petrop.*, Vol. 14 (1744–46), pp. 273–75.
[113] Schwaiger, *Ephem. Soc. Meteorol. Palat. anni 1783* (Mannheim, 1785), p. 300.
[114] His given name seems not to be be known even to the Académie, of which he was a member.
[115] *Hist. Acad. R. Sci. 1666–1699* (11 vols., Paris, 1733), Vol. 2, pp. 133–35.

natural water surfaces than small vessels freely exposed to the atmosphere. Most people, however, were quite used to statistics in which the annual evaporation was greater than the rainfall, even in western Europe; when observations along the Canal de Bourgogne after 1839 by its chief engineer, Delaporte, showed the contrary, F. Vallés was sure the results must indicate an exceptional regime.[116] The engineer Tarbe explained that the measurements had been made in tanks 2.5 meters square and 0.4 meters deep, built of masonry and lined with sheet zinc.[117] There was a scale on one of the interior faces. The bottom of the tank was at ground level, and a gently sloping bank was graded up to its edge. At Dijon, Ruinet installed a tank only 30 cm square for comparison, and found that it could evaporate a depth 50 per cent greater than the large one.[118]

Shortly before this the ideal measuring device for use with evaporation tanks had been invented by Uriah A. Boyden, "an eminent hydraulic engineer of Massachusetts,"[119] for use in experiments on the flow of water. This was the hook gauge, an utterly simple apparatus in which a point is brought up from beneath the surface of the water; the coincidence of the point with the surface was easily observed by looking at the reflections of nearby objects. In still water a precision of a few hundredths of a millimeter is readily obtained. The hook gauge is much better in liquids that will wet it than is a point lowered from above. It had become the usual equipment with large evaporation tanks at least by 1889, when such a tank six feet square was taken as a standard in comparative experiments at Stratfield Turgiss, Hampshire.[120] The hook gauge was placed in a small compartment that communicated with the main body of water.

It would be of little interest to record the large number of different small atmometers built in various countries during the nineteenth century, some read by weighing and some volumetrically. There were several of the latter sort in which the water had to be taken out of the evaporator and into a graduated tube for each measurement. Some of these were very complicated,[121] and as W. Napier Shaw pointed out, the reading depends on how long the evaporating dish is allowed to drain.[122]

Other people, aware of the fact that the evaporation from a dish depends on how far the water level is below the rim, applied the familiar principle of the "chicken feeder," or birds' drinking glass.

[116] Vallès, *Annales des Ponts et Chaussées*, Ser. II, Vol. 20 (1850), pp. 383–93.
[117] *Ibid.*, Ser. III, Vol. 3 (1852), pp. 249–52.
[118] *Ibid.*, Ser. III, Vol. 20 (1860), pp. 150–60.
[119] James Bicheno Francis, *Lowell Hydraulic Experiments* [etc.] (Boston, Mass., 1855), p. 18.
[120] G. J. Symons, *British Rainfall* (1889), pp. 18–43.
[121] Examples: J. Prinsep, *Asiatic Researches*, Vol. 15 (1825), Appendix, P. xiii; A. A. Mühry, *Ann. Phys.*, Vol. 113 (1861), pp. 305–8; R. von Vivenot, *Repert. phys. Techn.*, Vol. 1 (1865), pp. 103–30; J. von Lamont, *Zeits. österr. Ges. Met.*, Vol. 4 (1869), p. 81–86.
[122] Shaw, *Quarterly Weather Report for 1877* (London, 1885), pp. [35]–[36].

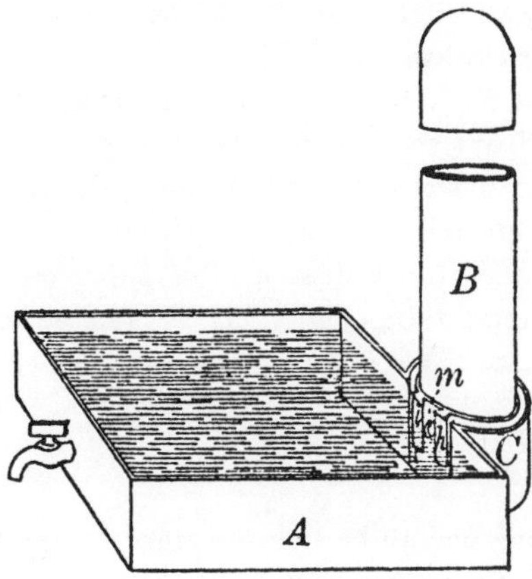

Fig. 4.31 Prestel's atmometer, 1864.

The best example is that of M. A. F. Prestel,[123] shown in Figure 4.31, in which the supply vessel *B* is a graduated glass and serves also as a measuring device. We need not linger over the elaborate automatic valve that permitted the tube to be filled and inverted, for the question that must arise is how rain is to be taken account of.

Evaporation takes place not only from the surface of bodies of water and the soil, but also from the leaves and other parts of plants, which are exposed to the air on all sides. The last sort of atmometer with which I shall deal attempts to take this fact into account. The first of these was that of John Leslie[124] (Fig. 4.32), in which

> A thin ball [*A*] of porous earthen-ware, two or three inches in diameter, with a small neck, to which is firmly cemented a long and rather wide tube [*B*], bearing divisions, each of them corresponding to an internal annular [*sic*] section, equal to a film of liquid that would cover the outer surface of the ball to the thickness of the thousandth part of an inch . . . to the top of the tube is fitted a brass cap [*b*] having a collar of leather, and which, after the cavity has been filled with distilled or boiled water, is screwed tight. The outside of the ball now being wiped dry, the instrument is suspended out of doors, and exposed to the free action of the air.

A few years later Angelo Bellani of Milan described another atmometer in which porous pottery was used, but which differed from Leslie's in that the evaporating surface was flat and horizontal, closing the mouth of a brass funnel.[125] The narrow measuring tube extended horizontally from the funnel, and was open at the end. The funnel

[123] Prestel, *Die Regenverhältnisse des Königreichs Hannover* [etc.] (Emden, 1864), cited by E. Gelcich in *Zeits. Instrum.*, Vol. 10 (1890), p. 52.
[124] Leslie, *A short Account of Experiments and Instruments depending on the Relations of Air to Heat and Moisture* (Edinburgh, 1813), p. 81.
[125] Bellani, *Giornale di Fisica* (Pavia), Decade II, Vol. 3 (1820), pp. 166–77.

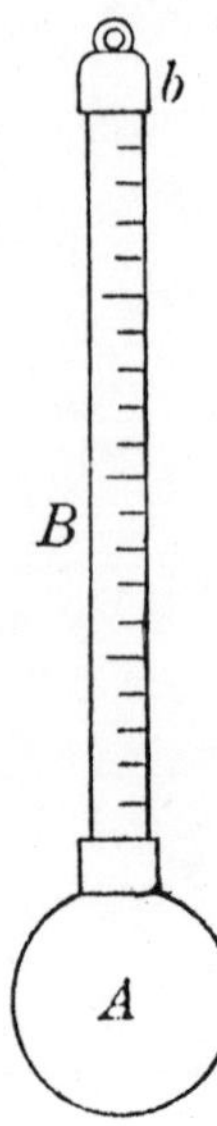

also communicated, through a stopcock, with a supply vessel from which the apparatus could be refilled.

Leslie's instrument could simulate the evaporation from a peach or an apple, Bellani's from the upper side of a leaf. In 1872, Albert Piche of Pau, in the Pyrenees, told his colleagues about an "atmismomètre" that could well represent a leaf in its entirety, for the evaporation took place from a disk of filter paper covering the open lower end of a graduated tube (Fig. 4.33).[126] The paper, which had a pinhole in the middle, was to be held in place by the pressure of the air, but a brass spring was, very wisely, provided for safety.

Fig. 4.32 Leslie's porous-sphere atmometer, 1813.

The Piche atmometer has been and is extensively used by agriculturists and foresters, as have various modifications of Leslie's instrument.

[126] Piche, *Bull. Assoc. sci. France*, Vol. 10 (1872), pp. 166–67. This author's name is frequently misspelled Piché.

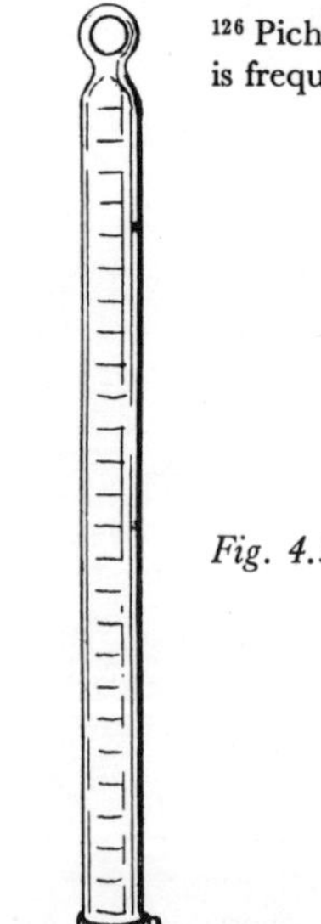

Fig. 4.33 The atmometer of Piche, 1872.

V

The Windvane and the Anemometer

1. Introduction. The use of simple instruments to indicate the direction of the wind is a practice of some antiquity, but the measurement of its force or speed is a much more recent accomplishment. The graphical recording of these elements was first done about 300 years ago, but practical and durable apparatus for this purpose first appeared half-way through the nineteenth century. To fulfill the program of this book we must therefore trace the history of wind-measuring devices for some 2,000 years.

A few words should be said about the specification of wind direction, although this is not strictly an instrumental matter.[1] Only four winds are mentioned in Homer and in the Bible, but these references are poetical, and Hellmann argues plausibly that the recognition of intermediate directions is not thereby excluded. The Babylonians, at least, had the eight-point wind rose, and combined the names of the four cardinal points to name the intermediate directions.[2] In the Greek and Roman periods, however, and as late as the sixteenth century in Italy, there were often twelve winds, in groups of three centered on the four principal directions;[3] there was also an eight-point wind rose, as in the "Tower of the Winds" at Athens.

With the introduction of the mariner's compass, the card of which was from the beginning divided into eight, sixteen, and thirty-two equal divisions, the twelve-point wind rose became less popular. In the Mediterranean each of the eight winds had its own Italian name—Tramontana, Greco (or Grego), Levante, Scirocco, Ostro, Garbin, Ponente, and Maestro; according to Thompson the earliest compass card bearing "the initials of the Frankish names of the winds—N., NNE., NE., etc.," dates from 1536. This nomenclature finally prevailed, but some of the old names survive to this day in popular use.

In the remainder of this chapter I shall deal first with the windvane,

[1] See S. P. Thompson's scholarly article in *Proc. British Acad.*, Vol. 6 (1913), pp. 179–209; also G. Hellmann, *Abh. preuss. Akad. Wiss., phys.-math. Kl.*, (1920), pp. 48–56.
[2] Gustav Hellmann, *Quart. J. Roy. Meteorol. Soc.*, Vol. 34 (1908), p. 224.
[3] Thompson, p. 184.

next with the anemometer, and last with instruments that combine the two devices; though because some kinds of anemometer have to be faced into the wind by a windvane, it will not always be possible to be entirely systematic. A selection will have to be made from the very large number of recording instruments, some of great ingenuity, but not all by any means sufficiently durable for practical use.

2. The Windvane. Andronicus of Cyrrha, who flourished about 100 B.C., "built at Athens an octagonal marble tower, and, on the several sides of the octagon, had representations of the winds carved opposite their several currents. And above the tower he caused to be made a marble upright, and above it he placed a bronze Triton holding a rod in his right hand. He so contrived that it was driven round by the wind, and always faced the current of air, and held the rod as indicator above the representation of the wind blowing."[4] Thus Vitruvius informs us about the earliest windvane of which we have any knowledge. It should be added that the primary purpose of this "Tower of the Winds," which still exists without its windvane, was to house a water clock that would supplement readings of the sundials on several of its eight faces.

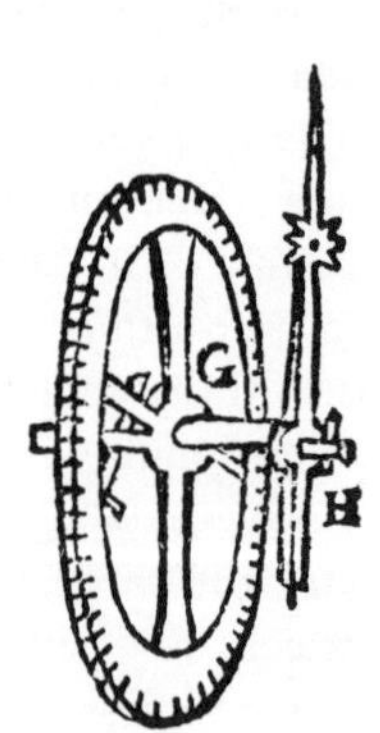

Fig. 5.1 Danti's "vertical instrument," 1578.

Only a little later, Marcus Terentius Varro had a windvane on his farm, apparently arranged so that the direction of the wind could be observed on a dial on the ceiling of the room below: "In the middle of the same hemisphere, there is a globe with the eight winds, as on the dial at Athens, which Cyrrhestes made, and an index is moved from the pole to the globe, that it may shew what wind blows, so that you may know this in the inside."[5]

[4] Marcus Vitruvius Pollio, *De architectura libri decem*, I, vi, 4; trans. Frank Granger. (Cambridge, Mass., and London, 1955), p. 57. (Loeb Classical Library.) Quoted by permission.

[5] Varro, *De re rustica*, III, v, 17; tr. T. Owen in *The three Books of Marcus Terentius Varro concerning Agriculture* (Oxford, 1800), p. 214.

The simple windvane, a flat sheet of metal cut in the shape of the silhouette of some creature or artifact, became common in the Middle Ages, especially on the towers of churches. Many of these vanes were very elaborate in design, and their decorative value was much greater than their scientific utility, for most of them must have required a fresh wind to make them turn.[6] The next improvement after that of Varro seems to have been made by Egnatio Danti, a professor of astronomy at Bologna; it dates from 1578.[7] The improvement was to have the wind direction indicated on a vertical dial, built into a wall, and Danti had the apparatus made for Cardinal Paleoto at Florence. He also knew that a windvane must be well above all obstructions, so as not to receive winds "reflected"[8] from nearby buildings; and thus he put the vane on a shaft 30 *braccia* (about 17 meters) long (Fig. 5.1). The lower bearing was a tempered steel point running in soft metal, and there were several bearings attached to the wall, to keep the shaft straight. The exact nature of the gearing is not shown clearly in this figure.

A further improvement was made by Jacob Leupold—the lower pivot ran in oil, and had a dust-cover.[9] A crown gear and two lantern gears transmitted the rotation of the vane to dials on two gables of a house. As Leupold usually acknowledges his sources, and cites none here, the oil bath was probably his own invention; at any rate, he had built such a vane on the country house of a well-known bookseller. It is most unlikely that he had heard of Danti.

A flat vane does not develop very much turning moment at small angles to the air flow. In 1797 G. F. Parrot described the first real improvement in the vane itself, the splayed vane (Fig. 5.2),[10] in which the single vane was replaced by two plates diverging from the axis. Parrot's vane was small, the "wings" were only seven Paris inches long and the counterweight was four inches. The counterweight itself was important, because few vanes in the eighteenth century were balanced. He suggested, without sufficient evidence, that the plates should have a cycloidal curvature, but most later splayed vanes seem to have had flat plates.

The splayed vane held the field until the twentieth century, when the progress of aviation led to the adoption of airfoil sections, as in the "R.A.E." vane, developed in England.[11] For some purposes it may be

[6] Drawings of more than 200 vanes dating from the late middle ages to the present time will be found in A. Needham, *English Weathervanes* [etc.] (Haywards Heath, 1953).

[7] Danti, "La fabbrica dello anemoscopio, strumento verticale," in his *Primo volume dell'uso et fabbrica dell'astrolabio, et del planisferio* [etc.] (Florence, 1578), pp. 273–81.

[8] "*Riflessi*" (this was probably an optical analogy).

[9] Leupold, *Theatri statici universalis. Pars III, sive theatrum aerostaticum* [etc.] (Leipzig, 1726), p. 300.

[10] Parrot, *Mag. für den neuesten Zustand der Naturkunde*, Vol. 1 (1797), pp. 144–58.

[11] See Fig. 5.22, p. 203. E. Gold, *Quart. J. Roy. Meteorol. Soc.*, Vol. 62 (1936), p. 177, says that this improvement is due to Professor G. I. Taylor.

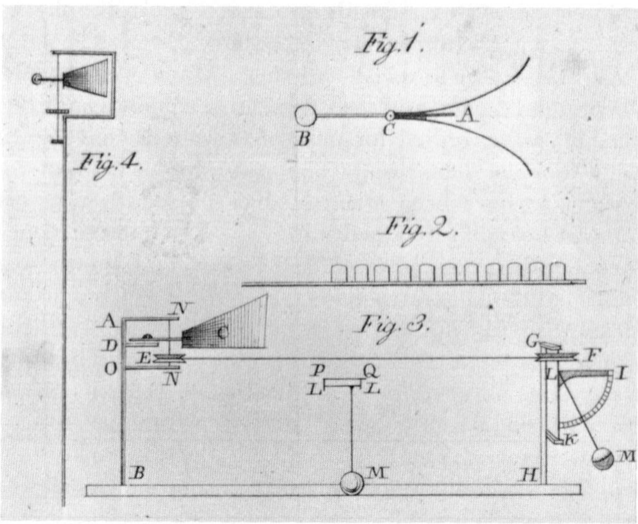

Fig. 5.2 Parrot's splayed vane, 1797.

desirable to have a vane that is insensitive to the rapid changes of wind direction that occur in gusty winds, and that will give a mean direction over a period of time. Such a device, suggested but not used by William Whewell,[12] formed part of Beckley's recording anemometer, to be described later.[13] It was an adaptation of the "fan tail" used to face windmills into the wind, an invention patented in 1745;[14] it consisted of a small subsidiary windmill at right angles to the main one, so that if the main windmill were faced into the wind, the small one would be stationary. A change in the wind direction will cause the fan to rotate in such a sense as to turn the whole mill into the wind again through suitable gearing. Beckley used two "fans" with a worm between them, working against a fixed worm-wheel.

In the last quarter of the eighteenth century there was a flurry of interest in vanes that would indicate the inclination of the wind to the horizontal as well as its direction, and a greatly exaggerated estimate of the utility of such measurements to the meteorology of the time. Franz Carl Achard of the Berlin Academy,[15] J. F. Benzenberg,[16] and Hyacinthe Carena[17] all made very similar suggestions—a balanced vane on a horizontal axis attached to the tail of an ordinary windvane. Such vanes, but much more delicately made, have been of use in micro-meteorological investigations in recent times, but the techniques of stereophotogrammetry are usually brought in to observe them.

[12] Whewell, *Trans. Cambridge Phil. Soc.*, Vol. 6 (1838), p. 302.
[13] See p. 216.
[14] Edmund Lee, British Patent 615, Dec. 9, 1745.
[15] Achard, *Mém. Acad. R. Sci., Berlin* (1788), pp. 103–4.
[16] Benzenberg, *Ann. Phys.*, Vol. 8 (1801), pp. 240–44.
[17] Carena, *Mém. phys. math. R. Acad. Sci. Turin*, Vol. 18 (1809), pp. 92–97.

Instruments to record wind direction are very numerous, but most of them are combined with recorders of wind speed, and will be dealt with later. Here we shall notice a few devices that record the direction alone.

For the earliest of these, attributed to Sir Christopher Wren, the reader must be referred to Chapter 7, as it formed part of a meteorograph. I shall pass on to 1789, when Christian Gotthold Herrmann, a pastor at Kämmerswalde in Saxony, described an ingenious machine which shows that the twelve-point wind rose had survived in some places until then.[18] In his apparatus a wheel carrying twelve sector-shaped compartments is rotated by a windvane. Once an hour a hammer operated by a clock ejects a numbered cube from a magazine into the appropriate box, depending on the wind direction at the

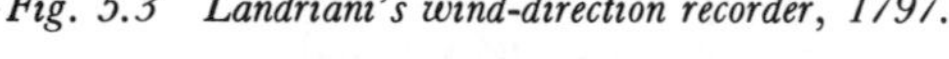

Fig. 5.3 Landriani's wind-direction recorder, 1797.

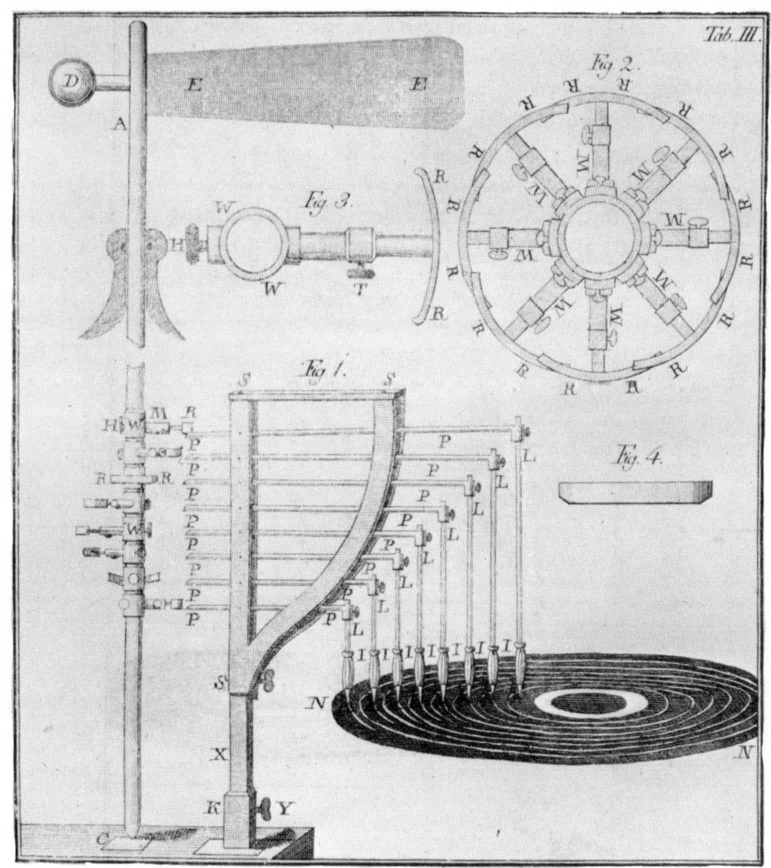

[18] Herrmann, *Mechanisch verbesserter Wind-, Regen-, und Trockenheits-Beobachter* (Freiburg i.S., 1789), pp. 18–31.

time, so that by looking at the numbers on the cubes in each compartment, the observer could tell what the direction had been at each hour.

In the following decade a recording windvane (Fig. 5.3) was set up at Milan by Count Marsiglio Landriani.[19] Cams *R* on the shaft of the vane displace one of eight pencils that otherwise would all trace concentric circles on a horizontal disk *N* rotated by a clock.

In 1837 James H. Coffin of Ogdensburg, New York, who built his own instruments and accumulated a great deal of statistical information about the weather, designed a "sand windvane," of which a small portion is preserved in the Museum of History and Technology in Washington, D.C.[20] In this instrument a hopper attached to the foot of the vane shaft distributed a continuous stream of dry sand among thirty-two receptacles arranged in a circle, and the time that the wind had blown from each direction was obtained by weighing these at the end of some fixed period. By 1849,[21] realizing that the diurnal variation might be important, he had substituted the rather improbable device shown in Figure 5.4, in which a clock-driven mechanism drops a card, numbered with the hour, into one of thirty-two boxes attached to a ring revolved by the vane. In principle this is exactly the same as Herrmann's instrument.

To turn now to recorders that give a detailed record of wind direction as a function of time (those that are not confined to four, eight, etc., directions), the first separate instrument since that of Wren was

Fig. 5.4 Coffin's second recording windvane, 1849.

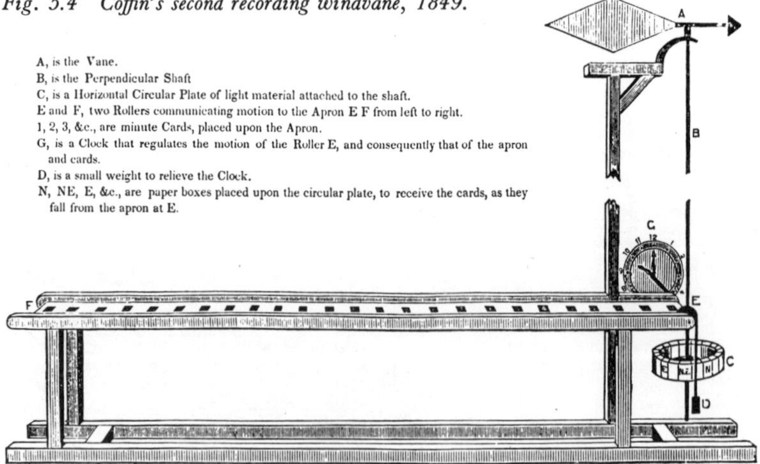

[19] Landriani, *Mag. Neuste Phys. Naturgeschicte*, Vol. II, part 3 (1797), pp. 93–102. This publication must not be confused with the one cited in note 10 above.
[20] Inv. no. 247,698.
[21] Coffin, Proc. A.A.A.S., Vol. 2 (1849), pp. 386–89.

described—I doubt very much whether it was ever built—by Jacob Leupold.[22] This solved the problem of transforming a rotary motion, that could make any number of turns in either direction, into a linear motion confined to the width of a chart. The trick was to use a chain or belt on two equal pulleys, with the chain carrying at least three uniformly spaced styli (Leupold used four). The circumference of either wheel is exactly equal to the distance between two adjacent styli, and if the chart is this wide, one stylus will leave it just as another enters.[23]

The next one actually built was that of Karl Kreil of Prague in the nineteenth century.[24] Like his other recording instruments, this one made its record on a flat chart, and the windvane operated a pencil by moving a nut on a fast screw. There was room for about two complete revolutions of the vane, after which the pencil could go no farther. "I have not yet," wrote Kreil, "been able to think of a way of overcoming this defect completely."[25] His insistence on a flat chart made the problem more difficult. If he had used a drum, he would have had another degree of freedom, for there are two ways of using a drum for this purpose. One can let the windvane turn the drum, probably on its own vertical axis, and move the stylus axially by clockwork; or one can turn the drum with a clock, and move a pen or pens axially by means of some mechanism attached to the vane. The first way is easy; the second poses a problem. On the other hand, if you want a continuous record of both speed and direction on one chart, you have to use the second configuration.

The drum rotated by the vane was used by F. Pfeiffer in 1868,[26] by Daniel Draper about 1880,[27] and by Joseph Baxendall at the turn of the century.[28] It is worth noting that Draper's wind-direction recorder is the only one of his instruments that does not use the large flat chart for which he had a predilection.

Windvanes indicating electrically at a distance began to appear about 1850, mostly in connection with anemometers, and will receive attention below. Here I shall refer only to the ingenious instrument of A. Moritz,[29] which used just three connecting wires to indicate 64 points. This depended on one set of contacts being made during motion in one direction, and another set during the reverse motion. The receiver had two electromagnets and two escapements. Unfortunately there was no feedback, with the result that a missed contact would affect all subsequent indications.

[22] Leupold, *Theatri statici universalis*, pp. 303–4 and plate XXII.
[23] See Fig. 10.2, p. 321. There are technical advantages to using only three styli.
[24] Kreil, *Entwurf eines meteorologischen Beobachtungssytems für die österreichische Monarchie* [etc.] (Vienna, 1850), pp. 201–2.
[25] *Ibid.*, p. 202.
[26] See p. 195 below.
[27] *Amer. Meteorol. J.*, Vol. 1 (1884–85), pp. 535–39.
[28] Baxendall, *Quart. J. Roy. Meteorol. Soc.*, Vol. 25 (1899), pp. 326–29.
[29] Moritz, *Repert. für phys. Technik*, Vol. 10 (1874), pp. 457–59.

3. Classification of anemometers. Although anemometers are of many kinds, and hundreds of designs have been imagined, they can be classified into three groups, somewhat as follows: (1) those in which the dynamic pressure of the moving air produces an observable deflection of some indicator (pressure anemometers); (2) those in which some mechanical system is continuously turned on its axis by the air stream, and the speed of its motion measured or the number of revolutions counted (rotation anemometers); and (3) those depending on the cooling power of the air. The third class is not ordinarily used in meteorology, and will not be discussed here.

Each of the above classes may be subdivided: 1*a*, Swinging-plate anemometers and analogous instruments; 1*b*, normal-plate anemometers; 1*c*, "bridled" anemometers; 1*d*, pressure-tube anemometers; 2*a*, rotation anemometers with horizontal axes; 2*b*, rotation anemometers with vertical axes; 3*a*, the Kata-thermometer, and analogous instruments; and 3*b*, hot-wire anemometers. Among these types, the "bridled" or restrained anemometers partake of the characteristics of both classes 1 and 2; technically they are rotation anemometers prevented from rotating continuously, and for this reason they will be dealt with last.

Of course the speed of the wind can be measured with no apparatus other than a piece of thistledown and some means of measuring time. This method was recommended by Edme Mariotte in 1686,[30] and was the one preferred by William Derham in 1708.[31]

4. The swinging-plate anemometer. The first anemometer of any kind that we know about was a swinging-plate instrument, and was described and illustrated by Leon Battista Alberti at some time near 1450. It is in a little work called "On the pleasures of mathematics," first printed in Italian in 1568,[32] and I have reproduced the Figure (Fig. 5.5) showing the little swinging board faced into the wind by a vane, and provided with an arc on which its deflection could be read. The idea may have arisen from watching a sign swinging in the wind, or some sheets drying on a clothes line.

On the strength of the well-known figure in the *Codex Atlanticus*, fol. 249^{v}.a, Leonardo da Vinci has usually been credited with the invention of this instrument, but Leonardo was born in 1452, and there is a passage in *Ms. G.*, fol. 54^{r},[33] in which he refers vaguely to an anemom-

[30] Mariotte, *Oeuvres* (2 vols., Leyden, 1717), II, 406.

[31] Derham, *Phil. Trans.*, Vol. 26 (1708), p. 30.

[32] Alberti, *Opuscoli morali . . . tradotti . . . da Cosimo Bartoli* (Venice, 1568), pp. 225–55. The anemometer is described on p. 253. According to the *Dizionario biografico degli italiani*, Vol. I (Rome, 1960), the *Ludi mathematici* was written between 1450 and 1452; G. Mancini, *Vita di Leon Battista Alberti*, 2nd ed. (Florence, 1911), p. 281, says that it was dedicated to Meliaduso d'Este in 1450.

[33] Cited by Ivor B. Hart, *The Mechanical Inventions of Leonardo da Vinci* (London, 1925), pp. 25–26.

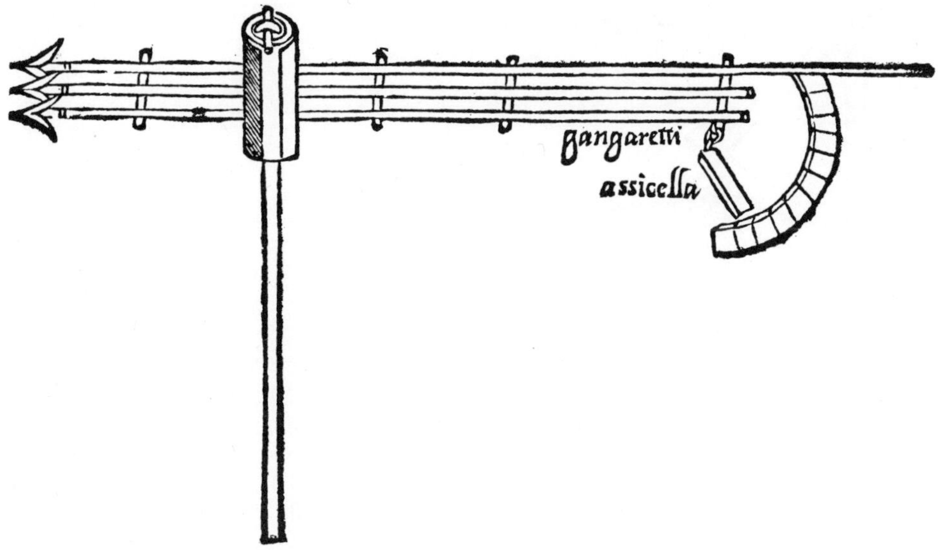

Fig. 5.5 Alberti's swinging-plate anemometer, ca. *1450.*

eter by Alberti. In view of this it can scarcely be maintained that Leonardo reinvented it independently. It is a little more likely that Robert Hooke did. On October 12, 1664, at The Royal Society, "it was mentioned, that there was an instrument made for measuring the swiftness of the wind."[34] This was probably the instrument (Fig. 5.6) described anonymously three years later[35] as part of some instructions for keeping "a register of all changes of wind and weather at all hours."

We next hear of this sort of anemometer in 1684, when Leibniz adumbrated a recording instrument based on it.[36] Like many of Leibniz' technical ideas, it was not developed at the time.[37]

The "anemoscope" of Roger Pickering[38] was very similar to that of Hooke but much larger, with the addition of a ratchet-and-pawl mechanism to indicate the maximum wind.[39] He doubted whether the construction of this 4¼-feet-high instrument was strong enough to stand very bad weather for long.

At about the end of the eighteenth century the Abbé Antonio Maria Vassalli-Eandi made the swinging plate into a recording instru-

[34] Birch, *History*, I, 475.
[35] *Phil. Trans.*, Vol. 2 (1667), pp. 444–45.
[36] E. Gerland, ed., "Leibnizens nachgelassene Schriften physikalischen, mechanischen und technischen Inhalts." *Abh. Gesch. Math.*, Vol. 21 (1906), pp. 1–256. The reference is to pp. 195–96.
[37] See also p. 39 above.
[38] Pickering, *Phil. Trans.*, Vol. 43 (1744), No. 473, pp. 9–11. (No. 473 is erroneously numbered from p. 1.)
[39] See End Papers.

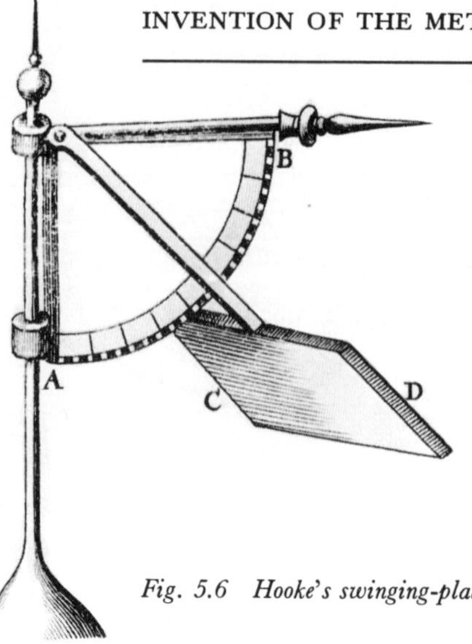

Fig. 5.6 Hooke's swinging-plate anemometer, ca. *1664.*

ment.[40] His plate was one meter square, and was faced into the wind by an enormous vane with an area of two square meters. A cord attached to the pressure plate traversed a red pencil axially along a drum rotated once every thirty hours by a clock. The vane also turned a large stepped cam that caused one of thirty-two black pencils to press on the drum 180° away from the red one that made the wind-speed record.

The swinging body need not be a plate. During a storm on December 13, 1747, Michael Christoph Hanov of Danzig thought of hanging a lead ball from the center of an astrolabe, and observing how far it was blown aside by the wind.[41] He calibrated this by drawing it aside by means of a thread passing over a smooth rod to a series of weights. Knowing the cross-section of the ball, which was only ⅔ of a Paris inch (about 18 mm) in diameter, he boldly extrapolated the wind's force to larger areas. He was aware that it is proportional to the square of the speed, "other things being equal."

The idea of a sphere blown aside by the wind had an ingenious development a century later at the hands of Sir Charles Wheatstone, well known for clever applications of electricity, who, according to Theodose Du Moncel,[42] came to Paris on a visit and suggested a distantly recording anemometer of great simplicity, a "parabolic roof" with five or six conducting rings, and a ball hanging by a wire from the apex (Fig. 5.7). The topmost ring was to be divided into eight or sixteen parts, to give the wind direction. The instrument was to work a suitable chronograph, but I know of no evidence that it was ever made.

The swinging-ball anemometer of Parrot has already been illus-

[40] Vassalli-Eandi, *Mem. Accad. Sci. Torino*, Vol. 12 (1800–1803), pp. 426–44.

[41] J. D. Titius, ed. *Michaelis Christophori Hanovii . . . Opuscula* (2 vols., Halle, 1761), Vol. II, pp. 369–77.

[42] Du Moncel, *Exposé des applications de l'électricité* (5 vols., Paris, 1856–62), Vol. II, pp. 339–400.

trated in Figure 5.2, and it will be seen that the vane rotates it into the wind by way of pulleys and a belt. I shall mention one more application of the resistance offered to the air by a sphere, the anemometer of S. B. Howlett.[43] As shown in Figure 5.8, a sphere *d* having a cross-sectional area of 0.2 square feet is supported in the wind by a brass tube *b* working in gimbals, and more than counterbalanced by a weight *e*. A pencil slides in the tube and marks the deflections on a circular chart, showing the maximum gust as a function of direction.

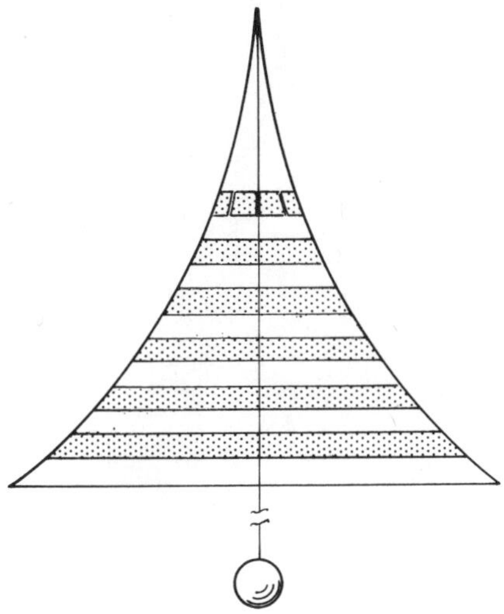

Fig. 5.7 Wheatstone's idea for an anemometer, ca. *1850.*

5. The normal-plate anemometer. An anemometer in which a surface is held normal, or nearly normal, to the wind is theoretically simpler but more difficult in practice than the instrument just referred to. The earliest anemometer of this kind was probably described by Santorio Santorre,[44] who played an important role in the history of the thermometer, as we saw in Chapter 2. His anemometer (Fig. 5.9) consists of a flat plate rising from the end of a steelyard; the illustration is self-explanatory if we assume that the piece of string joining the beam to the top of the page is a whim of the engraver. Santorio was a medical man, and the text reproduced with our figure shows why he was interested in anemometry—strong winds seem to be good for some people, bad for others. He also illustrated a current-meter with the plate turned downwards, but otherwise on the same principle.

On March 2, 1627, a patent was granted by the States-General of

[43] Howlett, British Patent 1,839 of 1865. See also *Proc. Meteorol. Soc.*, Vol. 4 (1868), pp. 161–64.

[44] *Sanctorii Sanctorii Iustinopolitani . . . commentaria in primam fen primi libri canonis Avicennae* [etc.] (Venice, 1625), Col. 245–46.

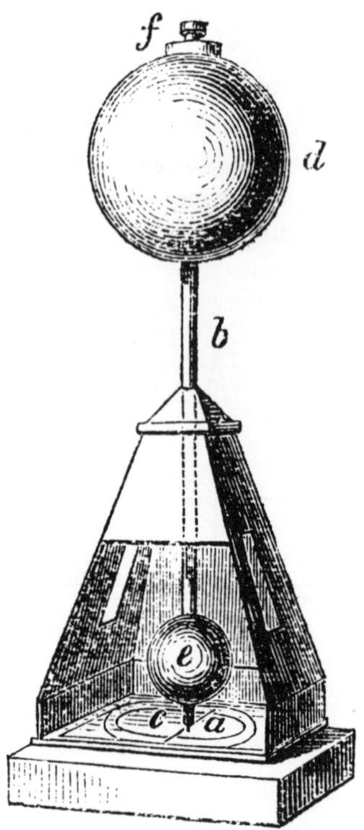

Fig. 5.8 Howlett's anemometer, 1868.

Holland to one Willem Douglas for various devices.[45] The tenth of fifteen claims is "an invention showing the increase and decrease of the winds proportionally." This should be an anemometer, but no further information is available, and I record it here only because its date is just lat r than that of Santorio's book.

It will not surprise us that this type of anemometer is represented in Leupold's compendium.[46] His normal-plate anemometer is shown in figures *I* and *II* of our Figure 5.10, and it will be seen that the plate, faced toward the wind by a vane, is mounted on a carriage provided with rollers. The resistance to its motion is provided by a weight acting on a spiral pulley, a device that turns up again and again in Leupold's plates.

A simple application of the pressure plate was made by Pierre Bouguer for measuring the wind at sea.[47] This hand anemometer was just a spring balance with a six-inch-square piece of cardboard pressing a square shaft into a square box against the resistance of a

[45] *Arch. Staten-Generaal* ('s-Gravenhaage), ms. 12303, fol. 225^{v}–226^{v}.

[46] Leupold, *Theatrum aerostaticum* (Leipzig, 1726), Tab. XLVIII.

[47] Bouguer, *Traité du navire, de sa construction, et de ses mouvemens* (Paris, 1746), pp. 359–60.

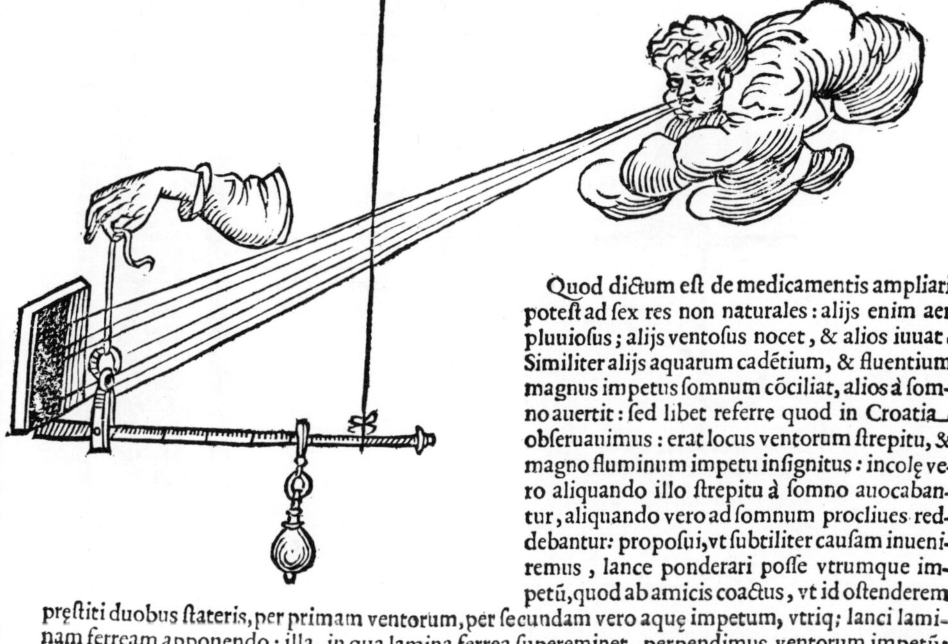

Quod dictum eſt de medicamentis ampliari poteſt ad ſex res non naturales: alijs enim aer pluuioſus; alijs ventoſus nocet, & alios iuuat. Similiter alijs aquarum cadētium, & fluentium magnus impetus ſomnum cōciliat, alios à ſomno auertit: ſed libet referre quod in Croatia obſeruauimus: erat locus ventorum ſtrepitu, & magno fluminum impetu inſignitus: incolę vero aliquando illo ſtrepitu à ſomno auocabantur, aliquando vero ad ſomnum procliues reddebantur: propoſui, vt ſubtiliter cauſam inueniremus, lance ponderari poſſe vtrumque impetũ, quod ab amicis coactus, vt id oſtenderem pręſtiti duobus ſtateris, per primam ventorum, per ſecundam vero aquę impetum, vtriq; lanci laminam ferream apponendo: illa, in qua lamina ferrea ſupereminet, perpendimus ventorum impetũ: cuius

Fig. 5.9 Santorio's anemometer, 1625.

spring. A scale on the shaft was established by calibration with weights. It is to be noted that Bouguer was interested in the pressure of the wind on the sails of the ship, and suggested that the plate could be held parallel to them. There was no calibration for wind speed.

The engineers began to be concerned with the relation between speed and pressure, and made experiments with whirling machines. Among the first to do this were Benjamin Robins, one Rouse, and the celebrated John Smeaton.[48] At St. Petersburg, Johann Ernst Zeiher made a whirling machine with which he calibrated a Bouguer anemometer,[49] and in 1783 Richard Lovell Edgeworth reported on his "experiments upon the resistance of the air,"[50] in which he established that the resistance offered by the air to a moving object depends upon

[48] Smeaton, *Phil. Trans.*, Vol. 5, part 1 (1759), pp. 100–174.
[49] Zeiher, *Nov. Comm. Petrop.*, Vol. 10 (1764), pp. 302–6.
[50] Edgeworth, *Phil. Trans.*, Vol. 73 (1783), pp. 136–43.

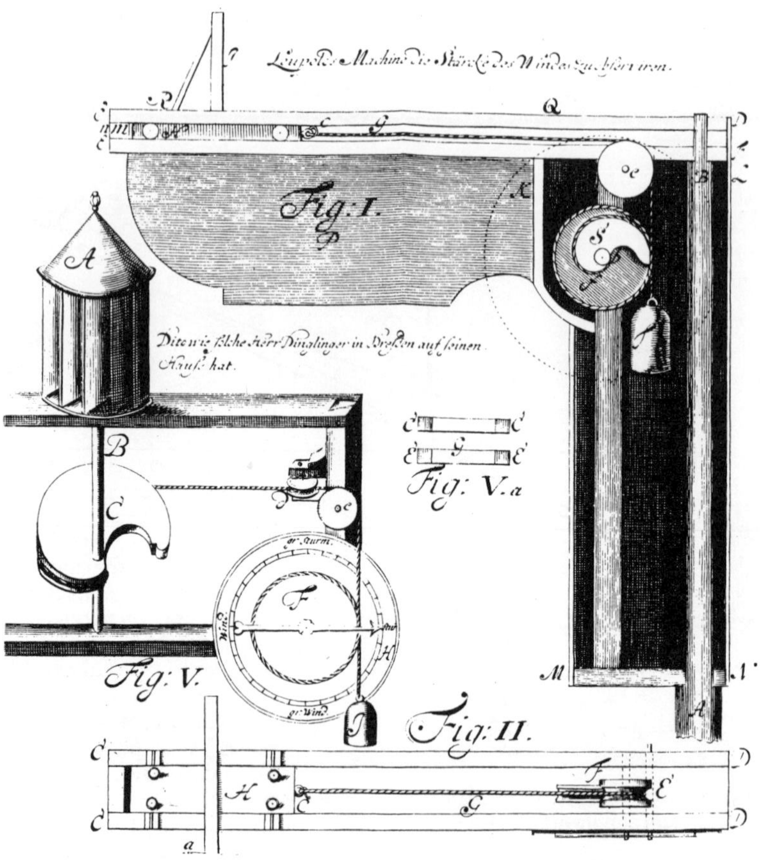

Fig. 5.10 Two anemometers, from Leupold's Theatrum aerostaticum.

its shape as well as its size, and that for surfaces of similar shape, it is not linear in area.[51]

It would be tedious to enumerate and more tedious to describe all the variations on the normal-plate anemometer that appeared in the century following Bouguer's book. These differ chiefly in the means of providing a calibrated resistance—a Roman balance (Dalberg),[52] a variable column of liquid (Wilcke),[53] a chain balance (Woltmann),[54] a fusee (Beaufoy),[55] or—to go back to Leupold—a spiral cam

[51] The history of such experiments to about 1840 is dealt with in great detail by Muncke in *Gehler's physikalisches Wörterbuch* (11 vols. in 20, Leipzig, 1825–45), Vol. 10, pp. 1779–1860.

[52] K.T.A.M. von & zu Dalberg, *Obs. sur la Phys.*, Vol. 17 (1781), p. 438. He also used a spring balance.

[53] Johann Karl Wilcke, *K. Vet. Akad., nya Handlingar*, Stockholm, Vol. 3 (1732), pp. 89–103.

[54] R. Woltmann, *Mag. für das Neueste aus der Phys. und Naturges.*, Vol. 11, part 3 (1797), pp. 106–8.

[55] Mark Beaufoy, *Ann. Philos.*, n.s., Vol. 2 (1821), pp. 431–33.

(Goudriaan).[56] Slightly later and unique is the design for an anemometer discovered by C. H. Gibbs-Smith among the papers of that forgotten genius Sir George Cayley, dating from 1849.[57] In this the pressure plate is at the end of a fairly heavy bar hung from a frame by two links forming a parallel motion. The deflection of one of the links is read on a quadrant. Damping is provided by a paddle projecting down from the bar into a trough of water.

Goudriaan's anemometer is interesting because it was surely one of the last instruments to incorporate a water clock, which provides the time coordinate. It was in fact a combined water-level or tide gauge, windvane, and anemometer, recording wind speed on a disk chart, wind direction and water level on two drums, one revolved by a vane and the other by a float. A normal pressure plate revolved the disk, over which the pencil was moved radially by the water clock.

With Goudriaan's instrument we have entered the field of recording anemometers. Although, as we have seen and shall see in what follows, a number of instruments to record the speed of the wind had been designed, the fact remains that in 1830 there was still no recording anemometer durable enough and cheap enough for continuous use at meteorological stations. This requirement was first met by Follett Osler in England with a combination pressure-plate anemometer and recording rain gauge,[58] which gave many years' service at more than one observatory.[59]

This instrument, shown somewhat diagrammatically in Figure 5.11, has a square pressure plate *T* faced into the wind by a large splayed vane *V*. The pressure plate is on appropriate guides and its motion, resisted by a spring *t*, is passed down the hollow mast *WX* by a cord to a stylus *2*. The vane also turns a pinion *p* that moves a rack *ef*, to which is attached a stylus *1*. These styli, and a third one connected to the mechanism of the rain gauge, make records on a flat chart traversed from left to right by a clock *C*. A problem that does not seem to have been solved is the one already mentioned—what happens if the vane makes more than one revolution in the same direction?

There were several later designs of normal-plate anemographs, of which the most original was that of Charles Cator in its various modifications.[60] In this instrument the back of the pressure plate was faired in as a cone (Fig. 5.12), and the mechanical parts were brought indoors as far as possible. In the first model the pressure of the wind was balanced by a weight acting on the edge of a spiral cam, but in

[56] A. F. Goudriaan, *Verh. Kon. Nederl. Inst., erste Cl.*, Vol. 3 (1817), pp. 1–25.
[57] Gibbs-Smith, *Sir George Cayley's Aeronautics 1796–1855* (London, 1962), pp. 239–40.
[58] Osler, B.A.A.S., Liverpool (1837), *Sections*, pp. 33–34. Regarding the rain gauge, see p. 155 above.
[59] One was installed at Greenwich in 1841, and was still giving good service in 1861. See J. Glaisher, *Proc. British Meteorol. Soc.*, Vol. 1 (1861), p. 22.
[60] Cator, *Proc. Meteorol. Soc. (London)*, Vol. 3 (1866), pp. 49–55; *ibid.*, Vol. 4 (1869), pp. 273–82; *ibid.*, Vol. 5 (1871), pp. 263–66.

the later one shown in our illustration, two curved weighted levers were substituted. Another novelty introduced by Cator was an attachment for measuring the "run of the wind" (the total number of miles) in 24 hours. In this device the movement of the pressure plate was made to shift the "regulator" of a special clock, so that the run of the wind was indicated by the number of minutes gained by the clock at the end of the period.

Fig. 5.11 Osler's combined anemograph and recording rain gauge.

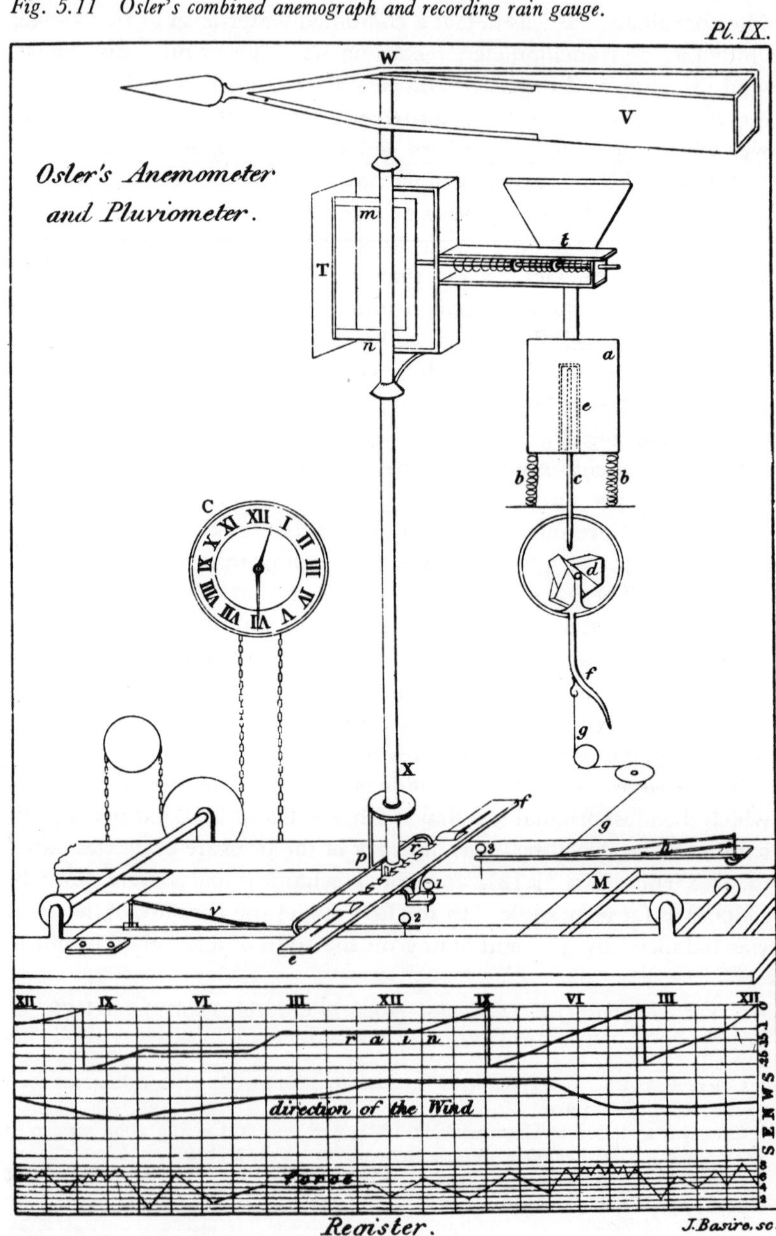

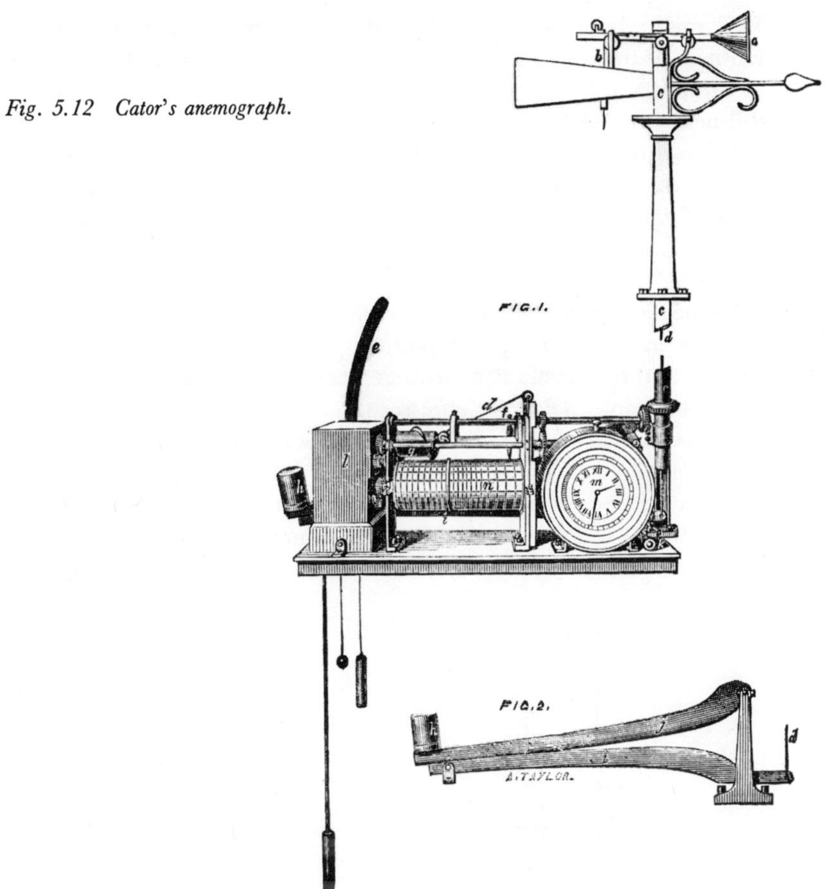

Fig. 5.12 Cator's anemograph.

6. The pressure-tube anemometer. Pierre Daniel Huet, Bishop of Avranches in Normandy, died in Paris in 1721. In the following year a miscellaneous collection of short essays was published,[61] among which is an account of his invention of a pressure-tube anemometer. It may be of interest to read his story of how this came about:

People have latterly succeeded in finding out exactly the properties of the air, its warmth, its humidity, and its weight, by means of the thermometer, the hygrometer, and the barometer, which is a weighing machine for air. But while we have sought to weigh the air, we have not thought at all about measuring the wind. I proposed this to Hubin the Englishman, an excellent maker of instruments of these sorts. He laughed about it, as a thing pleasant to think of, but impossible to carry out. I gave him the description of an instrument that I had contrived for doing this, and he was so pleased with it that he left me with the intention of making it as soon as possible; but his death prevented it.[62]

[61] *Huetiana, ou pensées diverses de M. Huet, Evesque d'Avranches* (Paris, 1722). Also printed in Amsterdam in 1723.
[62] *Ibid.*, pp. 55–56.

This might be expected to date the anemometer, but the date of Hubin's death is unknown, though he was one of the leading instrumentmakers in Paris, especially about 1675.[63] It is known that he was still active in 1688.

It is a little surprising the Hubin was unwilling to believe that wind could be measured, for, as we have seen,[64] he was acquainted with some of Robert Hooke's instruments. As to the date of Bishop Huet's idea, I am inclined to place it in the first decade of the eighteenth century.

In the form sketched by Huet (Fig. 5.13) the instrument seems very much a toy, as Sprung has pointed out.[65] The little windvane is absurd, and the mechanical arrangements at *I* are even more so. The wind blows into *AB* and depresses the mercury at *C*, raising it at *F*. Over the mercury in the tube *FK* there is some *eau seconde*, a green acid liquid obtained in the separation of the precious metals; while this would keep one surface of the mercury clean, it would not alter the sensitivity. I feel sure that if Hubin had made an anemometer of this

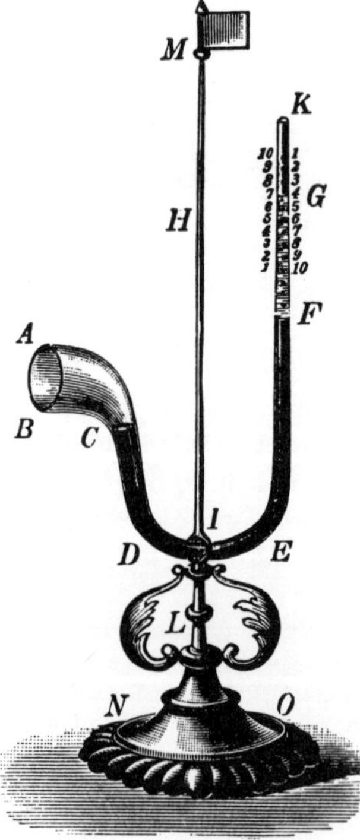

Fig. 5.13 Huet's anemometer.

[63] M. Maurice Daumas, the authority on French instruments and instrumentmakers of the period, is unable to date Hubin's decease (private communication).
[64] Page 26 above.
[65] *Zeits. Instrum.*, Vol. 12 (1892), p. 147.

sort, he would not have used the mercury; indeed, the deflection given by this instrument would have been very small in any ordinary wind.

Huet's elementary instrument antedated the celebrated "Pitot tube," intended solely for measuring the speed of the water in streams and described by Henri Pitot of the Académie des Sciences in 1732.[66] In 1743 Stephen Hales, who may well have seen Pitot's paper, invented an instrument for the measurement of artificial ventilation on the same principle as Huet's but using water as the liquid.[67] Nevertheless the first anemometer of this kind that can be considered as a satisfactory meteorological instrument was that of the Edinburgh physician James Lind.[68]

The construction of this simple instrument will be clear from Figure 5.14, with the notation that the elbow *F* was of brass. The main glass tubes *AD* and *BC* were of about 0.4-inch bore; the connecting tube *ab* was about 0.1 inch. The hole in the brass cap *G* was 0.2 inch in diameter. The *Z*-shaped appendage was to be put on at *F* to keep out rain! The scale of inches *HI* could be raised or lowered a little in order to bring its zero to the undisturbed level of the water in the tubes. It should be noted that there is no separate windvane, because Lind found that the anemometer, supported as shown, always turned into the wind. The instrument with a windvane at the top, illustrated in many textbooks, was probably a modification introduced by some nineteenth-century maker.[69]

Lind assumed that the pressure of the wind on an obstacle would be equal to that exerted on its base by the column of water sustained in the anemometer. He did not take into account the suction at *G*. Even with this added, it takes a breeze of about sixteen knots to give a difference of level of 0.1 inch. To make the instrument more useful for light winds, W. Snow-Harris made the interesting modification shown in Figure 5.15.[70] The small tube *cdefg* has a bore one-quarter that of the tube *ab*. If now the instrument, which is provided with a plumb-bob, is installed so that the segment *ef* of the tube is horizontal, the level in *av* will change only one-sixteenth as fast as the position of the meniscus in *ef*. Stronger winds are taken care of by the segment *fg*, which operates in the usual manner.

Other sorts of sensitive manometer have been used with the pressure-tube anemometer, but I shall confine myself to the development of the instrument into an anemograph. This began in 1836, when Richard Adie was given a silver medal by the Scottish Society of Arts for a pressure-tube anemometer with its head faced into the wind

[66] Pitot, *Mém. Acad. r. Sci. Paris* (1732), pp. 363–75.
[67] Hales, *A Description of Ventilators* [etc.] (London, 1743), p. 14.
[68] Lind, *Phil. Trans.*, Vol. 65 (1775), pp. 353–65.
[69] It seems to have become established by 1864. See Negretti & Zambra, *A Treatise on Meteorological Instruments* [etc.] (London, 1864), p. 117.
[70] Snow-Harris, *Nautical Mag. for 1858*, pp. 113–22.

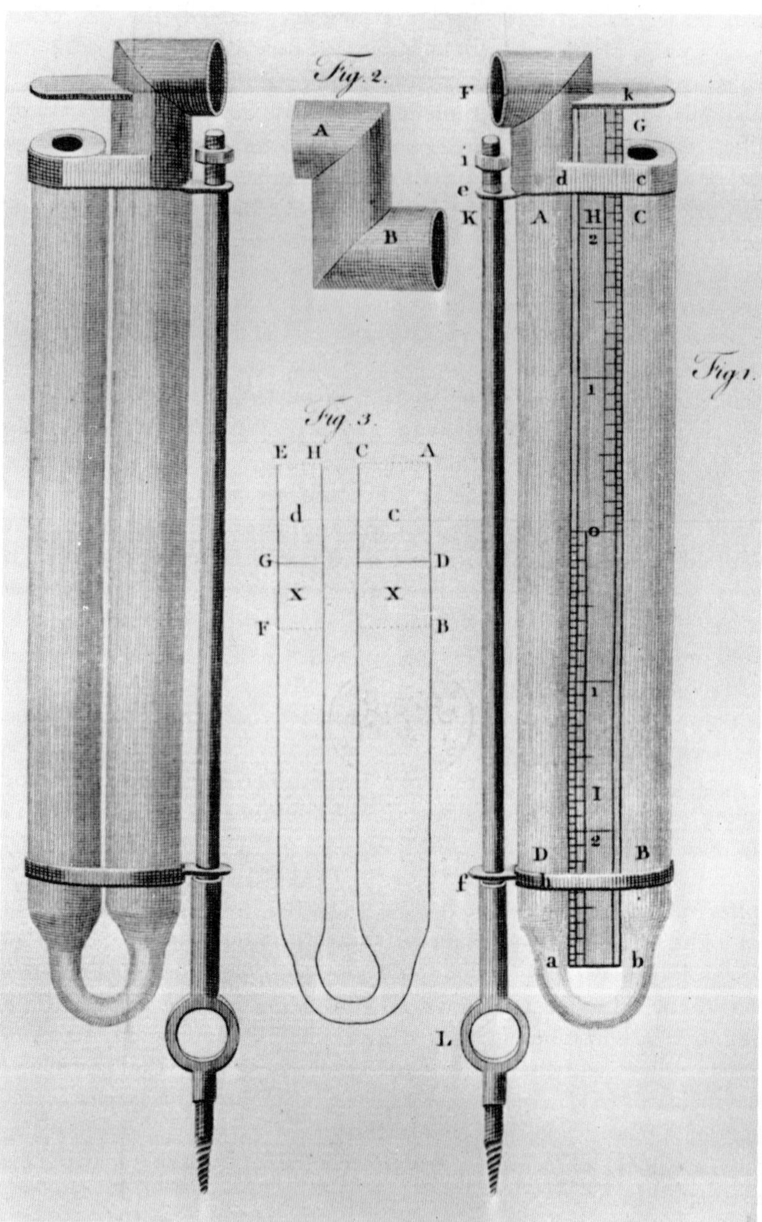

Fig. 5.14 Lind's pressure-tube anemometer, 1775.

by a vane.[71] A pipe from the head went to the space inside a floating cylinder; the upward force was aided by a weight acting on a spiral pulley in such a way that it straightened out the scale. This was not a recording instrument, and is mentioned here because it incorporated a hollow float, the essential element in future developments.

The earliest of these to be reported was the "Windautograph" of F. Pfeiffer, shown somewhat diagrammatically in Figure 5.16,[72] which needs little comment except to note the way in which a mercury seal is used at the bottom of the vertical pipe. This anemograph would

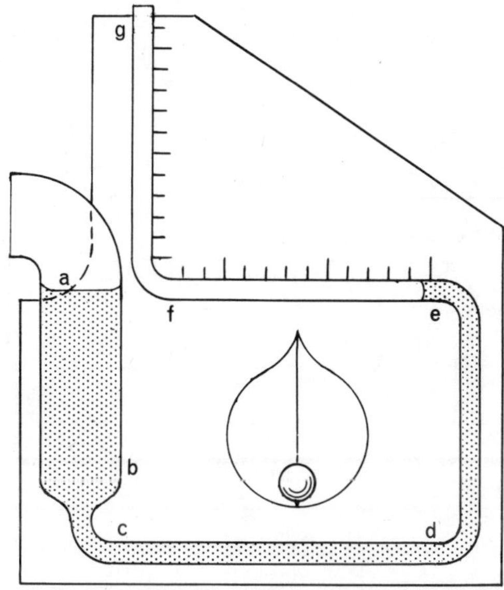

Fig. 5.15 Snow-Harris' modification of Lind's anemometer, 1858.

provide a record that is approximately linear in wind pressure, so that its deflections would be proportional to the square of the wind speed. The variations in upward force on the bell jar would be matched by opposite changes on the buoyancy of its walls.

The next instrument to appear was somewhat off the main line of development. This is the anemograph of R. Bourdon,[73] in which a small Venturi tube is mounted with its discharge end in the throat of a larger one; this is said to "multiply" the partial vacuum in the throat of the small one, and the vacuum is measured with a bell-jar manometer of constant section acting against a helical spring. The motion of the bell jar, which will be proportional to the square of the wind speed, is registered by a pencil moving radially over a circular chart turned by a clock. This instrument, which does not seem to have

[71] Adie, *Edinburgh New Philos. J.*, Vol. 22 (1836), pp. 309–13.
[72] See Ernst Mayer, *Zeits. österr. Ges. Meteorol.*, Vol. 3 (1868), pp. 409–12.
[73] See Léon Descroix, *Annuaire Soc. météorol. de la France*, Vol. 30 (1882), pp. 122–27.

been much used, is in the Conservatoire National des Arts et Métiers, Paris.[74]

Meteorologists naturally preferred to have the scale of an anemograph linear in wind speed, and if the pressure-tube anemograph was to compete with other types, some means of straightening out the

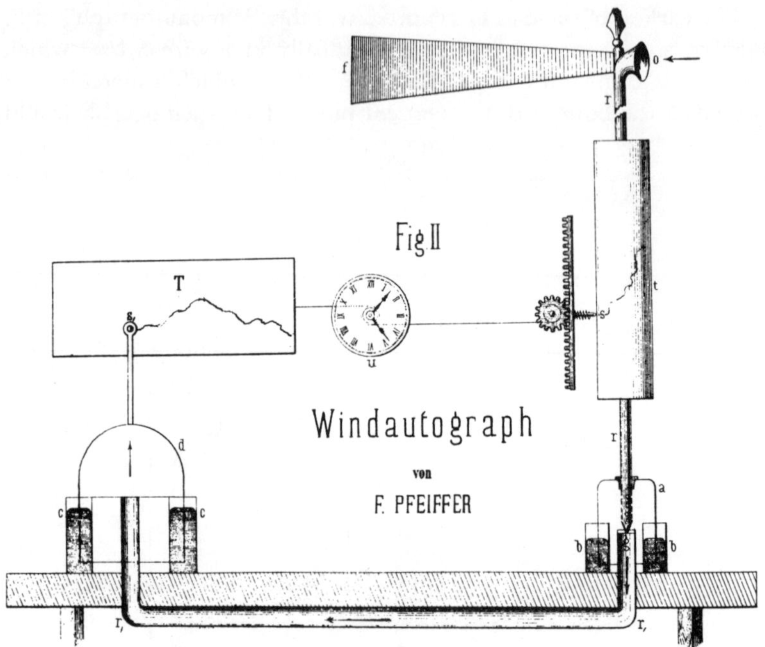

Fig. 5.16 Pfeiffer's anemograph and recording windvane.

scale had to be found. A brilliant solution to this problem was provided in the 1890's by William Henry Dines. Previous inventors had assumed that the pressure on one side of the manometer would be constant.[75] Dines knew that the pressure within a building exposed to the wind could fluctuate widely, and arranged things so that neither side of his manometer was open to the air indoors, as we shall see. In view of the importance of the Dines anemometer I shall go into some detail about its history. W. H. Dines got his degree from Cambridge in 1881, and after spending some years teaching mathematics, became interested in anemometry. He was soon "the most active member of a Wind-Force Committee appointed by the Meteorological Society in November, 1886."[76] It was in connection with the work of that committee that the first of his pressure-tube anemographs was described,[77]

[74] Inv. no. 10307.
[75] Cleveland Abbe knew that this was not so. See U.S. War Dept., *Ann. Rep. Chief Signal Officer for 1882*, p. 99.
[76] Obituary notice by Sir Napier Shaw, *Proc. Roy. Soc. A*, Vol. 119 (1928), p. xxiv.
[77] Dines, *Quart. J. Roy. Meteorol. Soc.*, Vol. 18 (1892), pp. 165–83. *Collected Scientific Papers of William Henry Dines, B.A., F.R.S.* (London, 1931), pp. (29)–(47).

but the advantages of such an instrument had been noted by Dines two years earlier, at the end of an account of some experiments with a whirling machine that he had had built in 1886.[78] At any rate, the instrument he described in 1892 had all the essentials of the final form of the pressure-tube anemometer. I reproduce his diagrams of the head (Fig. 5.17) and of the float (Fig. 5.18). The former needs the explanation that at *B* "twelve 1/12-inch holes are drilled in the outer tube in a ring, by which holes the annular space is connected with the outer air."[79] In this way a suction, independent of wind direction and varying as the square of the wind speed, was communicated through *E* (Fig. 5.18) to the space above the float, while the pressure from *A* is transferred through *F* to the space beneath it. The shape of the hollow chamber *N* of the float is so arranged that the change in the displacement of the float is proportional to the square of the distance it has risen from its original position of equilibrium. It is characteristic of Dines that he writes that "the chamber *N* is not exactly conical, but bulges out slightly in the central parts."[80] The rigorous theory of this anemograph was given only in 1936 by E. Gold.[81]

The first such anemograph was erected at Hersham in December, 1890, and moved to Oxshott in May, 1891. In this instrument the float was "connected with the pen carriage by a silk thread passing through a very small hole in the top of the zinc cylinder *H*."[82] It was, of course, necessary that the cylinder *H* should be very nearly airtight.

It is not certain who made this first instrument. Dines stated that it and two other anemometers were "obtained" in the autumn of 1890,[83] but his son, L. H. G. Dines, was sure in 1950 that it was procured from the London instrumentmaker R. W. Munro, who made the later ones for many years.[84] I shall let L. H. G. Dines continue with the story:

> In May 1891 we moved to Oxshott to a new house built to my Fathers own designs. The form of the roof had apparently been deliberately devised with anemometer comparisons in mind, for it had a flat leaded space 12 ft square which served no immediate purpose in relation to the accommodation inside the house . . .
>
> A great deal of work was carried out at Oxshott during the winter 1891–2 comparing anemometers.[85] The specimen of the Pressure Tube anemometer used in this case was the same as that previously erected on the tower at Hersham.

[78] *Ibid.*, Vol. 16 (1890), pp. 208–10. *Collected Papers*, pp. (26)–(28).
[79] *Ibid.*, Vol. 18 (1892), p. 169. *Collected Papers*, p. (33).
[80] *Ibid.*, p. 172. *Collected Papers*, p. (36).
[81] Gold, *Quart. J. Roy. Meteorol. Soc.*, Vol. 62 (1936), pp. 170–77.
[82] Dines, *Quart. J. Roy. Meteorol. Soc.*, Vol. 18 (1892), p. 192. *Collected Papers*, p. (36).
[83] *Ibid.*, p. 173. *Collected Papers*, p. (37).
[84] Letter from L. H. G. Dines to D. Brunt, Dec. 14, 1950 (Collection of the Poole family, Item 47, foll. 12–15). Quoted by permission. But Mr. J. S. Dines (private communication, Sept. 11, 1967) thinks that it is just possible that his father made it in his own workshop.
[85] See Dines, *Quart. J. Roy. Meteorol. Soc.*, Vol. 18 (1892), pp. 165–83.

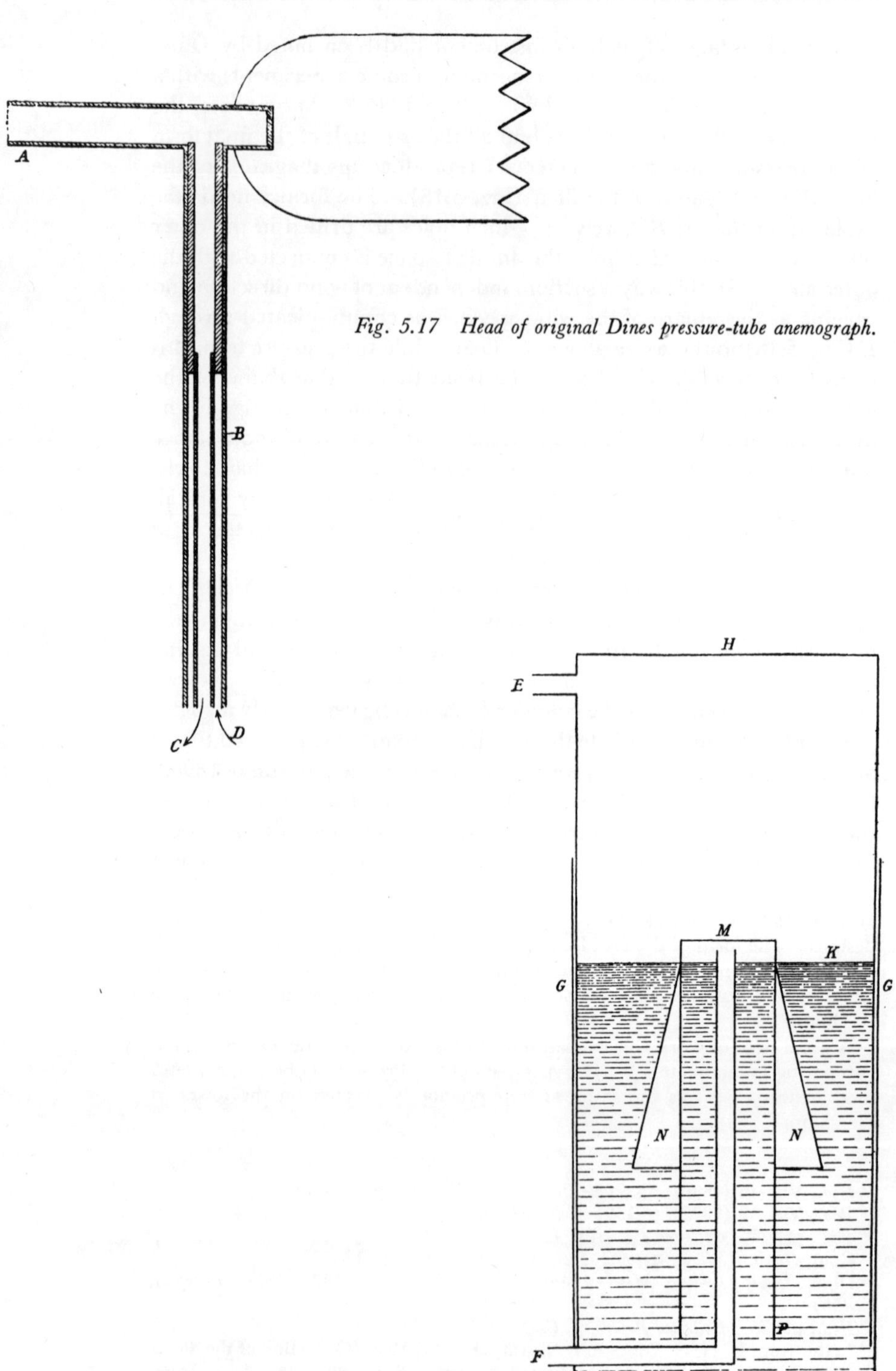

Fig. 5.17 Head of original Dines pressure-tube anemograph.

Fig. 5.18 Float of original Dines pressure-tube anemograph.

[F. J. W.] Whipple states[86] that on some (unspecified) date the Wind Force Committee requested Messrs Dines and Munro to prepare drawings of the P.T. Instrument. What I think certain is that sometime in 1893 a brand new instrument constructed by Munro was installed at Oxshott where it worked continuously for 14 years. This particular instrument remained in possession of the Dines family till 1938, when it was presented to the Science Museum.[87] In essential principles it followed its predecessor; the float was of the same type, but mechanical details of the recording mechanism were altered, and the whole instrument had the customary high finish of the product of a high class instrument maker. It was never fitted with a direction recorder in the modern sense.[88]

This instrument, of which at least four specimens were installed in various places, is shown in Figure 5.19. It should be noted that there are four rows of suction holes instead of one. In January, 1895, Munro's order book contained an entry about "new drawings and patterns with new shaped float and higher 10″ tank for 120 m.p.h."[89] This refers to the radically different and very sophisticated shape shown in Figure 5.20. This figure is a cross-section of the recorder in its modern form, including a wind-direction recorder, about which a few remarks are necessary because the original Dines anemometer had none. Nevertheless W. H. Dines, as early as 1898,[90] had imagined the ingenious double helical cam which puts one pen into action when the wind veers through north, while the other pen simultaneously goes out of action, while if the wind backs through north the roles of the pens are reversed. (I do not know how to explain this mechanism satisfactorily in the absence of the thing itself.) Dines's idea was to work the two pens through chains. He seems to have made such a recorder as a separate instrument and then given up the idea because, he told L. H. G. Dines, he thought that Baxendall's recorder[91] was so much better.[92] The combination of the two instruments came in stages. In 1903 Baxendall described a head for the Dines anemograph with a splayed vane that also revolved his direction recorder of 1899.[93] It was mechanically improved by F. L. Halliwell, and is shown in Figure 5.21.

J. S. Dines tells us about the next stages: "I found its remains [of W. H. Dines's original direction recorder] at Pyrton Hill when my job was to investigate wind currents and using Father's essential design fitted it into a P.T. and thereafter it came into fairly general use. This would have been about 1911, but I have no means of fixing

[86] Dines, Collected Papers, p. (11).
[87] Inventory no. 1937–177.
[88] L. H. G. Dines to D. Brunt, Poole Coll., item 47, fol. 13.
[89] Letter from R. C. Munro to J. S. Dines, July 12, 1967.
[90] L. H. G. Dines to J. S. Dines, Dec. 2, 1950 (typescript, signed "Lewen"). Poole Coll., item 47, foll. 5^r–6^r.
[91] Joseph Baxendall, *Quart. J. Roy. Meteorol. Soc.*, Vol. 25 (1899), pp. 326–29. This was exactly the same in essentials as Pfeiffer's recording vane (Fig. 5.16).
[92] L. H. G. Dines to J. S. Dines, Poole Coll., item 47, fol. 5^r.
[93] Baxendall, *Quart. J. Roy. Meteorol. Soc.*, Vol. 29 (1903), pp. 289–98.

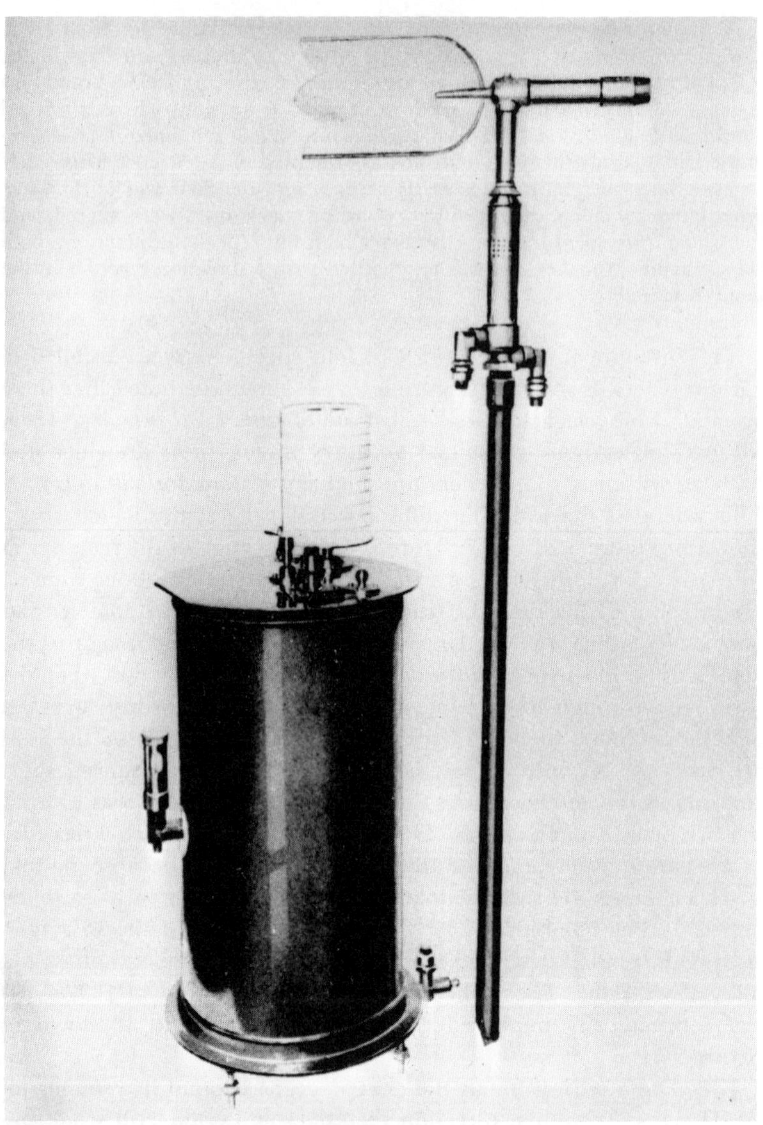

Fig. 5.19 The type of Dines anemograph made by Munro, 1893 to 1896.

the exact date. Stanley's predecessor at the M.O. modified it subsequently by fitting the long arms to replace the vertical slides, which made the thing cumbersome but also made it work better."[94] Whipple tells us[95] that "Stanley's predecessor" was J. H. James, a skillful mechanic who was in charge of the Meteorological Office

[94] J. S. Dines to L. H. G. Dines, Dec. 1, 1950 (typescript, signed "Sum"). Poole Coll., item 47, fol. 4.
[95] In Dines, *Collected Papers*, p. (12).

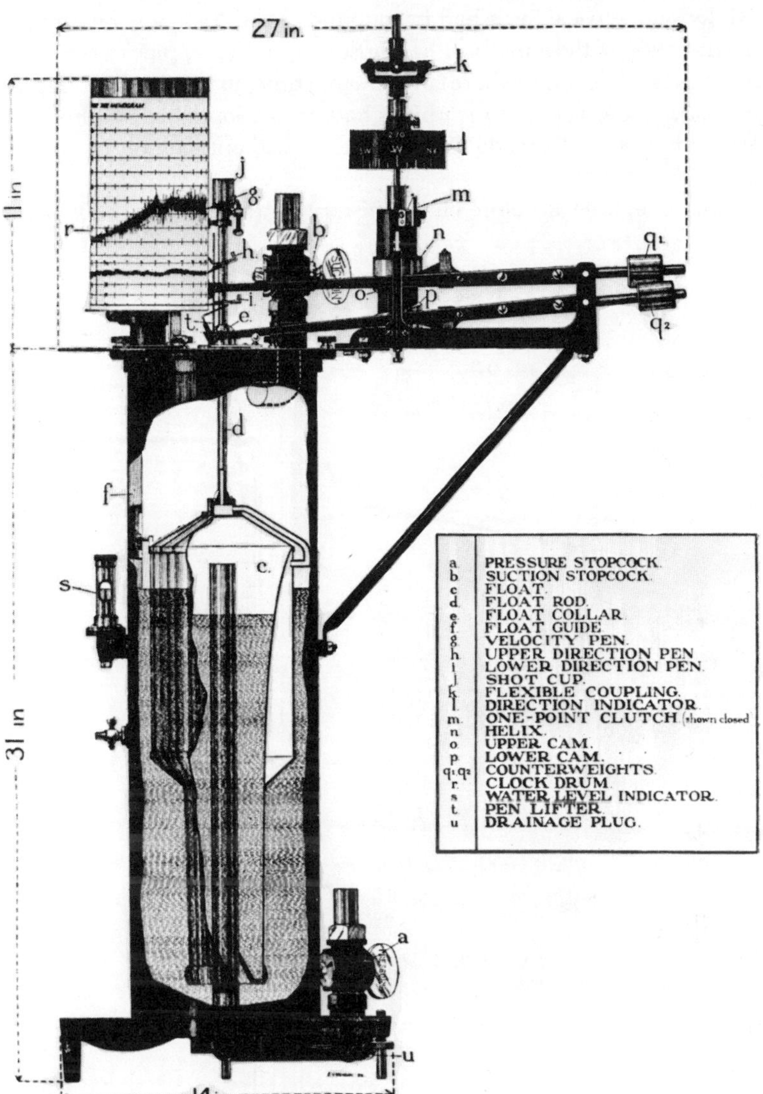

Fig. 5.20 Dines anemograph, modern form (from "The Observer's Handbook," 1934; by permission of the Controller, H. M. Stationery Office).

workshop, and that the date was 1918. The "long arms" are clearly indicated in Figure 5.20.

Also in 1918 the vane was transformed, at the suggestion of the famous mathematical physicist G. I. Taylor, into the tall airfoil section shown in Figure 5.22.[96] In 1927 the conical shield over the pipes was introduced in order to make the fixed part of the head

[96] E. Gold, *Quart. J. Roy. Meteorol. Soc.*, Vol. 62 (1936), p. 177.

axially symmetrical, as it had been found that the suction varied with the direction of the wind.[97] The pressure and suction pipes were made twice as large (one-inch bore) at the same time, to reduce the damping in gusts. However, one-inch pipes had been adopted at Southport in 1903 by Joseph Baxendall,[98] apparently without consulting W. H. Dines.

This is a brief account of the history of the Dines pressure-tube

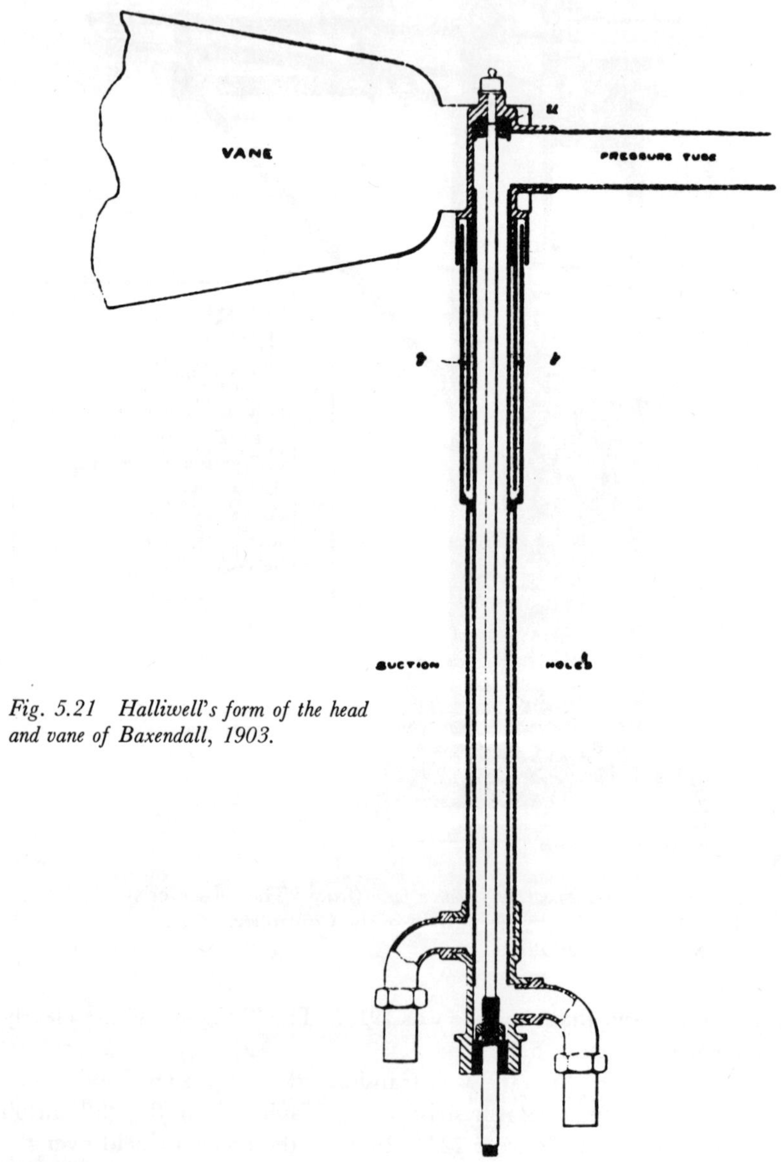

Fig. 5.21 Halliwell's form of the head and vane of Baxendall, 1903.

[97] *Ibid.*
[98] Baxendall, *Quart. J. Roy. Meteorol. Soc.*, Vol. 29 (1903), pp. 292–93.

anemograph up to about 1930. Apart from optional additions such as electrical gear for transmitting the records to distant points, it has altered little since then, and remains an instrument of remarkable convenience and accuracy.

7. Rotation anemometers with horizontal axes. The common European windmill dates from the twelfth century.[99] It was perhaps natural that when, in the seventeenth century, meteorological instruments began to be developed, the use of a miniature windmill as an anemometer was considered.

Robert Hooke considered and used the idea of a windmill in the "weather-clock" that was under construction in the winter of 1672/3, and was described by Hooke about December, 1678, according to Derham. This instrument will be dealt with in Chapter 7, and here we shall note only that "the strength of the wind [was] observed by the number of revolutions of the vane-mill, and marked by three

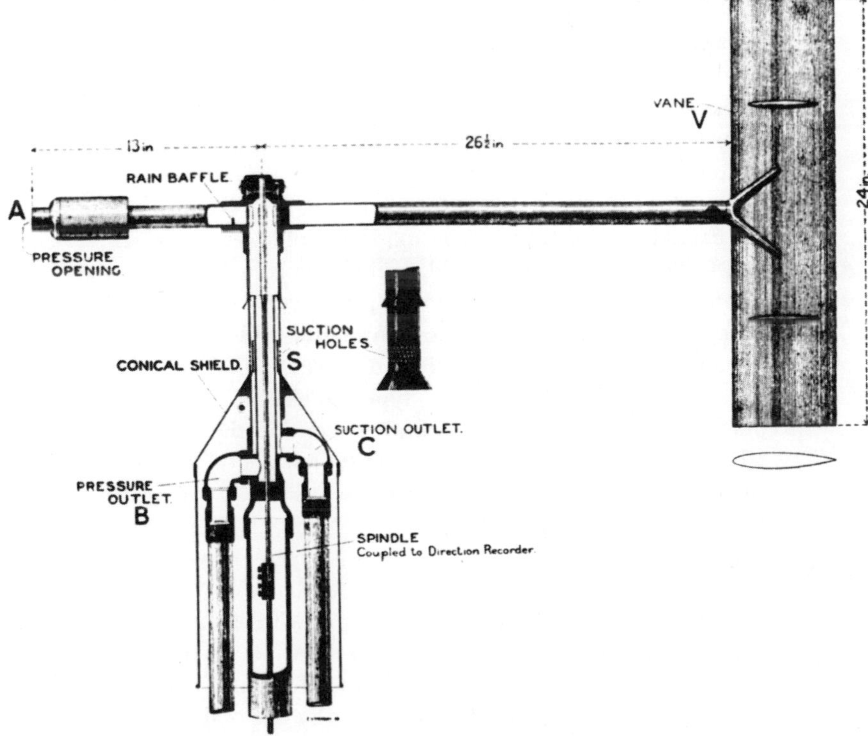

Fig. 5.22 Dines anemograph, head and vane, modern form (from "The Observer's Handbook," 1934; by permission of the Controller, H. M. Stationery Office).

[99] T. K. Derry and T. I. Williams, *A Short History of Technology* [etc.] (Oxford, 1960), p. 254.

punches; the first marks every 10000 revolutions, the second every 1000, and the third every 100."[100]

Even after the completion of the "weather-clock" Hooke seems to have made experiments along this line, for on

> November 14, 1683, Mr Hook shew'd an instrument to measure the velocity of the air or wind, and to find the strength thereof, which was by four vanes put upon an axis, and made very light and easy for motion; and the vanes so contriv'd, as that they could be set to what slope should be desir'd: It was several times try'd and examin'd in the long gallery in Gresham College; whereby it appear'd, that by walking from one end thereof to the other, and carrying the same above ones head, the doors and windows of the said gallery being shut, and so the air within it being not in motion but stagnant, the instrument made so many turns as there were circumferential lengths of the said vanes in the length of the gallery; and if by trial it were found to be more or less than the due measure of the circumferential lengths, then by setting the said vanes either flatter or sharper in respect of the way of its motion through and against the air, the same was easy to be adjusted; the use of which may be of very great consequence in the business of sailing and steering a ship upon the sea, and for examining the power and strength of the wind upon land in order to the theory of shipping for which it was designed.[101]

Hooke referred to this instrument again in a lecture on December 3, 1690, suggesting that it be mounted on a ship, and continued, "The instrument I design for this purpose, is but little differing from that which I long since contriv'd for measuring the velocity of the wind, and caused one of them to be made for the weather clock. It has been sufficiently seen and try'd, and therefore will not, for this time, need a more particular description; however, if any one designs to make trial thereof I shall not be wanting [he will be available] to give the workman, that makes it, sufficient instruction."[102] But alas! By the time these passages were set in type, Hooke was dead, and Waller confesses that this is all he can find concerning the instrument.

The next rotation anemometer with a horizontal axis was inspired not by a windmill but by an overshot water wheel. It was imagined by the important Russian physicist Michael Lomonosow, whose diagram is shown in Figure 5.23.[103] In a large narrow box *BCED* that forms part of a windvane, there is a paddle wheel *A*, formed of 16 vanes *a* 2 inches by 24 inches, braced with wires *c* and *g*. This wheel, of which only the top half is exposed to the wind, revolves the shaft *b*, carrying a single-tooth pinion *d*, so that for each revolution of the paddle wheel the large wheel *F* advances by one of its 800 teeth, being

[100] Hooke, in: W. Derham., ed., *Philosophical Experiments and Observations of the late Eminent Dr. Robert Hooke, S.R.S. and Geom. Prof. Gresh.*[am College] *and Other Eminent Virtuoso's in his Time* (London, 1726), pp. 41–42.

[101] Quoted from Royal Society, *Journal Book*, Nov. 14, 1683, in *The Posthumous Works of Robert Hooke* [etc.], ed. Richard Waller (London, 1705), p. 562.

[102] *Ibid.*

[103] Lomonosow, *Nov. Comm. Petrop.*, Vol. 2 (1749), pp. 128–33.

prevented from slipping back by the pawl *m*. The wheel *M*, with 400 teeth, is pressed by the spring *e* against the pinion *h*. As *M* revolves, a drum attached to it winds up a cord *lt* that goes down the hollow mast and is wrapped around the wheel *H*. This is in a narrow box *NN* and the whole assembly rotates with the vane over the thirty-two small boxes *z*. The wheel *H* has a scale on its rim, which can be read against the index *n*, and a tube *yu*, initially filled with mercury. As the wheel *H* is slowly revolved by the paddle wheel through the gearing and the cord, mercury spills out into the box *NN* and escapes at *k* into whatever box *z* corresponds to the wind direction at the moment. In theory, at least, the distribution of the run of the wind may be found by weighing the amount of mercury in the various boxes. The instrument is returned to zero by pulling the cord *qr*, which draws the wheel *M* away from its pinion against the resistance of the spring *e*, then rotating the wheel *H* counter-clockwise, and refilling the tube *yu* with mercury. Some reservations about this apparatus are legitimate, especially in view of its great size—the box *BCED* would be about 1.5 meters square—and the lack of a counterpoise for the huge vane, but at least it is unique.

To come back to the windmill, in 1752 C. G. Schober made experiments on the resistance of windmill sails, using a whirling machine driven by a weight.[104] He also devised a windmill anemometer that rang a bell after every six revolutions, so that the wind speed could be found by counting the strokes of the bell for some given time. This has led to the claim that he invented the windmill anemometer, a claim that has also been made for Reinhard Woltmann,[105] but we have seen that the priority belongs to Hooke. It may also be recorded that the great Euler produced an elaborate mathematical theory of the action of the wind on a windmill,[106] but he was thinking of wind power, not of measurement.

Omitting one or two anemometers of little novelty, we come to the instrument designed by William Whewell, the extraordinarily versatile Master of Trinity and later Vice-Chancellor of Cambridge, naturalist, mathematician, philosopher, mineralogist. This anemometer, described in 1836,[107] unfortunately revealed his lack of flair for instrument design, although a number were made and installed.[108] The working parts are demonstrated in Figure 5.24, which shows the rather exiguous eight-sailed windmill, faced into the wind by a vane that seems too small to turn the whole mechanism of the instrument. A lead screw *S*, revolved slowly by the windmill, lowers the pen *P* at a

[104] Schober, *Hamburgisches Magazin*, Vol. 9 (1752), pp. 115–44; 227–69.

[105] Woltmann, *Theorie und Gebrauch des hydrometrischen Flügels oder eine zuverlässige Methode, die Geschwindigkeit des Windes und der strömender Gewässer zu beobachten* (Hamburg, 1790).

[106] L. Euler, *Hist. Acad. R. Sci. Berlin* (1756), pp. 165–234.

[107] Whewell, B.A.A.S., Bristol, 1836, *Sections*, pp. 39–40.

[108] There is one in the Science Museum, London (Inv. no. 1893–166), which was at Greenwich Observatory.

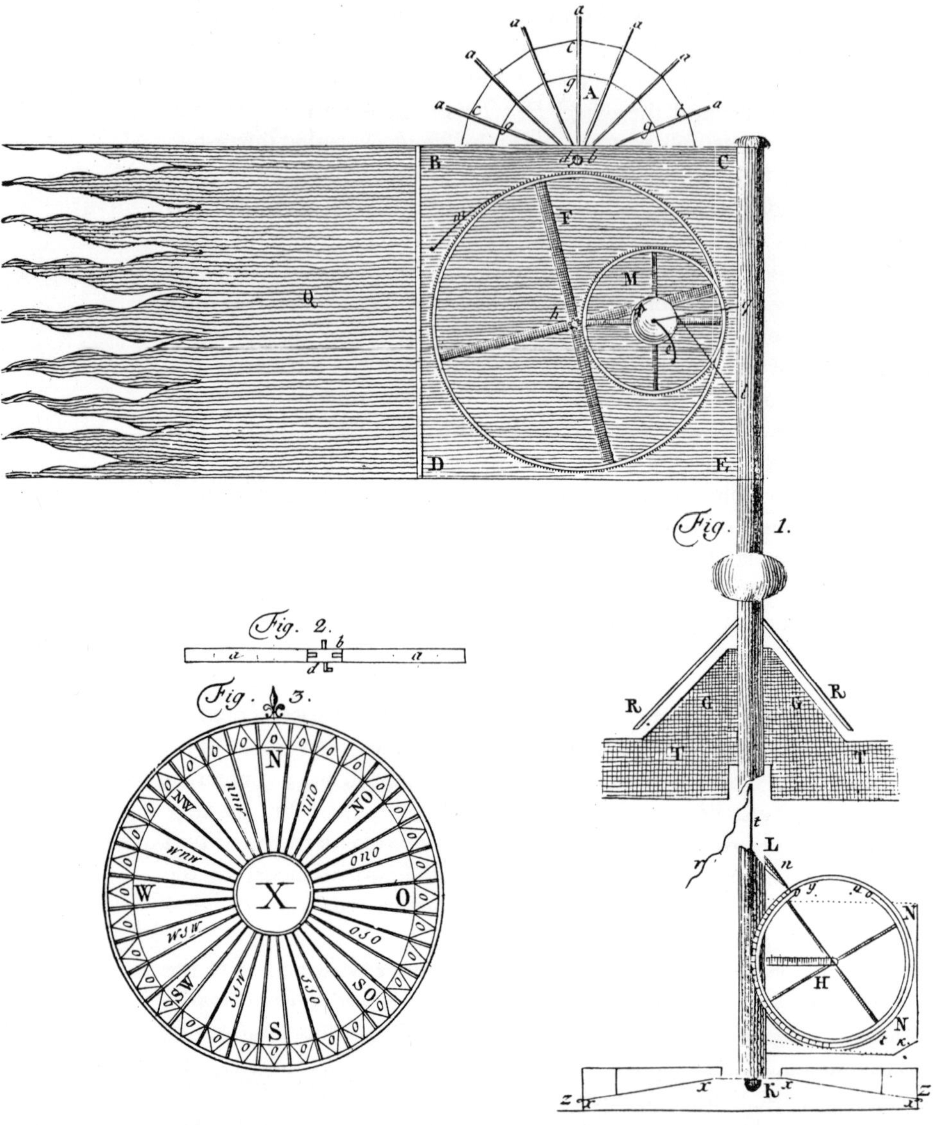

Fig. 5.23 The anemometer of Lomonosow.

rate proportional to the speed of the wind or, in other terms, through a distance proportional to its total run. At the same time, the pen is moved tangentially by the windvane around a stationary chart *C*. There is no clock, so that hourly values cannot be obtained.

Whewell himself soon realized that his engineering had not been of the best, and confessed in 1838 that "Various improvements in the instrument were suggested by using it; and as it had not been foreseen what strength of workmanship would be requisite to resist the weather, all the instruments were, at one time or another, disabled, so as to interrupt the observations."[109]

It is unlikely that the excellent instrumentmaker John Newman of Regent Street, who made the first ones,[110] had very much influence on the design. It was later severely criticised by W. Snow-Harris, who found it highly nonlinear,[111] and by the Astronomer Royal Sir George Airy, who "in his experience of this anemometer at Greenwich, thinks it cannot be depended upon for direction."[112] Surely Newman, just looking at it on his bench, must have had qualms.

We have followed the history of the windmill anemometer almost far enough, for the cup anemometer came to be preferred after the middle of the nineteenth century for general meteorological purposes. I shall mention only a remarkable form of rotation anemometer invented by W. H. Dines.[113] The idea of this instrument was to have a rotation anemometer entirely independent of friction, and therefore requiring no calibration. It is shown in Figure 5.25. Dines's brief description of its functioning could not be more concise:

> Briefly, a helicoid is fixed at the front, and a small pair of sails of variable pitch at the back of a steel rod, and just behind the helicoid is a light fan, which can turn on the same axis, but is independent of the helicoid and sails. If the rotation be too rapid, the fan turns in the same [relative] direction as the helicoid, and by its motion alters the pitch of the sails so that their motion is retarded; if on the other hand, the friction is increased, or from any other cause the motion becomes too slow, the fan is turned in the other direction [relative to the helicoid], and the rate is increased.[114]

Unfortunately, while this instrument behaved according to expectation, it was not robust enough for general use. It is interesting that it formed the subject of Dines's only patent, although at least two others were applied for and abandoned.

8. Rotation anemometers with vertical axes. The windmill with a vertical axis is older than the form to which we are accustomed. It was known in Persia in the seventh century A.D.,[115] but it found no favor in Europe, except in a form known as the Polish mill, which will be referred to later.

It so happens that, apart from an impracticable suggestion by Athanasius Kircher,[116] the first rotation anemometer to be described

[109] Whewell, *Trans. Cambridge Phil. Soc.*, Vol. 6 (1838), p. 302.
[110] *Ibid.*, Later ones were made by Simms of Fleet Street.
[111] Snow-Harris, B.A.A.S. York, 1844, *Report*, p. 245.
[112] *Ibid.*
[113] Dines, *Quart. J. Roy. Meteorol. Soc.*, Vol. 13 (1887), pp. 218–23. See also British Patent 1345 of 1886, Jan. 30.
[114] *Ibid.*, pp. 219–20.
[115] Derry and Williams, *A Short History of Technology*, p. 254.
[116] Kircher, *Magnes, sive de arte magnetica* [etc.] (Rome, 1641), pp. 512–13.

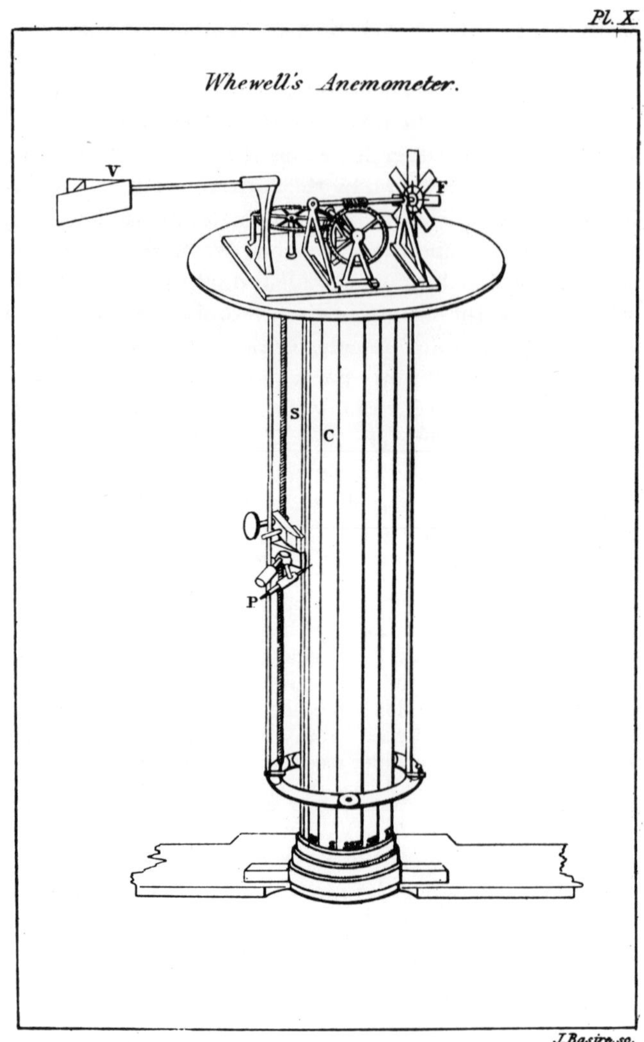

Fig. 5.24 Whewell's anemometer.

in print had a vertical axis.[117] It seems to have been the invention of a very aggressive Paris instrumentmaker called René Grillet, who was accused of plagiarism regarding another of his "curiositez,"[118] but may indeed have invented this anemometer and windvane, even though the rotating part bears a strong resemblance to one of the windmills in Faustus Verantius' book of machines.[119]

The anemometer (of which Grillet gives no figure) is a horizontal

[117] [Grillet, René]. *Curiositez mathématiques de l'invention* du *S*r *Grillet horlogeur à Paris* (Paris, 1673), pp. 9–11.
[118] See my *A History of the Thermometer*, p. 63.
[119] *Machinae novae Faustii Verantii siceni* [etc.] (Venice, 1616), plate 8.

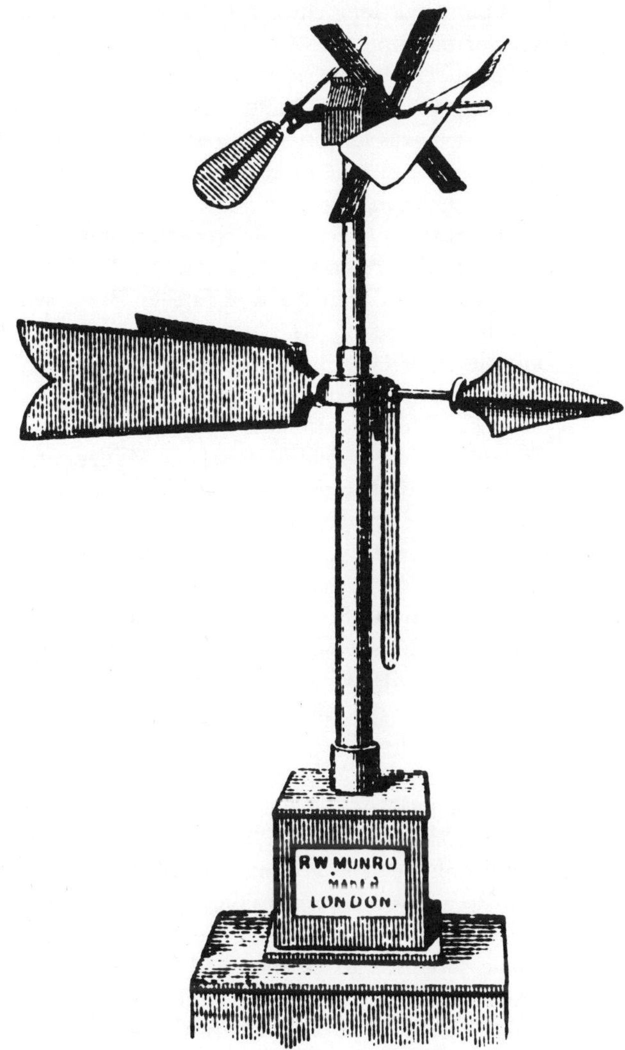

Fig. 5.25 Dines's helicoid anemometer.

windmill with four sheet-metal sails hinged to two pairs of wooden crosspieces one above the other. Each sail can swing as far as it will in one direction, but is prevented by a bar joining the crosspieces from swinging the other way. The mill will therefore revolve. At the bottom of the shaft that comes down through the roof is a worm that engages two wheels of 100 and 101 turns on concentric tubular shafts. These two shafts are brought out to appropriate dials, permitting the registration of 10,000 turns of the mill. The use of a timepiece makes it possible to measure the speed of the wind.

The vane is on a separate vertical shaft, geared to a pointer on a horizontal shaft inside the two concentric shafts already mentioned.

This pointer is long enough to indicate the wind direction on a wind rose surrounding the wind-speed dials.

The description is so circumstantial that I think one may assume that one was actually built. As it is not especially complicated, the fact that it was forgotten suggests that the windmill portion was not adequately resistant to the weather, but Grillet must be given full marks for originality.

The next and most famous application of this sort of windmill is found in the remarkable anemograph of Louis XV's Postmaster General Louis Jean Pajot d'Ons-en-Bray, which, to translate the descriptive title of his memoir,[120] "marks of itself on paper, not only the winds that have blown during 24 hours, and at what time each one began and ended, but also their different speeds or relative forces." This instrument, which is one of the treasures of the Conservatoire National des Arts et Métiers at Paris,[121] is, as far as I know, the oldest recording instrument that has been preserved, and is certainly one of the outstanding meteorological instruments of the eighteenth century. It is probably the "Grand anémomètre en cuivre" that forms item 49 in the catalogue of the Academy's instruments prepared after the French Revolution by J. A. C. Charles.[122]

Figure 5.26 is a photograph of this instrument, which is in reality two instruments, one on each side of a weight-driven pendulum clock. On the left is the wind-speed recorder, on the right the device that records the direction of the wind. Let us first deal with the wind-speed instrument.

D'Ons-en-Bray says nothing about the windmill except that it is of the sort commonly called "moulin a la polonaise." It is in fact a sort of reaction turbine, an outer set of stationary vanes channelling the wind to a revolving set inside, much as in Leutmann's anemometer, which will be dealt with in the next section.[123] This assembly is of course supposed to be mounted above the roof, and a shaft comes down to a 21-tooth pinion, which drives an 84-tooth gear *16* (fig. 5.27), on the shaft of which is a single-start worm engaging a 100-tooth worm gear, that revolves a "snail," a stepped spiral cam. This turns once for 400 revolutions of the windmill, and drops the weighted lever *29*.

A band of paper rubbed with finely-ground calcined hartshorn is unrolled from a spool *22*, over a roller *23*, on to a take-up spool *24* driven by the clock. There is a clever use of a weight acting through a fusee *37* to keep the tension of the paper almost constant. Every 400 turns of the mill, the weight *29* drops and makes a mark near the top

[120] Pajot d'Ons-en-Bray, *Mém. Acad. r. Sci. Paris* (1734), pp. 123–34 (with 6 plates).

[121] Inv. no. 5608.

[122] *Inventaire des Instrumens de Physique Optique, Géométrie, Astronomie &c. Du Cabinet de la cy dev! Academie des Sciences de Paris.* Institut de France, ms. 1,986, no. I [1793].

[123] See Fig. 5.41, p. 227. This sort of windmill was illustrated by Verantius, *Machinae novae*, plate 13.

of the band of paper. This paper will not move at a uniform speed because it is being rolled up on the spool *24*, so d'Ons-en-Bray has another lever *30* make a mark on it every half or quarter of an hour as desired.

The mechanism for moving the paper on which the wind direction is recorded is similar, except that the paper is much wider and the

Fig. 5.26 The anemograph of d'Ons-en-Bray, 1734 (courtesy of the Conservatoire National des Arts et Métiers).

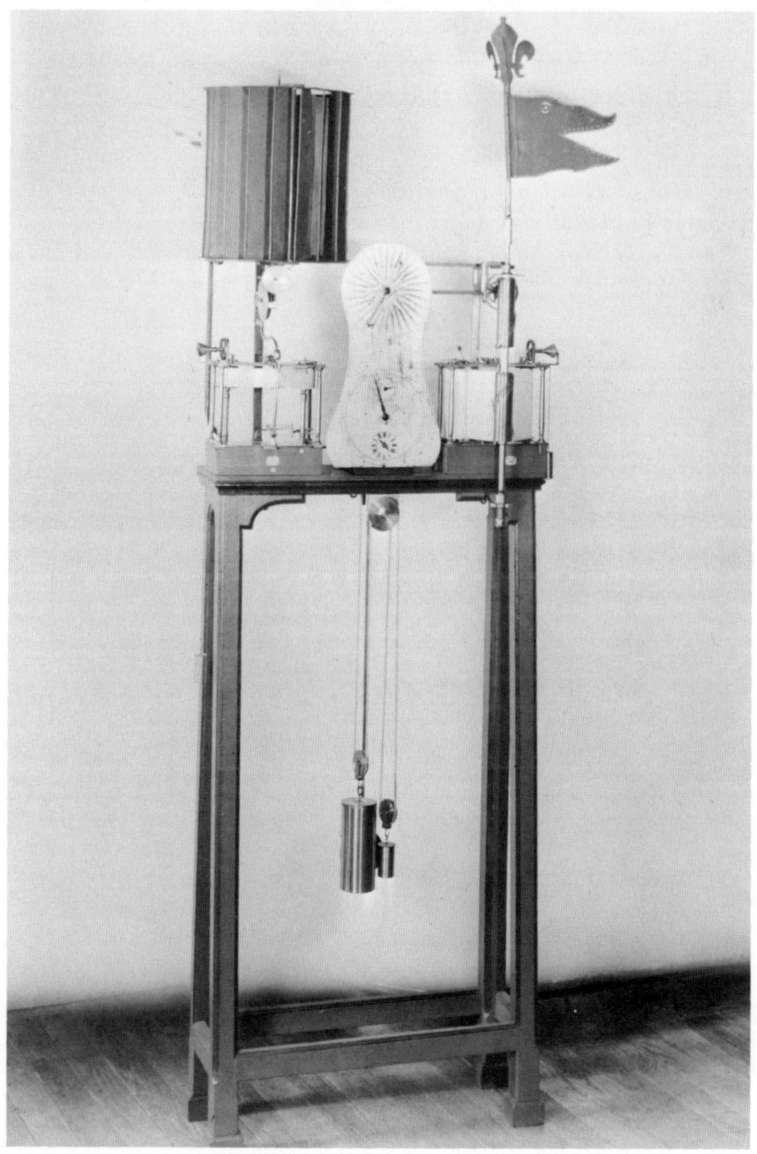

time mark is made near the top. The motion of the vane *5* turns the shaft *4* and also moves a pointer on the dial *B* by way of three pairs of gears. On the shaft *4* is a cylinder *6*, carrying thirty-two pins arranged in a one-turn helix, each pin on a spring. Figure 5.28 shows them more clearly. (It also shows another cylinder, not represented in d'Ons-en-Bray's drawings, the purpose of which I cannot understand.) Whatever the orientation of the vane, either one or two of the pins will be pressed against the paper, making a horizontal line as this goes by; thus the wind direction can be determined to one sixty-fourth of a circle, which seems almost luxurious. Indeed, the whole instrument is luxurious; its inventor even suggested that duplicate paper-transport mechanisms should be provided, to obviate loss of record while the paper was being changed. No doubt Louis XV's Postmaster-General could afford such things. This extravagance of design caused Georg Wilhelm Muncke, writing a century later in the second edition of *Gehler's Physikalisches Wörterbuch*,[124] to suppose that it was never built. It was, but it had no influence.

Even while Muncke was writing, the cup anemometer was being developed, entirely changing the situation as far as rotation anemom-

Fig. 5.27 D'Ons-en-Bray's anemograph.

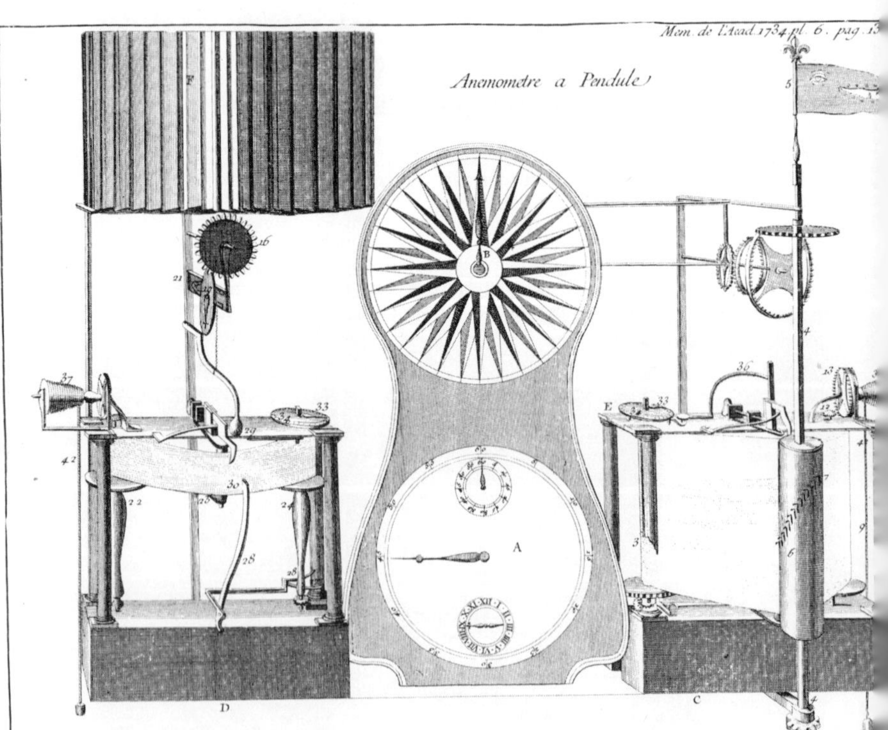

[124] Vol. 10, p. 2164.

Fig. 5.28 D'Ons-en-Bray's anemograph, detail (courtesy of the Conservatoire National des Arts et Métiers).

eters were concerned. This well-known instrument was invented by the Irish astronomer Thomas Romney Robinson, and announced to the British Association in 1846 as a "Modification of Dr. Whewell's anemometer for measuring the velocity of the wind,"[125] a title that at least shows in what awe the Master of Trinity was held. Robinson thought he had reason to believe that the centers of the cups move with exactly one-third of the speed of the wind.

Four years later he went into greater detail, but before considering that, I wish to draw attention to a passage and a diagram in a long report by W. Snow-Harris "on the working of Whewell's and Osler's anemometers at Plymouth."[126] This passage refers to an anemometer built by the Rev. W. Foster at "Sturbington" (certainly Stubbington) in Hampshire. A crude diagram (Fig. 5.29) is given, in which "*A* is a cross horizontal fly of three feet in diameter, having four vanes *a*, *b*, *c*, *d*; these vanes are six inches square, and are so contrived as to cause the fly to revolve in one direction only."[127] They also turn a plate *P*, across which a pencil *l* is moved radially by a clock; there is a wind-direction recorder *RS* on similar principles, and a recording rain gauge *MN* which is comprehensible only if we suppose that *M* is a tipping-bucket mechanism that in some way turns the plate *t*. The instrument, Snow-Harris reports, has withstood severe gales, and he gives a summary of a year's observations.

We can only suppose that the vanes *a*, *b*, *c*, and *d* must have been curved, and on this assumption, Foster's instrument is surely the precursor of the cup anemometer. Inquiry among his surviving descendents has elicited no further information. More might have been heard of it if, like Robinson, the Reverend Mr. Foster had belonged to the scientific Establishment.

Let us go on to 1850 and Robinson's second paper.[128] He begins with an interesting piece of history: the versatile Richard Lovell Edgeworth, whom Robinson, as a young man, had known well, constructed a windmill with a counter in 1793. It had had four light vanes driving an "endless screw" (a worm), which turned two gears having n and $n + 1$ teeth respectively, so that an index mark on the first passed one tooth of the second every n revolutions—a principle later widely used in the United States to count the revolutions of cup anemometers.[129]

[125] B.A.A.S., Southampton, 1846, *Sections*, pp. 111–12. The real ancestor of the cup anemometer is probably the "mola cum brachiis triangularibus" described by Verantius (*Machinae novae*, plate 9 and p. 6), in which each of the four sails consists of two sides of a horizontal triangular prism with a dihedral angle of about 35°.

[126] B.A.A.S., York, 1844, *Report*, pp. 241–66. The passage in question is on pages 250–52.

[127] *Ibid.*, p. 250.

[128] Robinson, *Trans. Roy. Irish Acad.*, Vol. 22 (1850), pp. 155–78.

[129] See U.S. Weather Bureau, *Circular D*, *Instrument Division*, 2nd ed. (Washington, 1900), pp. 9–23. Such instruments were introduced about 1871. We have seen (p. 208) that Grillet used this principle in 1673.

Robinson admitted that the cup wheel "was suggested to [him] many years ago by Mr. Edgeworth."[130] Edgeworth had died in 1817. The advantages of the cup wheel are that it has considerable motive power, that its movement is slow relative to the wind, that it needs no vane to face it into the wind, and that it can be duplicated easily, and various examples will give identical results.

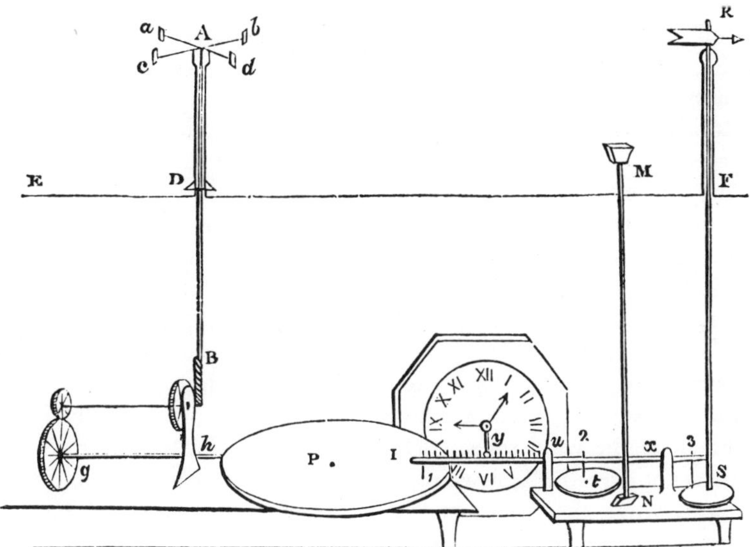

Fig. 5.29 Foster's anemometer, diagram.

Figure 5.30 shows some of the details of his anemometer and windvane. It is noteworthy that a ball journal bearing was provided for the upper end of the cup shaft, and we must remember that such things were not then commercially available. Separate records of wind speed and direction were made on circular charts; the pencils were moved radially by means of clockwork.

Robinson first tried to calibrate cup wheels by mounting them on a carriage, but calm enough days were rare, and later he used a whirling machine. Unfortunately he managed to deceive himself to the extent of producing too simple a result: "I think we are warranted in laying down the law, that in a horizontal windmill of this description, the centres of the hemispheres move with one third of the wind's velocity, except so far as they are retarded by friction."[131] This was at first taken for gospel, and for many years the official statistics in both the United Kingdom and the United States were computed on this three-to-one basis. But there were doubters, and in 1872 the Rev. Fenwick Stow published his comparisons of eight cup anemometers of various

[130] *Trans. Roy. Irish Acad.*, Vol. 22 (1850), p. 160.
[131] *Ibid.*, p. 168.

sizes, the results of which were "utterly irreconcilable" with any such law.[132]

Elaborate experiments were reported in 1874 by F. Dohrandt in St. Petersburg.[133] A whirling machine produces a circulation of its own, the so-called *Mitwind*, and Dohrandt paid a good deal of attention to this. He found that the wind speed v should be expressed as a function $a + bn + cn^2$ of the rate of rotation n; but $v = a + bn$ is a useful approximation. He measured Robinson's "constant" for eight cup anemometers by six different makers, obtaining values ranging from 2.09 to 3.08.

In his eighty-third year Robinson published a critical discussion of Stow's and Dohrandt's experiments.[134] While he objected to some of their methods, he recognized the importance of the matter and promised further investigations, which he reported some years later.[135] Meanwhile M. F. Thiesen had attempted to treat the subject purely theoretically,[136] a difficult matter for such a complicated shape, obtaining rather poor agreement, as he frankly confessed, with Dohrandt's results. Charles Chree discussed Thiesen's paper in 1895[137] and produced an elaborate theory of his own, concluding that there is no reason to suppose that the "anemometer constant" is really independent of the wind speed. He also found that various cupwheels should not agree in light winds, even if the friction is small, and that in gusty winds the cupwheel exaggerates the mean speed. This last effect poses a problem of which there is no obvious solution.

To return to the instrument itself, it was rapidly refined. One of the best of the early designs, which had a long innings, was that of Robert Beckley, the remarkable mechanical assistant at Kew Observatory. This was briefly described at Cheltenham in 1856 by John Welsh,[138] and more completely at Leeds two years later,[139] with the plates that are reproduced here as Figures 5.31 and 5.32. An important innovation was the use of coaxial shafts for the cupwheel and the windvane, both with ball bearings at the top. The vane was of the "fly governor" type. At first the wind-run recorder was an ingenious half gear and two racks (Fig. 5.32, upper left), giving a trace of fifty miles in one direction and fifty in the other. The direction recorder was a plane equiangular spiral with a sharp edge. "De la Rue's metallic paper" was used; this gave a trace by pressure with a brass edge or point. Later, sharp-edged helices were substituted (Fig. 5.32, bottom). Such an anemometer was exhibited at

[132] Stow, *Quart. J. Met. Soc.*, Vol. 1 (1872), pp. 41–49.
[133] Dohrandt, *Repert. Meteorol.*, Vol. 4 (1874), No. 5.
[134] Robinson, *Proc. Roy. Irish Acad.*, 2nd ser., Vol. 2 (1875), pp. 427-42.
[135] *Phil. Trans.*, Vol. 169 (1879), pp. 777–822; *ibid.*, Vol. 171 (1880), pp. 1055–70; *Proc. Roy. Soc.*, Vol. 30 (1880), pp. 572–74.
[136] Thiesen, *Repert. Meteorol.*, Vol. 5 (1877), no. 11.
[137] Chree, *Phil. Mag.*, Vol. 40 (1895), pp. 63–90.
[138] B.A.A.S., Cheltenham, 1856, *Sections*, pp. 38–39.
[139] B.A.A.S., Leeds, 1858, *Report*, pp. 306–7 and Plates 19, 20.

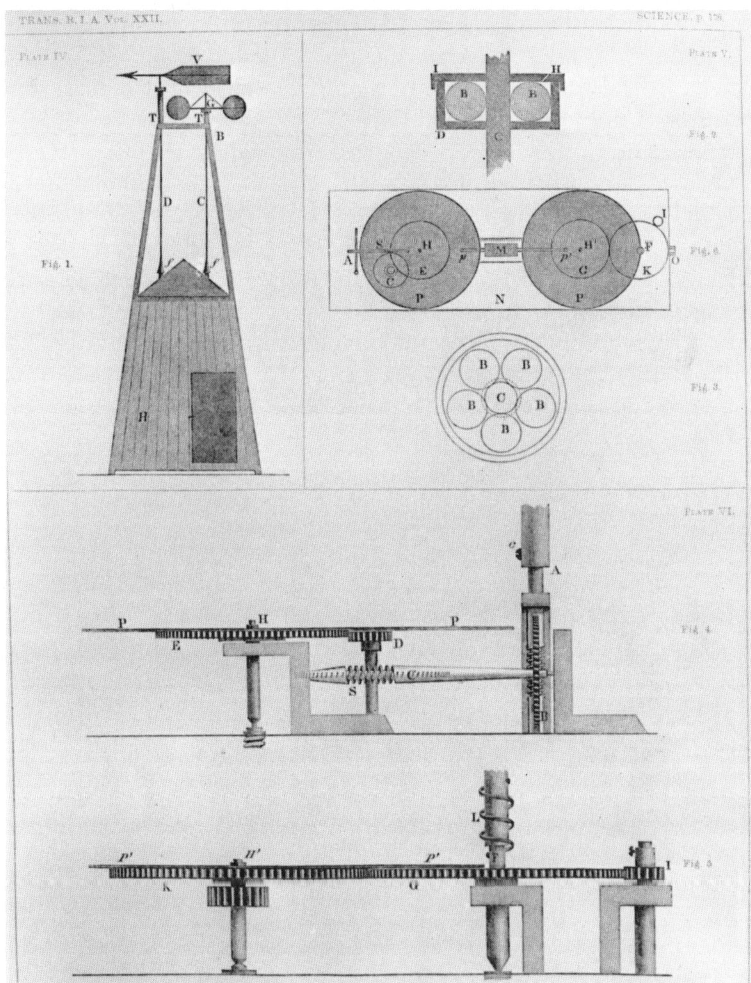

Fig. 5.30 Robinson's anemograph and recording vane, 1850.

the International Exhibition of 1862 by the Kew Observatory Committee of the British Association.[140]

In this instrument the two concentric shafts, one a rather large and heavy tube, had to be brought all the way down to the instrument, ending there in some elaborate gearing. It may have been the instrumentmaker James Hicks, who supplied these commercially,[141] who stood the anemometer on a gear box, bringing two separate light shafts down to the recording apparatus (Fig. 5.33), incidentally

[140] London, International Exhibition of 1862, *Reports by the Juries* [etc.] (London, 1863), p. XIII–39 (this is paged in "classes").

[141] *James J. Hicks' Illustrated & Descriptive Wholesale Catalogue of . . . . Meteorological Instruments, and all Kinds of Chemical and Philosophical Apparatus.* (London, James J. Hicks, 8, Hatton Garden, n.d.), p. 15. This was certainly printed after 1874.

avoiding problems of alignment. In this form, Beckley's anemometer was "adopted" by the Meteorological Committee of The Royal Society,[142] which later became the Meteorological Council, the precursor of the Meteorological Office.

It would be possible to fill a fair-sized book with descriptions of cup anemometers and the recorders, some of them extremely elaborate, designed for use with such instruments. It would also be very boring. I shall confine myself to a few, chiefly among the earlier arrivals.

The cup anemometer appeared on the scene just when people were getting to know how to use electromagnets. It seems very easy to cause a cupwheel to close a switch momentarily after a given number of turns—though it is not as easy as it looks to make such a mechanism reliable in all weathers—producing a pulse of current through an

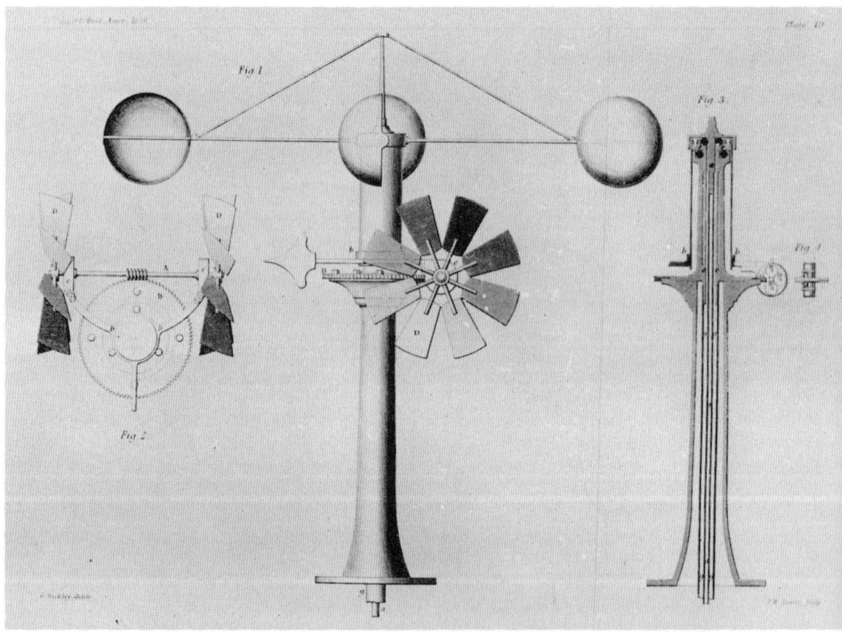

Fig. 5.31 Beckley's anemometer, outdoor part.

electromagnet, which can then be used to jog a pen, rotate a ratchet, operate a punch or even a printing wheel, or produce any other simple discontinuous motion. One of the first of these electrical anemographs was set up at the Observatory of the Jesuit College in Rome by the important astronomer Angelo Secchi,[143] who chose to make his electromagnet move a pencil in steps for one hour, after which another

[142] Royal Society of London, *Report of the Meteorological Committee for 1867*, p. 47.
[143] Secchi, *Atti nuovi Lincei*, Vol. 12 (1858–9), pp. 205–7.

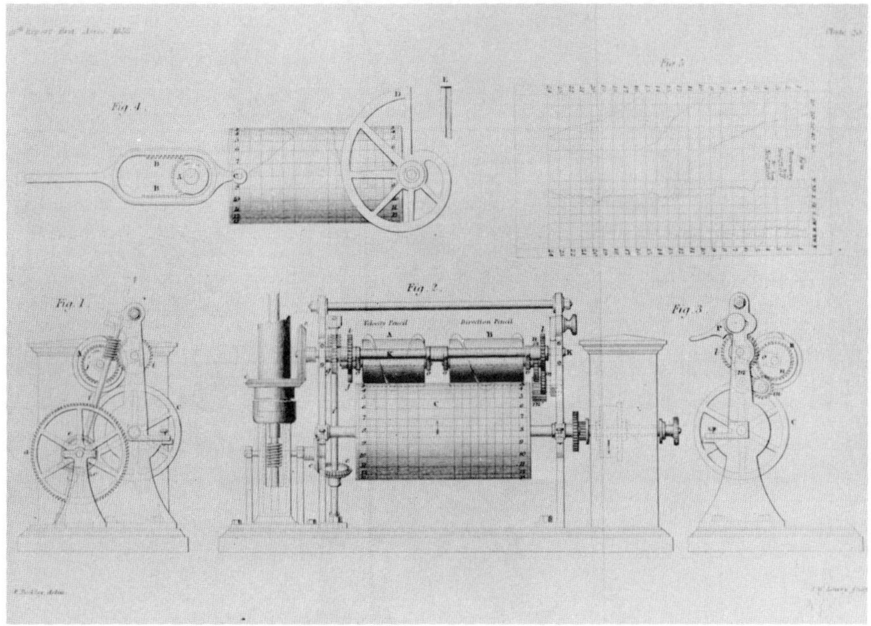

Fig. 5.32 Beckley's anemometer, recording mechanisms.

electromagnet, operated by a clock-driven switch, returned the pencil to its original position. At the same time the wind direction was recorded to eight points, a vane rotating two contacts (or one contact covering a sector of 22½ degrees) over four insulated quadrants. Secchi's way of recording the run of the wind has the great advantage that the average speed for each hour can be obtained immediately by inspection, as can be seen from Figure 5.34, which shows how this principle was employed thirty years later by the Paris firm of Richard Frères.[144]

A simpler idea is to have the switch on the anemometer make a dot on moving paper, as was done in the mechanically elaborate instrument made by another Paris maker, Jules Salleron, for the observatory at Modena.[145] Much of the complication (Fig. 5.35) is due to the use of a "fantail" windvane, which moves a wide contact over four insulated sectors. The cupwheel makes a contact every 200 revolutions; six wires connect all this to a simple drum chronograph with five electromagnets, recording wind direction to eight points every time the switch is closed by the cupwheel. A pushbutton permits the determination of the direction at any other time.

Essentially the same idea, except that the cup anemometer and the

[144] *Notice sur les instruments enregistreurs construits par Richard Frères* [etc.] (Paris, 1889), p. 35.
[145] See *Repert. phys. Techn.*, Vol. 5 (1869), pp. 304–14.

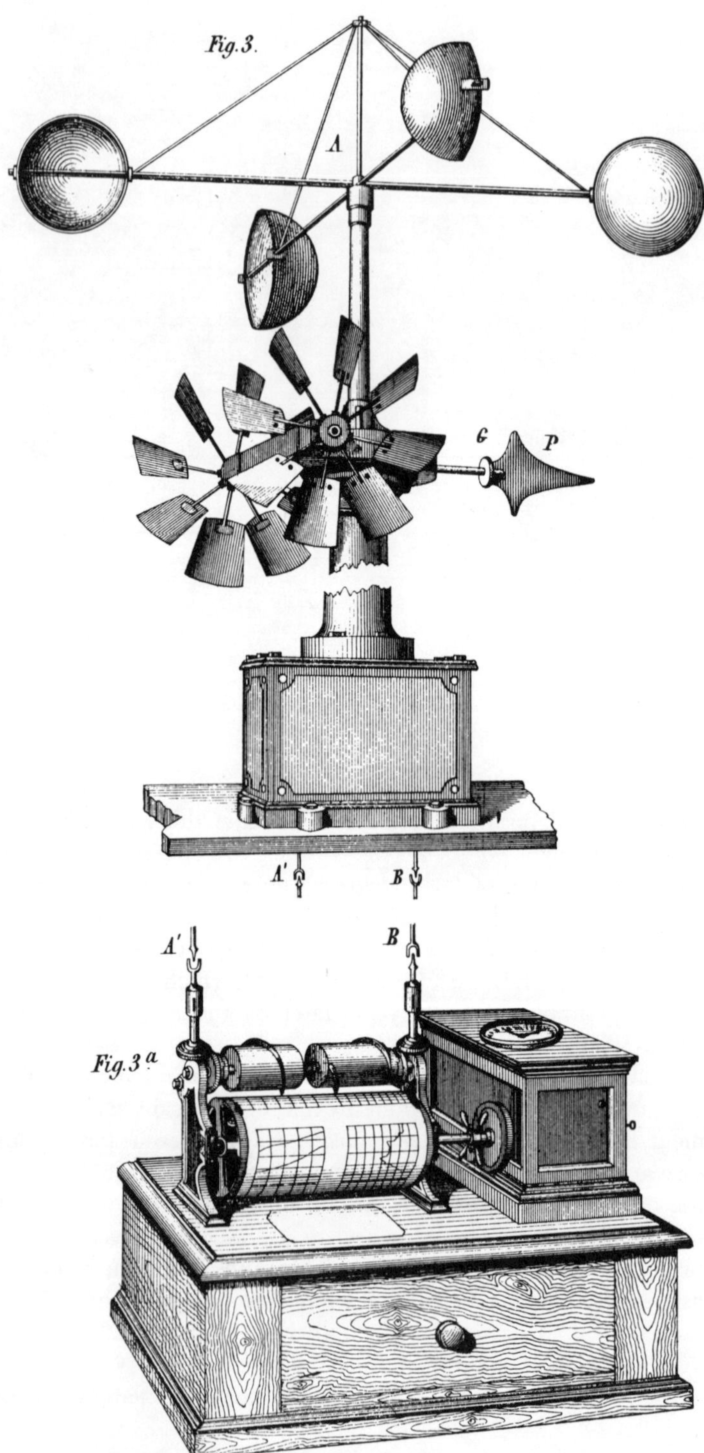

Fig. 5.33 Beckley's anemometer, later form, ca. *1867.*

windvane—a splayed vane—were separate instruments, was adopted in the United States about 1871 and used with little change for many years.[146] It is curious that the cup anemometers themselves (Fig. 5.36) continued to be provided with indicating dials as well as electrical contacts.

While electrical anemographs became more common, perhaps because of their greater flexibility of installation, completely mechanical ones continued to be designed. Beckley cooperated with the firm of Casella in an "embossing anemograph," in which a clockwork let fall a hammer at each hour to strike a paper ribbon onto a die in the form of an arrow that rotated with the windvane, and a type wheel advanced by the Robinson cups.[147] A simple subtraction gave the number of miles in the hour just past. This instrument was notable in that the only connection between the head and the recorder consisted of chains running on sprockets.

A pair of instruments, both completely mechanical, were announced by the Berlin maker R. Fuess in 1884.[148] Figure 5.37 shows the recording part of the simpler one, suggested by Richard Assmann.

Fig. 5.34 Richard's electrical wind-speed recorder, giving the mean hourly speed by inspection.

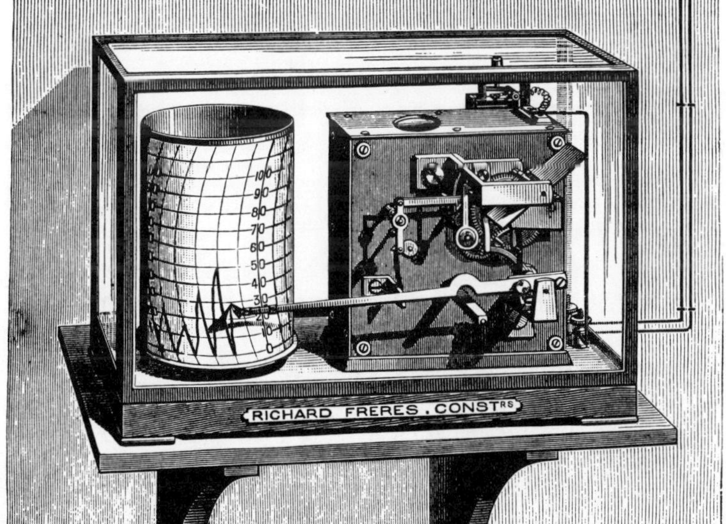

[146] U.S. Weather Bureau, *Circular D, Instrument Division*, in its various editions.
[147] *An Illustrated and Descriptive Catalogue of Surveying, Philosophical, Mathematical, Optical, Photographic, and Standard Meteorological Instruments, manufactured by L. Casella . . . 147, Holborn Bars, London, E.C.* (London, n.d. [1871]), p. 28.
[148] Fuess, *Zeits. Instrum.*, Vol. 4 (1884), pp. 297–303.

A paper tape and an inked ribbon are advanced by clockwork. At the end of every kilometer of wind the shaft *S* is caused to strike these by a large cupwheel, 1.06 meters in diameter. At the same time the shaft is turned by the windvane. The end of the shaft is in the form of an arrow, so that the wind direction is also printed on the tape for each kilometer.

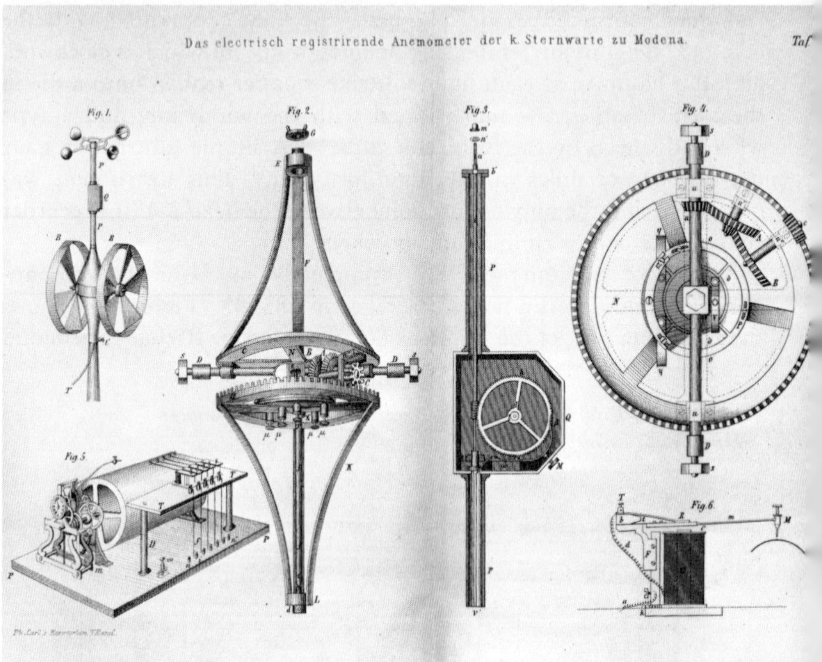

Fig. 5.35 Salleron's anemograph at Modena, 1869.

The second, the invention of Adolf Sprung, is more complicated (Fig. 5.38). In this the band of paper is moved by a roller geared to the cupwheel, and therefore with a speed proportional to that of the wind. The direction is recorded to sixteen points by one or two of eight pencils, let down by a cam geared to the vane. Another pencil is moved uniformly across the chart for an hour, and quickly back again, so that the wind-speed at any time is indicated by the slope of the line, and in addition, time marks are made every ten minutes. This elegant device must have been very costly.

An instrument operating on entirely different principles from any that have been mentioned so far is the "anemo-cinégraph" developed by Richard Frères,[149] which makes it possible to indicate or record the instantaneous speed of the wind with a cup anemometer. The mode of operation of the anemo-cinégraph may be described with reference

[149] *Notice sur les Instruments*, pp. 46–57.

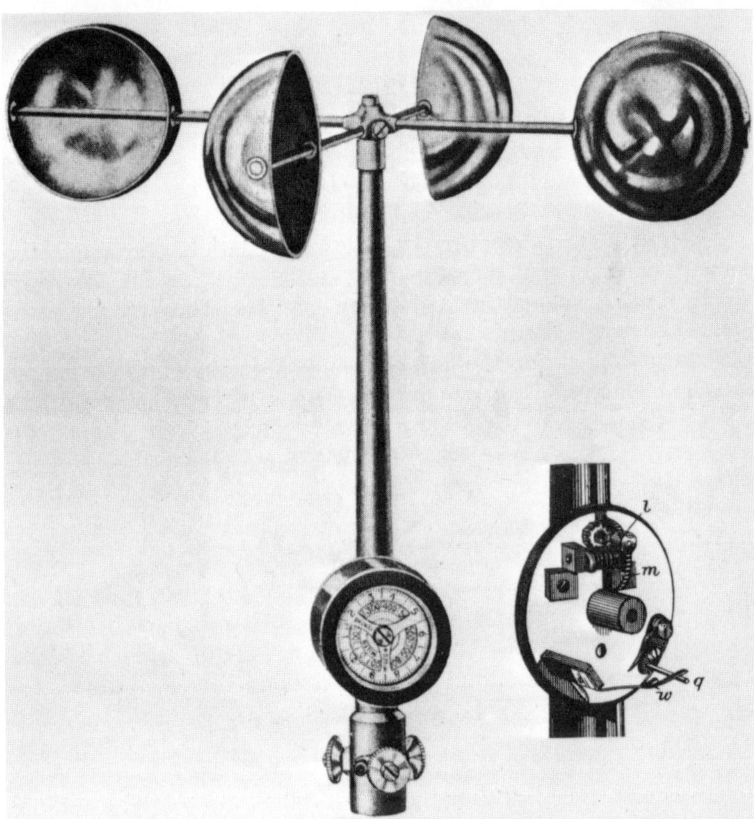

Fig. 5.36 Standard cup anemometer, U.S. Weather Bureau.

to Figure 5.39, in which *P* is a flat plate rotated at a uniform speed by a clock. A hardened steel wheel *Q* is set in rotation by friction at a rate proportional to its distance *r* from the center of the plate. The shaft *A* of *Q* can move freely to right and left, and carries a long screw *S*, which engages a toothed wheel *W*. This wheel is revolved, either mechanically or by an electrical mechanism, at a rate proportional to that of the cupwheel. If the radius *r* is such that the wheel *Q* and the screw *S* make one turn in exactly the time that *W* advances by one tooth, there will be no tendency for *S* to move to the right or left. If on the other hand the speed of the wind increases, so that *W* rotates faster, *S* and *Q* will be moved to the right until equilibrium is reestablished, and vice versa. A pointer or pen can be connected to the right-hand end of the long shaft in any suitable manner. This instrument will average the more rapid fluctuations in wind speed to an extent depending on the dimensions and angular speeds chosen by the designer.

Cupwheels became smaller and lighter as materials improved and better bearings became available, but they continued to have four

Fig. 5.37 The Assmann-Fuess recorder, 1884.

cups until in 1926 John Patterson of the Meteorological Service of Canada showed by extensive wind-tunnel experiments the advantages of using only three cups.[150] Traditionalists proved hard to convince, but eventually his conclusions were generally accepted.

9. Anemometers with restrained rotation. The last class of anemometers that will be discussed here comprises the so-called "bridled anemometers," a convenient term, in spite of its equine connotations, for instruments in which the rotation that would be produced by the wind is limited to an angular deflection.

The earliest such anemometer was described in 1709 by the great philosopher Christian Wolff, professor of mathematics and physics at the University of Halle, in a strange and stiff book[151] whose old-fashioned arrangement is betrayed by its title. The instrument is shown in Figure 5.40. A mill with four vanes *A*, *B*, *C*, *D* drives a

[150] Patterson, *Trans. Roy. Soc. Canada*, Sec. III, Vol. 20 (1926), pp. 1–54.
[151] Wolff, *Aerometriae elementa, in quibus aliquot aeris vires ac proprietates juxta methodium geometrarum demonstrantur* (Leipzig, 1709), pp. 320–23.

Fig. 5.38 The Sprung-Fuess recorder, 1884.

worm *EF*, and this engages a worm wheel *H* on a horizontal axle at right angles to that of the windmill. To this axle is clamped a bar *K* carrying an adjustable lead weight *L*, and also an index *N* moving over a quadrantal scale. The whole thing is supposed to be turned into the wind by a minuscule vane *OQRP*, so inadequate that if Wolff had not specifically told us so,[152] we might doubt whether it was ever built.

This instrument read not the wind at any moment but the maximum wind. This was why the worm gearing had been used: "For it is certain from mechanics that while a wheel can be driven by a screw, on the other hand a screw cannot be driven by a wheel; so that when the weight *L* has been raised to a height at which it is balanced by the force of the wind impinging on the sails *A*, *B*, *C*, *D*, then if the force of the wind ceases or becomes smaller, the weight cannot descend again."[153] The observer presumably had climb up and turn the sails back by hand.

[152] *Ibid.*, p. 321. "Dico, anemometrum esse constructum."
[153] *Ibid.*, pp. 323–24.

P
Q
A
S
W
r

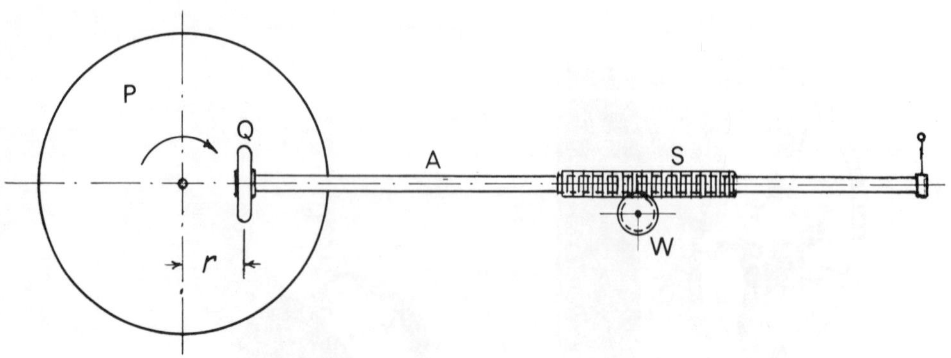

Fig. 5.39 The anemo-cinégraph.

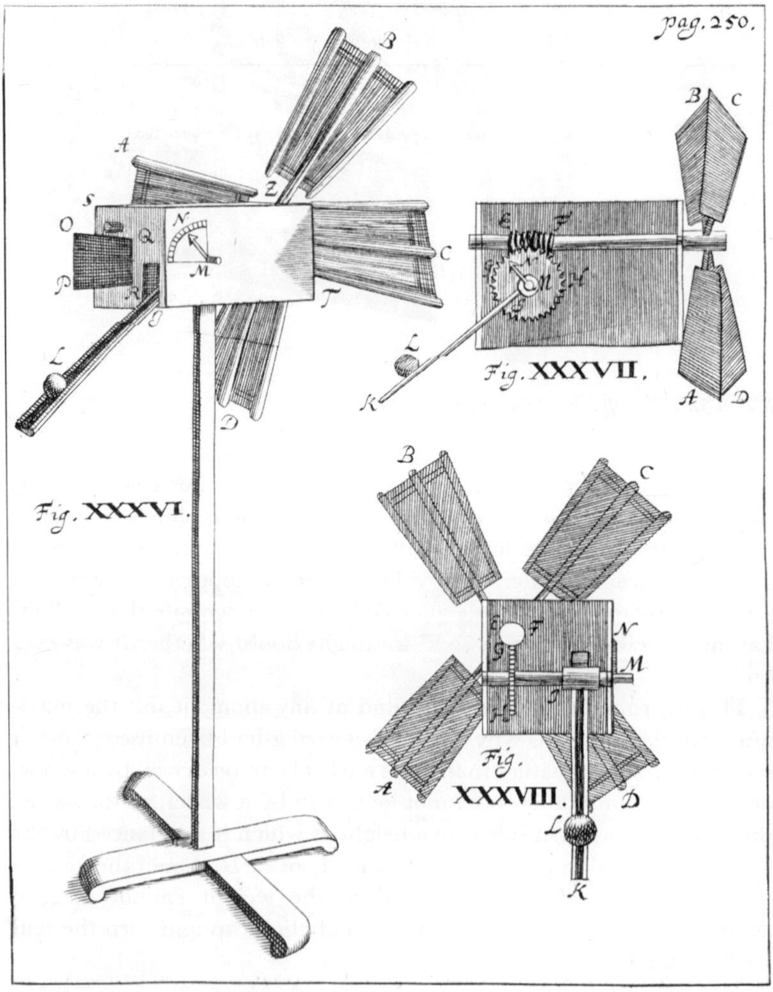

pag. 250.
Fig. XXXVI.
Fig. XXXVII.
Fig. XXXVIII.

Fig. 5.40 Wolff's anemometer, 1709.

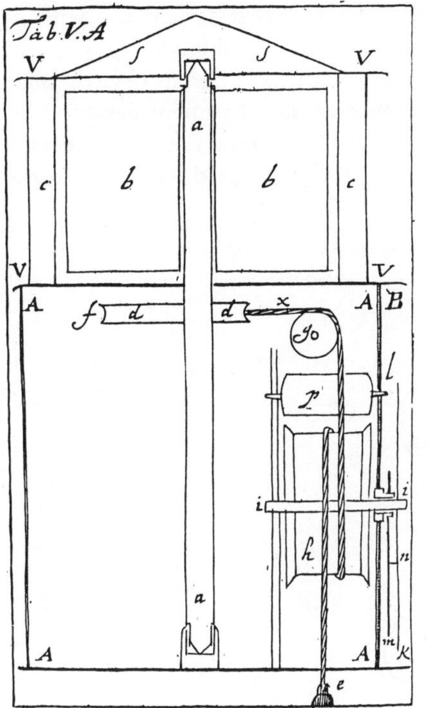

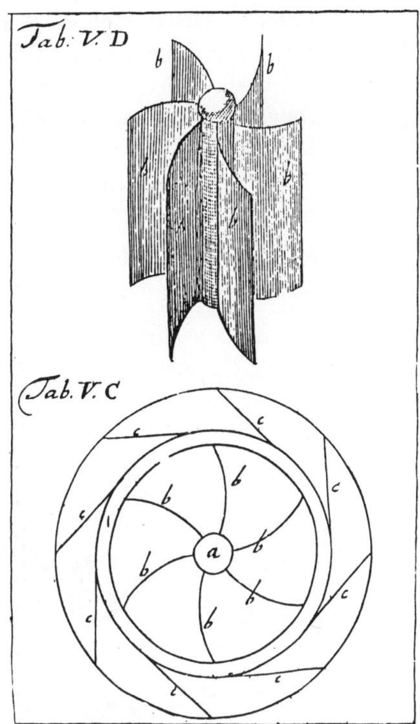

Fig. 5.41 Leutmann's anemometer, 1725.

In 1725 Johann Georg Leutmann published his surprising little book on his meteorological instruments.[154] His anemometer, Figure 5.41, uses the "Polish windmill" also employed by d'Ons-en-Bray, as we have seen. But Leutmann's mill is not allowed to rotate, being restrained by a cord *x* acting on a spiral pulley *d*, and passing over a roller *g* and past another one *p*, then wrapped around the drum *h*, which carries a pointer on its axis *ii*.[155] In Leutmann's anemometer the pointer would indicate the instantaneous wind, not the maximum.

In the following year Jacob Leupold published his *Theatrum aerostaticum*, to which I have referred several times. The reader is asked to turn back to page 188, where in Figure 5.10 he will see the anemometer "such as Herr Dinglinger has on his house at Dresden," differing only in minor respects from Leutmann's. In his text Leupold gives credit for the anemometer on Dinglinger's house to one Gärthner, about whom I have no further information.

Leupold describes and illustrates another bridled anemometer in which a propellor winds up a weight by means of a cord and a fusee.

[154] Leutmann, *Instrumenta meteorognosiae inservientia, das ist Werkzeuge, die Beschaffenheit der Luft zu erkennen* (Wittenberg, 1725). Anemometer, pp. 115 ff.
[155] The drawing is misleading, but the intention is clear.

Later, a Dr. Burton of Windsor made a similar instrument with a ratchet and pawl to hold it at the maximum reading.[156]

I shall neglect one or two later and not very different devices, and conclude this chapter with a reference to an instrument designed in 1880 by Sir G. G. Stokes for the purpose of recording the instantaneous maximum values in gusts. It was in use at Holyhead for many years,

Fig. 5.42 Stokes's bridled-cup anemometer, out-of-doors part.

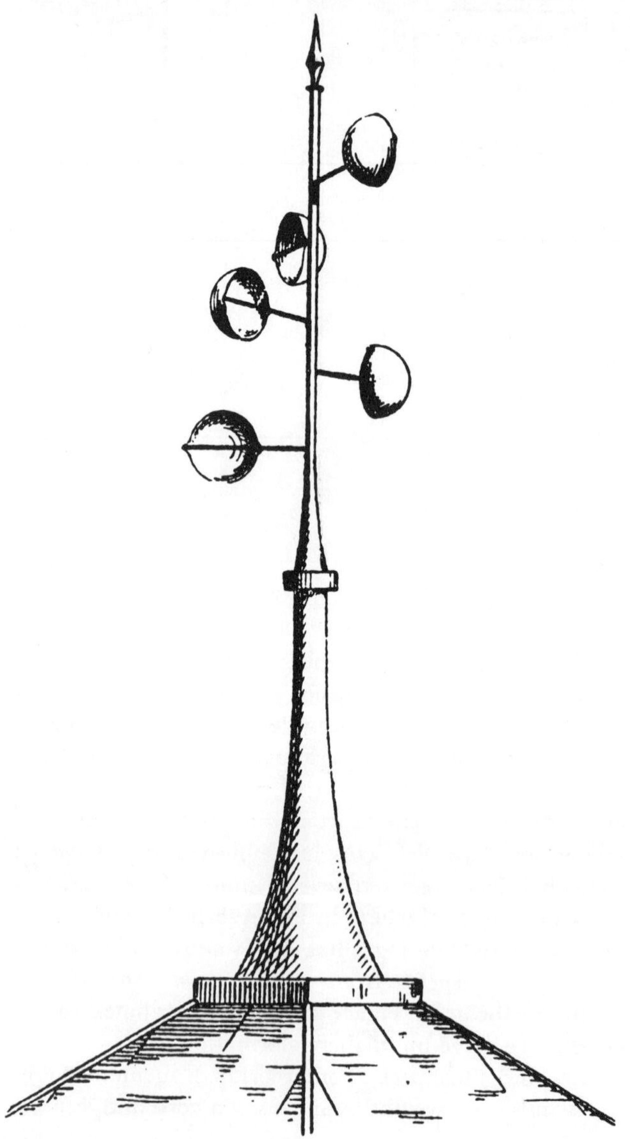

[156] See Benjamin Martin, *Philosophia Britannica* [etc.] (2 vols., Reading, 1747), Vol. II, p. 82.

but was not described until 1898.[157] It is now in the Science Museum, London.[158] The cups were originally in one plane but in 1890, at the suggestion of W. H. Dines, who had been asked by the Meteorological Council to make experiments on the instrument,[159] they were arranged helically, as in Figure 5.42. The motion of the cups is resisted by two weights acting through cords in the rims of two "snails" or spiral

Fig. 5.43 Stokes's bridled-cup anemometer, recorder.

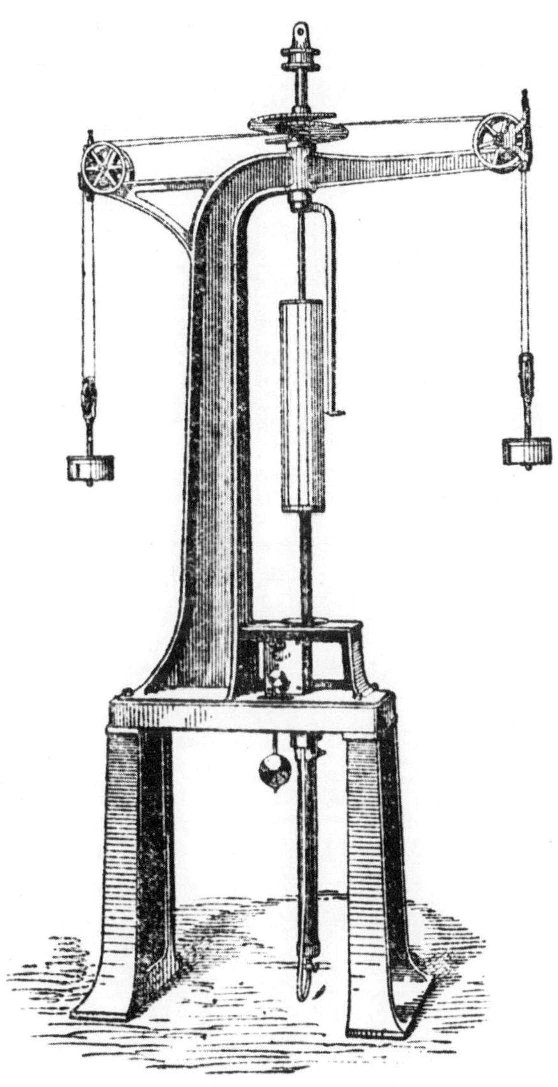

[157] R. H. Curtis, *Rep. Meteorol. Council* (1898), pp. 28–30.
[158] Inv. no. 1912–195.
[159] *Report of the Meteorol. Council for 1890* (London, 1891), p. 21.

pulleys, as shown in Figure 5.43. These pulleys have a profile in which the radius is proportional to the square of the angle, so that the deflection is linear in wind speed, but the instrument is not intended to measure light winds. The pen is on an arm attached to the spindle. The drum is drawn down axially by the clock, without revolving.

VI

The measurement of the duration of sunshine

---

1. Introduction. The measurement of the duration of direct sunshine is more an attempt to quantify human feelings of well-being than a serious contribution to meteorological theory. We all like the sun to shine, at least in the higher latitudes, and it is necessary to have statistics of sunshine in order that we may be promised more sunshine than usual or threatened with less, and of course the competition of rival holiday resorts must be considered.

Sunshine recorders can be divided into two classes, according to whether they make use of the photochemical changes produced by sunlight or utilize its thermal effects. The second class can be split into two, one depending on the principle of the burning-glass, the other on the differential thermometer. I shall deal with these classes in the order of their introduction. Meanwhile it may be well to point out that there is another possible classification into image-forming and non-image-forming instruments. The thermometric recorders are alone in the latter class. Since no image is formed, it is impossible for these recorders to distinguish between radiation from the sun and that from the rest of the sky.

The other two classes seem to give roughly the same over-all results, but the photographic ones tend to be less sensitive near sunrise and sunset, and to give higher values on days with cirrus cloud.[1]

2. Photographic sunshine recorders. In January, 1839, H. Fox Talbot made public his process of "photogenic drawing" on paper,[2] and Mandé Daguerre's method of photography on metal plates was presented to the Académie des Sciences by François Arago.[3] At this time the secretary of the Royal Cornwall Polytechnic Society was an exceedingly imaginative and energetic man called T. B. Jordan.[4]

[1] See F. Albrecht in *Handbuch der meteorologischen Instrumente* (E. Kleinschmidt, ed., Berlin, 1935), p. 186.

[2] Talbot, *Proc. Roy. Soc.*, Vol. 4 (1839), pp. 120–21 (Jan. 31); *ibid.*, pp. 124–26 (Feb. 21).

[3] *Compt. Rend.*, Vol. 8 (1839), pp. 4–7 (Jan. 7).

[4] Not to be confused with James B. Jordan, who comes into this story quite independently 50 years later.

Within three weeks of the date of Talbot's announcement to The Royal Society, he discussed with a committee at Penzance the use of photography to register the readings of meteorological instruments, and within less than two months he exhibited some barometric records.[5]

He also made a sunshine recorder, which he called a "heliograph." This "consists of a light cylinder of metal, with a similar cylinder revolving round it on a screwed axis once in 24 hours. The inner and fixed cylinder is covered with a sheet of the prepared paper, and the outer or revolving one has a small hole in it, through which the light shines on the paper.

"The axis of the cylinder and the position of the hole is so adjusted, that the hole shall at all times be opposite to the place of the sun . . . ."[6] It passes over the paper "in a spiral [*sc.* helical] line, every point of which will be illuminated for an equal time."[7] In this way one piece of paper may contain the record for a week or longer.

In the following year he described "a new arrangement of the heliograph"[8] in which he sacrificed the weekly record by substituting a long, narrow triangular slit, resulting in a photographic illumination meter or recording actinometer, which falls outside the scope of this book, but certainly presaged the later instruments of Bunsen and Roscoe.[9]

In 1856 the French physicist C. S. M. Pouillet reported on a photographic sunshine recorder, which he called (to translate the title of his paper) an "Actinograph, an instrument that records the times of day at which the sun shows or hides itself, and the duration of its appearances or disappearances."[10] In this instrument there was no clock. A metal box 20 cm square and 10 cm high had its axis—the line joining the centers of its square sides—set parallel to the polar axis of the earth. The other sides of the box were oriented toward north, east, south, and west. In the east, south, and west sides were thin brass plates, each with a small hole. Inside the box there was a movable piece 2 cm thick, with three edges cut out to form parts of three cylinders 6 cm in radius, centered on the three holes, and these cylindrical segments carried a band of photographic paper. The movable piece could be fixed at a suitable height each day, corresponding to the declination of the sun.

This apparatus, which is seldom referred to, must have been a fairly satisfactory sunshine recorder. The next one to appear was the invention of J. T. Goddard of Hounslow, and was exhibited at the

[5] T. B. Jordan, *Royal Cornwall Polytech. Soc., Annual Report for 1838*, p. 184, note. The Society's year seems to have extended into 1839, and this has caused some confusion.
[6] *Ibid.*, p. 185.
[7] *Ibid.*, p. 186.
[8] *Royal Cornwall Polytech. Soc., Report for 1839*, pp. 115–16.
[9] R. W. Bunsen and H. E. Roscoe, *Phil. Trans.*, Vol. 153 (1863), pp. 139–60.
[10] Pouillet, *Compt. Rend.*, Vol. 42 (1856), pp. 913–17.

London Exhibition of 1862. We read that "Mr Goddard . . . exhibits an instrument to which he has given the name of sunshine recorder; it works by letting the sun's rays pass through a narrow slit, and fall on photographic paper wound round a barrel moved by clockwork; the paper being changed daily, and the photographic impression developed and fixed in the usual manner."[11] I have found out nothing else about this instrument, but as far as I know Goddard was the first to use the term "sunshine recorder." His instrument was inferior to T. B. Jordan's, because it was its inner barrel that revolved, so that the hole faced the sun only at midday.

The whole subject aroused much interest in the United Kingdom after about 1875, largely on account of the activities of Robert H. Scott, who administered the affairs of the Meteorological Council, so that by March, 1885, several different sunshine recorders where shown at an exhibition organized by the Royal Meteorological Society.[12] Two of these were photographic, and these are both shown in Figure 6.1. That of Professor H. McLeod "consists of a glass sphere silvered inside and placed before the lens of a camera, the axis of the instru-

Fig. 6.1 (a) *McLeod's sunshine recorder;* (b) *James B. Jordan's first sunshine recorder.*

[11] London, International Exhibition of 1862, *Reports by the Juries* [etc.] (London, 1863), p. XIII–40 (this is paged in "classes").
[12] *Quart. J. Roy. Meteorol. Soc.*, Vol. 11 (1885), pp. 242–50.

ment being placed parallel to the polar axis of the earth."[13] It can easily be seen that the track of the sun's image will be a circular arc, with a uniform angular time scale throughout the day. The instrument was made by James Hicks, while that of James B. Jordan[14] (Fig. 6.1*b*) was constructed by Negretti & Zambra. It is simplicity itself, consisting merely of a tube placed with its axis parallel to the polar axis of the earth, and having two very small holes 90 degrees apart. The sun shines through one of these in the morning, through the other in the afternoon, and through both near midday, producing two curves on the ferroprussiate paper (blueprint paper) with which most of the tube is lined. There was a scale of latitude.

A few months later Jordan and Gaster, one of the assistants at Kew, reported on a slight improvement (Fig. 6.2), in a general paper about

Fig. 6.2 Jordan's improved recorder, 1886.

sunshine recorders.[15] The improvement was a sort of roof which reduced the overlap in the record.

In 1888 Jordan described a much better photographic recorder in

[13] *Ibid.*, p. 243. The rough prototype is in the Science Museum, London, inv. no. 1885–112.

[14] Jordan had attained some degree of celebrity by constructing glycerine barometers. See Middleton, *The History of the Barometer*, pp. 369–71.

[15] J. B. Jordan and F. Gaster, *Quart. J. Roy. Meteorol. Soc.*, Vol. 12 (1886), p. 21–25.

which he had divided the tube into two semicircular tubes, each with a hole in the flat side (Fig. 6.3).[16] In addition to giving an almost uniform time scale—the curves would be straight lines at the equinoxes when the paper was flattened out—this design also had the enormous practical advantage that a paper for the afternoon could be put in at any time during the morning, and vice versa. But it could not be used in the arctic summer, in spite of the latitude scale extending to 90 degrees.

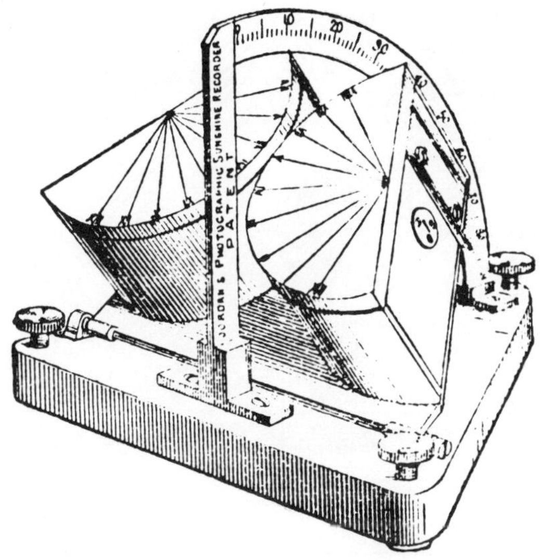

Fig. 6.3 Jordan's double sunshine recorder, 1888.

In the same year J. Maurer of the Swiss weather service proposed a modification of the first Jordan recorder in which the tube holding the blueprint paper terminated in a horizontal plate with one fine hole in the middle instead of the two holes in the tube.[17] It would have had to be made specially for a given latitude.

Charles F. Marvin in the United States designed a photographic sunshine recorder, also in 1888 according to D. T. Maring.[18] It is shown in Figure 6.4. Fundamentally it is like James Jordan's first type, except that 31 days' record can be made without replacing the paper. Two pieces of paper are used, one for morning, the other for afternoon, held by wire clips against sectors of cylinders back to back, before being slid into the main tube. At about 45° from the top of the tube are fastened dovetail slides in which move brass bars, each with a small central hole and a scale reading 1 to 31. These are moved each day, and are held by a spring catch. Recorders of this pattern were

[16] Jordan, *Quart. J. Roy. Meteorol. Soc.*, Vol. 14 (1888), pp. 212–17.
[17] Maurer, *Meteorol. Zeits.*, Vol. 5 (1888), pp. 110–12.
[18] Maring, *U.S. Monthly Weather Review*, Vol. 25 (1897), p. 487.

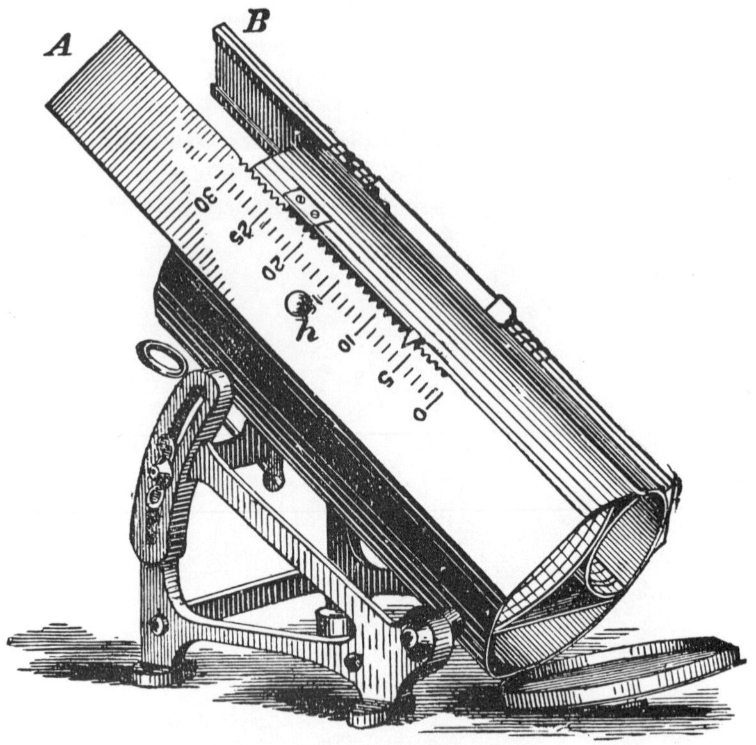

Fig. 6.4 Marvin's photographic sunshine recorder, 1888 (courtesy of the Smithsonian Institution).

still being used at a few stations twenty years later. There is one in the Museum of History and Technology in Washington, D.C.[19]

3. *Sunshine recorders making use of a burning glass.* The second kind of sunshine recorder is essentially a recording sundial. As early as 1646 Athanasius Kircher, professor of mathematics and Hebrew at the Collegio Romano, seems to have realized the general shape that such a device might take with his "horlogium helio-causticum, sive solare-ustorium."[20] The lower part of Figure 6.5 shows a crystal ball, or in default of this, a spherical flask full of liquid, placed centrally in a hemispherical shell. "And, thirdly, a sundial is drawn in the shell, together with things for burning and making sounds." Kircher suggests that small holes should be made on the bowl, and fuses put in them leading to little bombs to announce the hours. On the outside of the bowl is written:

[19] Inv. no. 314,539.
[20] Kircher, *Ars magna lucis et umbrae* [etc.] (Rome, 1646), pp. 790–92.

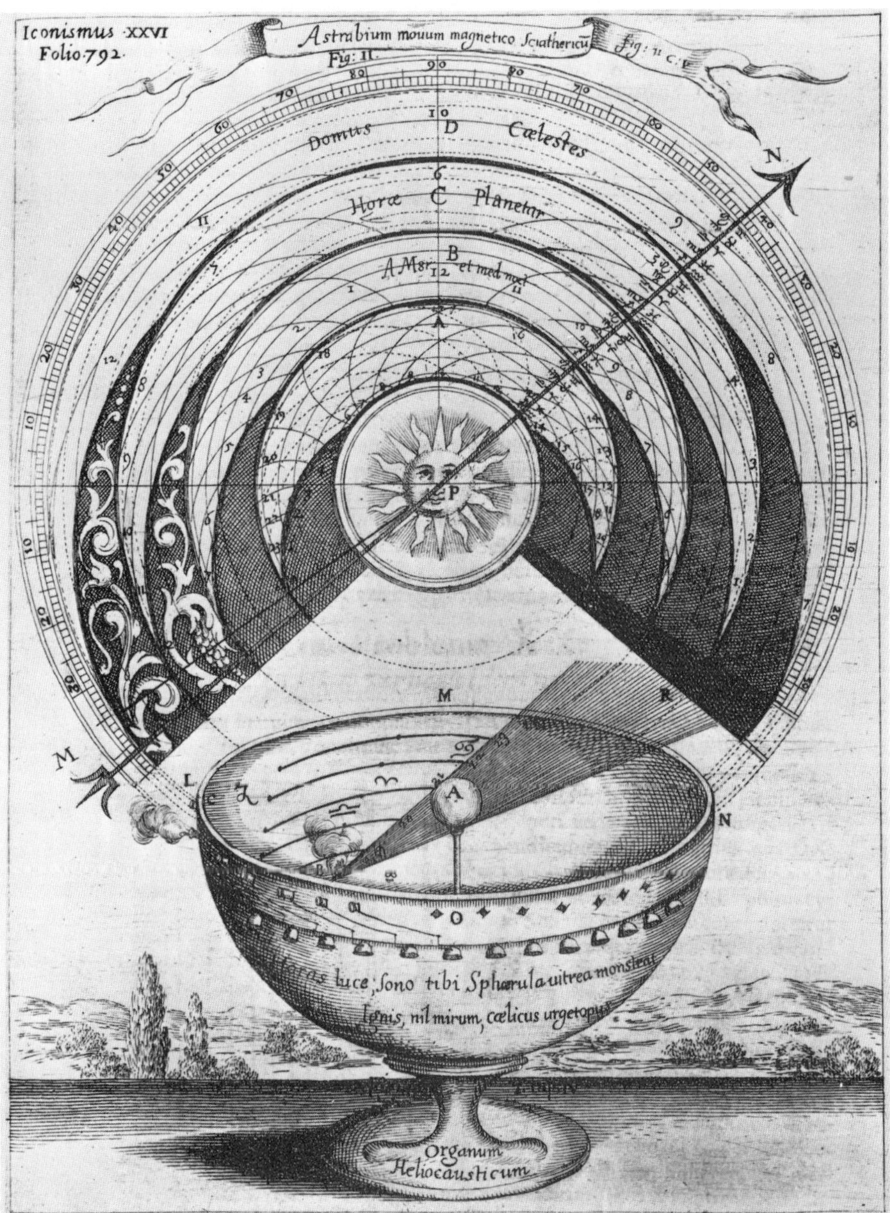

Fig. 6.5 An idea from Kircher, 1646.

With light and sound the glassy sphere shows thee the hours;
Truly it is the work of the heavenly fire.

It is not known whether J. F. Campbell of Islay, near London, an employee of the Board of Health, had seen Kircher's book when, in 1853, he designed a "self-registering sundial" on these principles and began regular observations in Whitehall.[21] Campbell's cross-section of the apparatus is shown in Figure 6.6. It is amusing that neither

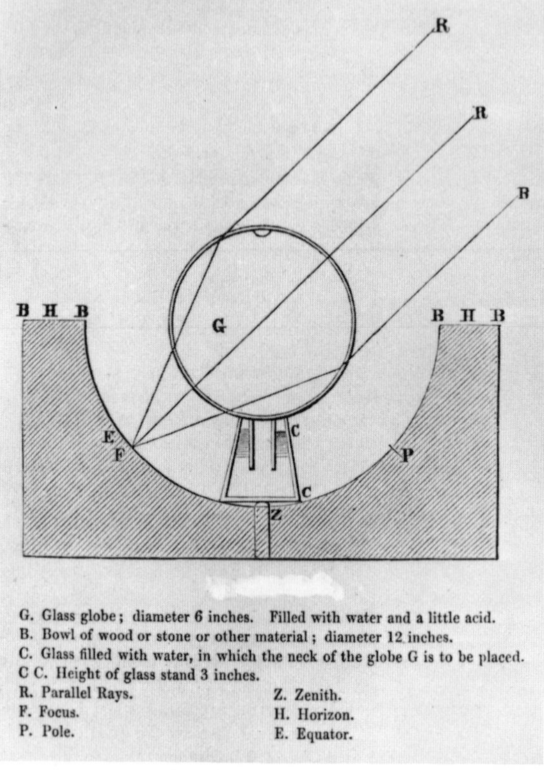

Fig. 6.6 Campbell's original sunshine recorder, 1853.

Kircher nor Campbell drew an accurate picture of the course of the solar rays, which of course suffer two refractions. Campbell recognized that the best scheme, although a troublesome one, would be to fasten a paper ribbon to the bowl, and suggested that rubber cement might be tried; but he decided to use a wooden bowl, which was charred by the heat when the sun was shining. These bowls, which were of fine-grained mahogany, were replaced only twice a year, at midsummer and midwinter, and showed little more than the difference in the heat of the direct sunshine for the two half-years.

Just when his paper was being published, Campbell obtained a solid glass sphere,[22] which was probably more accurate, and at any rate would not be damaged by freezing. In 1875 or 1876, according

[21] Campbell, *British Meteorol. Soc., Rep. of Council for 1857*, pp. 18–26.
[22] *Ibid.*, p. 23, note.

to Richard H. Curtis, he was able to get a really true sphere, made of good glass.[23]

In 1875 H. R. Roscoe and B. Stewart tried to quantify the method by taking the forty-eight bowls on which records had been burned, weighing each one, and then filling up the burned-out traces with a mixture of beeswax and olive oil until the whole surface appeared smooth, and weighing it again.[24] The results were scarcely worth the effort.

Campbell's original suggestion of a paper band was revived in 1877 by R. H. Scott,[25] who recognized that the bowl need not be a complete hemisphere, and indeed that there need not be a bowl at all. We do not know exactly what the improved device looked like, but

> The principle of the instrument is that the strip of paper is held in clips along the inner surface of a ring which is concentric with the spherical lens, and whose radius is equal to the radius of the lens plus its focal length. This ring is attached to a vertical circle, along which it is moved to correspond with the varying declination of the sun.
>
> The material employed for the record is thin millboard. It is obvious that a time scale could be printed on the millboard.[26]

In 1878 the two forms of recorder, the one with the mahogany bowl and the other with the strip of millboard, were both in use at Kew.[27] The results furnished by the latter were so satisfactory that in the following year the Council ordered twenty four-inch glass spheres to be used in sunshine recorders.[28] Scott had further ideas, and it was proposed that both a straight and a curved piece of millboard should be supplied, the former to be used when the sun's declination is between $-8°$ and $+8°$, the latter when it is outside these limits.[29] Installed in the instrument, the straight one will become part of a cylinder, the curved one part of a right circular cone. Scott goes on: "The bowl for the slips may be a mere skeleton, presenting four fiducial edges forming the interior [?] edges of four arcs of annuli in planes parallel to the plane of the equator, the centres of the annuli being in a line passing through the centres of the sphere and parallel to the axis of the earth."[30]

In the year in which this was published, Sir George Stokes, a member of the Meteorological Council, contributed to the *Quarterly Journal of the Meteorological Society* a "Description of the Card Supporter for Sunshine Recorders adopted at the Meteorological Office."[31]

[23] Curtis, *Quart. J. Roy. Meteorol. Soc.*, Vol. 27 (1901), p. 63.
[24] Roscoe and Stewart, *Proc. Roy. Soc.*, Vol. 23 (1875), pp. 578–82.
[25] Scott, *Quart. J. Meteorol. Soc.*, Vol. 3 (1877), pp. 18–19.
[26] *Ibid.*, p. 19.
[27] *Rep. Meteorol. Council to the Roy. Soc. for 1878* (London, 1879), pp. 15–16.
[28] *Report . . . for 1879* (London, 1880), p. 16.
[29] *Ibid.*, pp. 24–25.
[30] *Ibid.*, p. 25.
[31] Stokes, *Quart. J. Meteorol. Soc.*, Vol. 6 (1880), pp. 83–94.

This was the well-known slotted holder into which strips of cardboard, straight or suitably curved, could simply be slid, and which has been known as the Campbell-Stokes sunshine recorder ever since. It does not seem quite fair that R. H. Scott's name is not remembered in this connection, in view of what had been going on; of course it is possible that Stokes had really been the moving spirit behind these developments. This is, however, rendered less likely by his reference near the beginning of his paper to "Mr Scott's modification of the instrument, which substitutes a strip of card for the wood."[32]

The first instruments (Fig. 6.7) were made for a fixed latitude, small adjustments being allowed for by tilting the slate base of the instrument, but within a few years the recorder was being made with a latitude adjustment (Fig. 6.8), and G. M. Whipple had suggested to the firm of Casella an elaborate instrument (Fig. 6.9) that was not only usable in any latitude, but also needed only straight cards, having an adjustment for the declination of the sun.[33] This does not seem to have found favor.

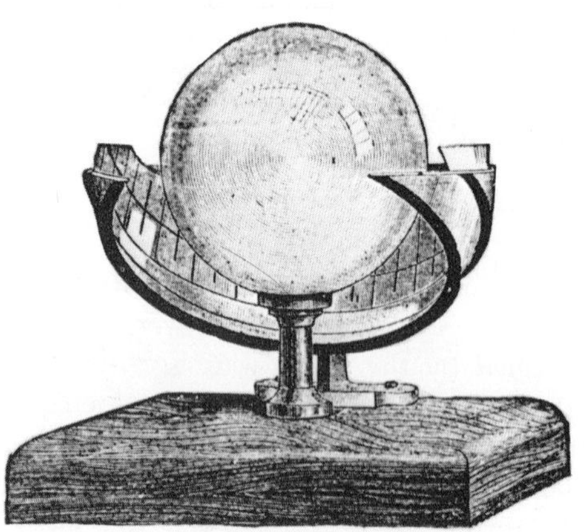

Fig. 6.7 Sir George Stokes's modification, 1880.

After some trials, cards printed in Prussian blue ink were adopted, but the ink, Stokes said, should be only moderately dark, because otherwise the fainter parts of the burn are hard to see. The supply of suitable cards became a sore point with continental meteorologists, and as early as 1883 we find Assmann, at Magdeburg, complaining of their high price, and of the "secret process" by which they were made,

[32] *Ibid.*, p. 83.

[33] Figures 6.8 and 6.9 are from *Quart. J. Roy. Meteorol. Soc.*, Vol. 11 (1885), plate 7.

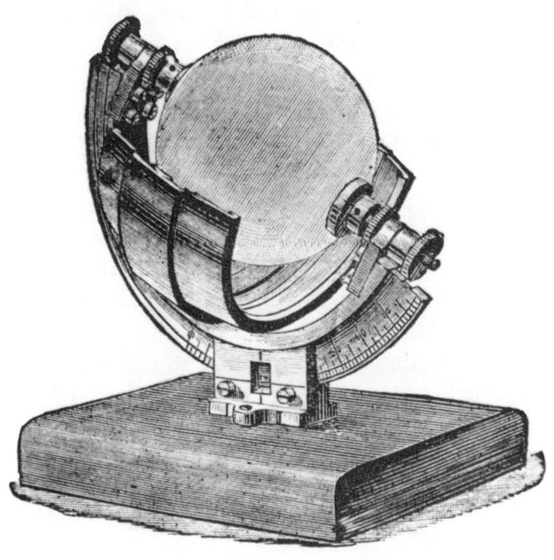

Fig. 6.8 Campbell-Stokes sunshine recorder with latitude adjustment.

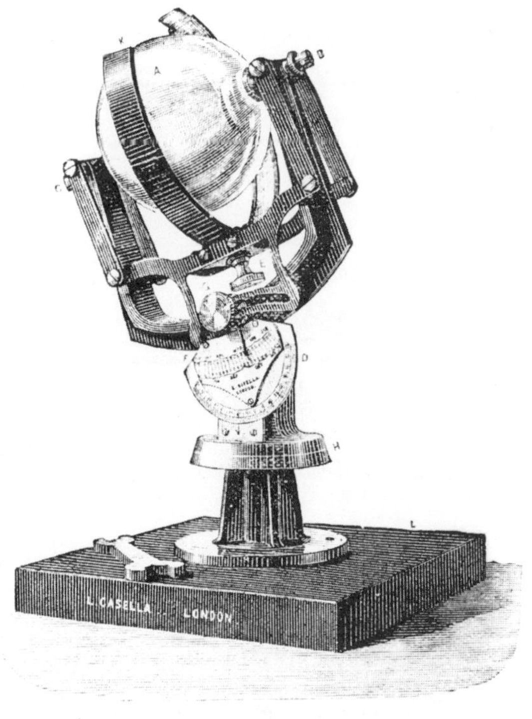

Fig. 6.9 The Whipple-Casella sunshine recorder.

although he praised the instrument.[34] Nevertheless, the Campbell-Stokes apparatus quickly superseded photographic sunshine recorders in most countries, in spite of the argument that the photographic and the burning recorders must be considered to complement each other.[35]

If the Campbell-Stokes recorder is to work properly, the glass ball must be of the right size and must be accurately centered in the card-holder. The importance of uniformity in the glass spheres was emphasized in 1898 by R. H. Curtis,[36] who also exhibited a model of an improved mounting for the sphere that permitted it to be adjusted in three directions. This was described in detail later.[37] The design is still used in temperate latitudes.

I shall end this section by referring to a somewhat different burning recorder, designed by J. Maurer after he had become the director of the Swiss Meteorological Service.[38] Maurer objected to the short time-scale of the Campbell-Stokes recorder. His solution (Fig. 6.10),

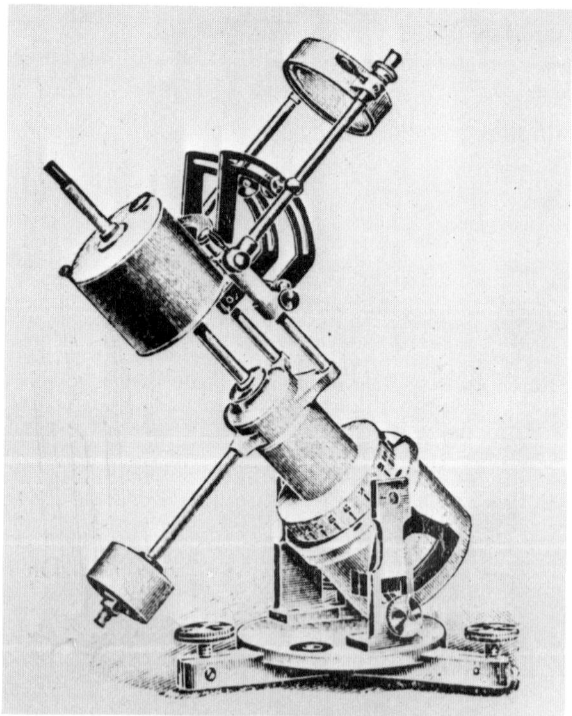

Fig. 6.10 Maurer's Sonnenscheinchronograph.

[34] Assmann, *Zeits. Instrum.*, Vol. 3 (1883), pp. 301–8.
[35] Jordan and Gaster, *Quart. J. Roy. Meteorol. Soc.*, Vol. 12 (1886), pp. 23–24.
[36] Curtis, *Ibid.*, Vol. 24 (1898), pp. 1–30.
[37] Curtis, *Ibid.*, Vol. 27 (1901), pp. 63–68.
[38] Maurer, *Meteorol. Zeits.*, Vol. 31 (1914), pp. 193–96.

which he called the *Sonnenscheinchronograph*,[39] had an achromatic lens 80 mm in diameter and 150 mm in focal length, rotated by clockwork to face the sun. This lens made an image of the sun on a cylinder covered with blue cardboard, rotated once an hour and moved axially so that the record formed a helix. Maurer later reported[40] that this instrument agreed with the Campbell-Stokes to about one per cent over all, but on days with small broken clouds it recorded much less sunshine. It did not, at any rate, arouse enthusiasm among other meteorologists.

4. Sunshine recorders based on the differential thermometer. The third class of sunshine recorder is based on the so-called photometer of John Leslie, a differential air thermometer with one bulb coated with a heat-absorbing layer.[41] If such a thermometer is fitted with some sort of a switching arrangement instead of a scale, it is easy to record electrically on a chronograph the periods when the radiation reaching the instrument is above and below some arbitrary limit.

The use of this principle seems to have been suggested in 1891 by D. T. Maring of the United States Weather Bureau,[42] and was first described by C. F. Marvin two years later.[43] It is therefore somewhat unfair to refer to the resulting instrument as the Marvin sunshine recorder, as is commonly done. It is a differential air thermometer with a clear bulb A and a black bulb B, mounted so that the axis of the instrument is approximately parallel to that of the earth (Fig. 6.11).

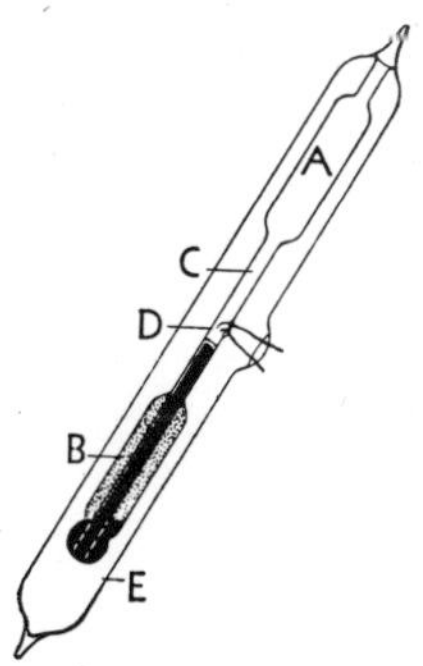

Fig. 6.11 The U.S. Weather Bureau sunshine recorder, diagram.

[39] In German the ordinary sunshine recorder is called *Sonnenscheinautograph*.
[40] *Meteorol. Zeits.*, Vol. 37 (1920), pp. 88–89.
[41] Leslie, *J. Nat. Phil., Chem. & Arts*, Vol. 3 (1799), pp. 461–67.
[42] Maring, *Monthly Weather Rev.*, Vol. 25 (1897), p. 487. This statement would not have got past the editor, Cleveland Abbe, if it had not been justified.
[43] Marvin, "Description of Instruments," in U.S. Dept. of Agriculture, *Report of the Chief of the Weather Bureau* (Washington, D.C., 1893), pp. 17–19.

The two bulbs are joined by a tube *C* with two platinum wires sealed through it at *D*. Parts of *B* and *C* are filled with mercury, the rest of the space with air. An evacuated shield *E* protects the instrument from air currents.

After some experience, Marvin reported, a little alcohol had been put into the instrument at the suggestion of C. B. Tuch. This "acts as a lubricator for the mercury . . . and its vapor plays an important part in the thermometric action of the instrument."

When radiation falls on the instrument, the black bulb will absorb more than the clear one, the air in it will expand, the vapor pressure of the alcohol will increase, and the mercury will rise in *C*, closing the electrical circuit at *D*. This circuit usually includes one pen magnet of a chronograph which also records wind, rainfall, and sometimes other quantities.

Another instrument of this type, in which ether was the operating liquid, was invented by W. H. Dines.[44] In this, the greater vapor pressure in the blackened end of the instrument shifted mercury until the tube tipped over and a circuit was established between two platinum wires. This device does not seem to have been used, except by its inventor.

Much later, R. Strutz substituted bimetallic strips, one blackened and the other bright, for the bulbs of the air thermometer.[45] Differential bending of the two strips caused two contact points to come together; one can easily imagine that such an instrument might be more easily adjustable, and reproducible, than the earlier devices of this kind.

One thing that should be made entirely clear is that no meaningful comparison can be made between the results obtained with the differential-thermometer type of recorder and with the other types. The latter are supposed to record when the sun is visible; the former will respond to some particular flux-density of radiation, no matter whether it comes from the sun or from the rest of the sky.

[44] Dines, *Quart. J. Roy. Meteorol. Soc.*, Vol. 26 (1900), pp. 243–45.
[45] Strutz, *La Météorologie* (1932), pp. 68–72.

VII

Early meteorographs

It is a platitude that the desire to make life easier is one of the most powerful incentives to mechanical invention. Reading a meteorological instrument with unfailing regularity at equal intervals of time can certainly become a most boring and confining occupation, so that it is not surprising that very soon after the invention of meteorological instruments attempts should have been made to have them produce a record by automatic means. What is rather surprising is that almost the very first devices of this sort which we know about should have been designed to record the readings of several meteorological instruments at once. They were what we should nowadays call "meteorographs."[1, 2]

The first ones were invented by the young man who later became Sir Christopher Wren, mathematician, astronomer, and—as almost everyone knows—the architect of St. Paul's Cathedral in London and many other famous buildings. Time has only strengthened the judgment of his contemporaries that he was nothing less than a genius, and in things mechanical his competence developed early, for when he was fifteen he wrote from Oxford to his father that one of his inventions, "a weather clock namely, with revolving cylinder, by means of which a record can be kept through the night," was being constructed at the expense of Dr. (later Sir) Charles Scarburgh, with whom he was staying.[3] This "weather clock" was most probably an air thermograph with a thermometer in the form of the "perpetual motion" invented by Drebbel and described, very inadequately, by Thomas Tymme.[4] At least, the "revolving cylinder" would suggest such an interpretation. A thermograph, even an air thermograph, was remarkable enough in 1647. However, it seems likely that Wren only combined two or more recording instruments into a meteorograph

[1] Some passages in this chapter are reproduced from W. E. K. Middleton, *Physis*, Vol. 3 (1961), pp. 213–22. I have also been helped by a paper by H. E. Hoff and L. A. Geddes, *Isis*, Vol. 53 (1962), pp. 287–324.

[2] In the eighteenth century the word "meteorograph" was used for any single recording meteorological instrument.

[3] Lena Milman, *Sir Christopher Wren* (London, 1908), p. 19.

[4] See p. 48 above.

years later, for he wrote to Lord Brouncker, the president of the new Royal Society, a letter mainly about meteorological observations, including the passage:

But this is not yet enough, for many changes may happen when the observer is absent or asleep. I might seem to promise too much, should I say, an Engine may be fram'd, which if you visit your Chamber but one half Hour in the day, shall tell you how many Changes of Wind have been in your Absence, though these were Twenty, and at what Hour every Change happen'd, and whether it were soft, stiff, or vehement. Neither shall the Instrument be subject to be out of Tune, or if it be, your own hand may rectify it.

Neither shall the Thermometer need a constant observance, for after the same Method may that be made to be its own Register.[5]

This reads as if Wren had promised to record the force and direction of the wind as well as the temperature. I am not aware that he did anything about the force, but by June, 1663, at Oxford, he had a meteorograph which registered not only the temperature and the direction of the wind but also the rainfall. In this month he was visited by a French traveler, Balthasar de Monconys of Lyon, who journeyed over a good deal of Europe in 1645–46 and again in 1663–64, making a point of visiting "philosophers" but not forgetting to record other things as well in his copious notebooks. On his return to France de Monconys published an account of his travels from which much can be gleaned about the state of natural philosophy in the various countries he visited.[6]

De Monconys went to Oxford especially to see Mr. Renes [*sic*]

a great mathematician, although small of stature, but one of the most civil and open I found in England, for although he did not wish that his ideas be divulged, he did not neglect to speak most freely about his weather-clock, which moves a ruler upon which is a stylus that marks on concentric circles corresponding to the hours all the changes in the wind indicated by a wind-vane that turns them [the circles]; and in the same way the rain, hail, and snow by vases attached to this wheel, which pass at each hour under a funnel into which it can rain, snow, or hail; and the heat and cold are registered by a thermometer that raises or lowers a tablet, against which a pencil on the above-mentioned ruler, lengthened as much as is needed for this purpose, marks the hours crosswise, while the tablet indicates the changes vertically. The following figure [Fig. 7.1] is a rough sketch of this.[7]

Hoff and Geddes discovered a drawing of Wren's at the Royal Institution of British Architects (R.I.B.A.) which may be a slightly

[5] Christopher Wren, *Parentalia: or, Memoirs of the family of the Wrens; viz. of Matthew, Bishop of Ely, Christopher Dean of Windsor & e., but chiefly of Sir Christopher Wren* (London, 1750), p. 224. The date of the letter is not given but I suggest January, 1663.

[6] Balthasar de Monconys, *Journal des voyages de Monsieur de Monconys* (3 vols., Lyons, 1775–66).

[7] *Ibid.*, Vol. II, p. 53.

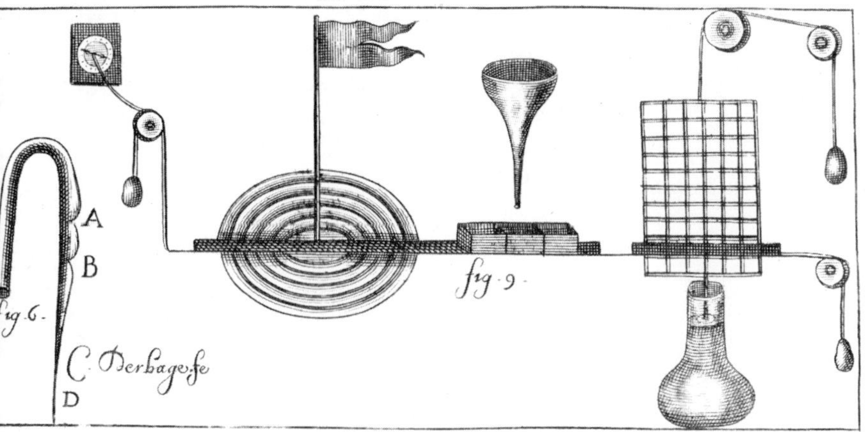

Fig. 7.1 Monconys' sketch of Wren's meteorgraph.

later version of this meteorograph (Fig. 7.2). The thermometer is both different and more comprehensible; it is an air thermometer, probably with some mercury in the U-tube, but the float on water or possibly oil, and an entirely reasonable mechanism for the pencil. There seem to be two rows of boxes for the rain, and I do not understand this. If we may assume that the little series of marks near the bottom represents a foot divided into inches, this was a large instrument; the "bulb" of the thermometer must have been about fifteen inches in diameter.

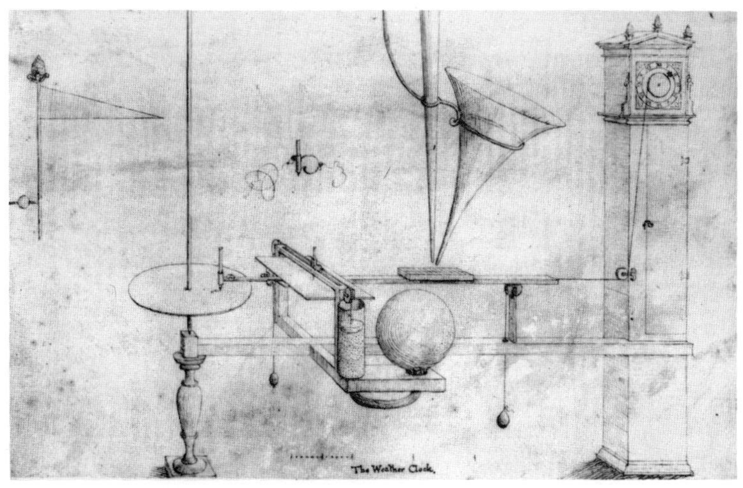

Fig. 7.2 A sketch by Wren (reproduced by permission of the Royal Institute of British Architects).

It must be remembered that all this had been going on at Oxford, and it appears that The Royal Society had heard of it,[8] for at a meeting on September 16, 1663,[9] "It being again[10] mentioned, that Dr. Wren's scheme for the observation of all the changes of weather should be sent for by Dr. Wilkins, the latter alleged, that he thought it was left with Mr. Boyle, whom the Secretary was therefore desired to speak to concerning it."

Presumably Oldenburg (the secretary) spoke to Boyle, and presumably Boyle denied that he had the document, because at the following meeting,[11] "Dr. Wilkins was desired to write to Dr. Wren, for his scheme of the instruments for observing all kinds of weather." Wilkins then wrote to Wren, who was at Wadham College, Oxford; but it was the 26th of November before Wren gave him an answer, as follows:[12]

"Wadh: Coll:
Nov: 26 1663

"Hon.^d S.^r

If you will pardon me for being a little late in observing your Commands, you will find heer enclosed the Designe I promised of the Weather Clock, changed a little into a more convenient forme, consisting of two winges wch may be added to pendulum Clock. A little description will make it fully intelligible.

"Upon the Index wheele of the Clock within the plate, is fixed the litle wheele A, wch mooves the Rack B, annexed by the joynt C to the crooked peece HCDEG, wch being made thin & light passeth through the Basis of the Clock at D & E, & through slittes in the pillars of the outward frame FG. at the ends it holds two black-lead pencills, upon the pointes of w.^ch (& not in the slittes) the crooked peece depends. The pencill H restes & drawes lines on the cylindricall surface of the Tumbrill K, wch is the Weather-Wheele mooved by Quicksilver after the manner I have formerly shown. The pencill I restes on the flat of the Wind-Wheele L mooved by a Vane without. on both surfaces are described Circular lines for the Houres according to the motion of the Rack, & crosse to them streight lines, shewing degrees of weather in the one, and Rumbes in the other; amongst these permanent lines, the pencills describe irregular lines compounded of the motions of the Rack & wheeles, much like the motion of the ship described among the longitudes & latitudes of the Chart; & from these tracks of the pencills may be collected the changes of wind & weather that have been in the 12 houres last past. these surfaces may be printed papers slightly stuck on with mouth glew, using bread to efface the old tracks. Or they may be a proper ground of whiting into wch the durable lines are stained the others being to be wiped out with a spunge. or they may be Box or Ivory, or unburnished Silver; if the lines engraven be soe filled, that the pencills stick not in them to hinder the Motion.

[8] Wren had written to Lord Brouncker about his thermometer on July 30, 1663 (*BM*, Sloane ms. 2903, fol. 104^r–105^v).

[9] Thomas Birch, *History of The Royal Society*, 4 vols., (London, 1756–7), Vol. 1, p. 304. Also *Journal Book* of The Royal Society for that date.

[10] I have not been able to identify the previous occasion.

[11] September 23, 1663. Birch, *History*, Vol. 1, p. 305.

[12] Royal Society, *Letter-book* W. 3, item 4, one folio, holograph. Copyright by The Royal Society: quoted by permission.

"I have willingly in this last contrivance omitted the Boxes, because I thinke they may be better disposed by themselves; & I doubt too whether they would not be drie, ere the observer comes to looke in them. I doe not esteeme this Instrument as a thing for common use, but for his Ease who shall undertake to give a Journal of wind & weather, wch without such an instrument is impossible to be allwaies don. S.^{r} if you please you may in any thing else command.

"Your most humble servant Chr. Wren."

The "Designe" referred to[13] is reproduced in Figure 7.3. On December 9, 1663, according to the *Journal Book:*

Dr. Wren's description of his Weather-clock was read, and ordered to be entered,[14] together with the Scheme thereof. Upon some debate, it was referred to the Council, to consider of the Expences and the most convenient way of reducing this Engine into practice;[15] as also of Additions to be made thereunto, whereof some were mentioned by Mr. Hook.

Mr. Hook made mention, that a common Seal'd Weather-glasse might be made applicable to this Weather-clock; and he was desired, to give proof thereof to the Company at their next Meeting.

Fig. 7.3 Wren's third meteorograph (copyright by The Royal Society; reproduced by permission).

[13] Copyright by The Royal Society; reproduced by permission. It is now attached to page 322 of Royal Society *Register Book II.* Presumably it is an original drawing of Wren's.

[14] I.e., in the *Register Book* of the Society. Only the long second paragraph was copied out; *Register Book II*, pp. 321–22.

[15] There is no record in the *Minutes* of the Council of any discussion of this.

One can imagine Hooke, who had a mind of extraordinary quickness, "thinking out loud" about how this "engine" might be improved. It would be interesting to know whether he saw at once how, in principle, a barometer could be adapted to it. He seems to have spoken too soon about the "sealed weather glass," which is, of course, the ordinary liquid-in-glass thermometer. I can find no record of his having given the desired "proof," either at the next meeting or later.

The "Weather-Wheele mooved by Quicksilver" is explained in the letter to Lord Brouncker already referred to, dated July 30, 1663.[16] The relevant passage is the following:

> I have pleased myself not a little with the play of the weather wheele (the Onely true way to measure Expansions of the Ayre) & I fancy it must needs give others satisfaction if it were once firmely made which I suppose may be donne if the circular pipe which cannot be truly blown in glasse were made of Brasse by those who make Trumpets & Sackbuts (who wire draw their pipes through a hole & equal, & then filling them with melted Lead turne them round into what flexures they please) but the inside of the pipe must be vernished with China vernish (which Gratorex hath) to preserve it from the quicksilver, & the glasses must be fixed to the Pipe with Varnishe, which I take to be the best cement in the world, for thus the Chinese fix glasse & mother of Pearle in their workes."[17]

In fact, this was Drebbel's "perpetual motion," with mercury instead of water as the liquid. On June 12, Wren had told Monconys about a somewhat similar "thermometer with a drum," but with water in it, and Monconys correctly identified it as Drebbel's machine.[18] The interesting thing to notice is that between the middle of June and the end of July Wren had again changed his ideas about the best sort of thermometer for a meteorograph. There also remains the open question of the relative dates of the instrument seen by Monconys and the one in Wren's sketch preserved by the R.I.B.A. I do not think we can quite exclude the possibility that they were the same instrument, for many of Monconys' drawings are somewhat schematic.

As his letter to Wilkins shows, Wren had second thoughts about his recording rain gauge for two excellent reasons, of which the second alone is more than sufficient.

From this time onward the development of the "weather clock" seems to have been taken over by Robert Hooke. The next reference I have been able to find to it is in Hooke's diary[19] under the date February 19, 1672/3, when he was "at home writing an account" of it.

[16] *BM*, Sloane ms 2903, fol. 104ʳ–105ᵛ. Quoted by permission.

[17] *Ibid.*, fol. 105ʳ. "Gratorex" refers to Ralph Greatorex, a well-known instrument-maker who had his shop on the Strand (see E. G. R. Taylor, *The Mathematical Practitioners of Tudor and Stuart England* [Cambridge, 1954], p. 229). "China vernish" is presumably shellac.

[18] Monconys, *Voyages*, Vol. II, p. 54. See also p. 246 above.

[19] Henry W. Robinson and Walter Adams, editors, *The Diary of Robert Hooke, M.A., M.D., F.R.S., 1672–1680* (London, Taylor and Francis, 1935).

Progress is reported on February 26, and on March 12 he read a lecture about it, most probably to The Royal Society. It seems to have been of continuing interest, because three years later he records "... read my Lecture about Weather clock. A pretty good auditory ..." This was not to The Royal Society, for the same entry[20] contains the notation "Society sat not."

In the years between there is an interesting entry, dated January 2, 1674/5: "With Tompion[21] to Sir W. Petty's. Directed his clock for weather." This sounds as if one was being built for the famous political economist. This is the earliest evidence that I have found that the idea had taken bodily form, but there is much more to come. In 1678, on September 11, "Crawley gave pattern for wheels to the founder for Weather clock." On September 13 "Crawley made wheeles" (fancy getting the castings back in two days!). And on December 5: "At Society ... shewd Weather clock," as also on January 9, 1678/9. There may have been an accident, for on February 4, 1678/9, Hooke "Made glasses for weather clock." Finally on May 26, 1679, "Removed Weather Clook [*sic*] to Mr. Henshaw."

We know that this particular instrument was being made at least partly at the expense of The Royal Society, for on December 26, 1678, the Council ordered that Hooke should be reimbursed for the £ 5 he had paid to Crawley "for workmanship about the weather-clock."[22] The *Journal Books* contain more details of great interest. On December 5, 1678:

> Mr Hook produced a part of his new weather Clock which he had been preparing which was to keep an Account of all the Changes of weather which should happen, namely the Quarters and points in which the wind should blow 2!y the strength of the Wind in that Quarter. 3!y The heat and cold of the Air. 4!y The Gravity and Levity of the Air. 5!y the Dryness and moisture of the Air. 6!y The Quantity of Rain that should fall. 7!y The Quantity of Snow or Hail that shall fall in the winter. 8!y the times of the shining of the Sun. This he was desired to proceed with all to finish he hoped to doe within a month or six weeks.[23]

As far as I know, the sunshine recorder is referred to nowhere else, and does not appear in Hooke's own description of the instrument published by William Derham in 1726, long after Hooke's death, in a collection of unpublished snippets, mainly by Hooke.[24] It is ascribed

[20] *Ibid.*, under the date April 20, 1676.
[21] Thomas Tompion was a famous member of the Clockmakers' Company, and one of his specialties was barometers. See E. G. R. Taylor, *The Mathematical Practitioners*, p. 247.
[22] Birch, *History*, Vol. III, p. 450.
[23] This and four later extracts are from the manuscript copy of the original *Journal Books* at The Royal Society. Quoted by permission.
[24] W. Derham, ed., *Philosophical Experiments and Observations of the late Eminent Dr. Robert Hooke, S.R.S. and Geom. Prof. Gresh.*[am College] *and Other Eminent Virtuoso's in his Time* (London, 1726).

by Derham to about the same date:

Dr. Hook's Description of his Weather-Wiser; about Dec. 5.1678.

The Weather-Clock consists of two parts. First, that which measures the time, which is a strong and large pendulum-clock, which moves a week, with once winding up, and is sufficient to turn a cylinder (upon which the paper is rolled) twice round in a day, and also to lift a hammer for striking the punches, once every quarter of an hour.

Secondly, Of several instruments for measuring the degrees of alteration, in the several things to be observed. The first is the barometer, which moves the first punch, an inch and half, serving to shew the difference between the greatest and least pressure of the air. The second is, the thermometer, which moves the punch that shews the differences between the greatest heat in summer, and the least in winter. The third is, the hygroscope, moving the punch, which shews the differences between the moistest and driest airs. The fourth is, the rain-bucket, serving to shew the quantity of rain that falls; this hath two parts or punches; the first, to shew what part of the bucket is fill'd, when there falls not enough to make it empty itself; the second, to shew how many full buckets have been emptied. The fifth is, the wind vane; this hath also two parts; the first to shew the strength of the wind; which is observed by the number of revolutions in the vane-mill, and marked by three punches; the first marks every 10000 revolutions, the second every 1000, and the third every 100; The second to shew the quarters of the wind, this hath four punches; the first with one point, marking the north quarters, *viz.* N: N. by E: N by W: N.N.E: N.N.W: N.E. by N. and N.W. by N: N.E. and N.W. The second hath two points, marking the east and its quarters. The third hath three points, marking the south and its quarters. The fourth hath four points, marking the west and its quarters. Some of these punches give one mark, every 100 revolutions of the vane-mill.

The stations or places of the first four punches are marked on a scrowl of paper, by the clock-hammer, falling every quarter of an hour. The punches, belonging to the fifth, are marked on the said scrowl, by the revolutions of the vane, which are accounted by a small numerator, standing at the top of the clock-case, which is moved by the vane-mill.[25]

Although the original of the description given by Derham is not in the archives of The Royal Society, there is a holograph account of the instrument, to which the date 1689 has been assigned:

The designe was first suggested by Sr Christopher Wren, as feasible to make a watch to mark the quarters of the wind and the Degrees of Heat & Cold in the air. But at the Desire of the Society I have made one by other contrivances to keep an account not only of the mutations of those two, but also of Divers others. to witt of the strength or swiftness of the wind as well as of the quarter of it; of the Drynesse and moysture of the air, of the quantity of the rain that falls and the times of its falling, of the pressure of the air and its variety, of the heat and cold of the air adjusted to the standard of freezing, and if the place were convenient of the sunshine and cloudinesse of the Sky and could have added some others if the watch had had a place adapted for their observation.[26]

[25] *Ibid.*, pp. 41–42. I have not been able to find the original manuscript of this.
[26] *Classified papers* of The Royal Society, Vol. XX, no. 80.

But before this, Nehemiah Grew had described the instrument with annoying brevity in his fantastic catalogue of The Royal Society's collections.[27] Among the "Instruments relating to Natural Philosophy" there was

A weather clock. Begun by Sir Chr. Wren . . . to which other motions have since been added, by Mr. Robert Hook Professor of geometry in Gresham-Colledge. Who purposes to publish a description hereof. I shall therefore only take notice, that it hath six or seven motions; which he supposeth to be here advantagiously made altogether. First a pendulum clock, which goes with ¾ of a 100 lib. weight, and moves the greatest part of the work. With this, a barometre, a thermometre; a rain-measure, such an one as is next described; a weather-cock, to which subserves a piece of wheel-work analogous to a way wiser; and a hygroscope. Each of which have their regester, and the weather-cock hath two; one for the points, the other for the strength of the wind. All working upon a paper falling off of a rowler which the clock also turns.

Let us endeavor to reconstruct in imagination this astonishing instrument.

The clock-and-roller mechanism was probably rather similar to that in some modern recorders, the drum or roller being in this instance either directly coupled to the hour-shaft of the clock, or geared to it at a ratio of one to one. A brilliant innovation was the use of "punches" to make the record, instead of the pencils employed by Wren. This device, which makes the motion of the meteorological instrument concerned just as frictionless as if there were no recording gear at all, is usually ascribed to Changeux[28] and dated a century later. The paper seems to have been in a long strip which unrolled as the clock ran, and was presumably held taut by a weight attached to the free end. Unless it was joined, this strip could not have been longer than the length of the biggest mold, as all paper was mold-made in the seventeenth century.

Because the axis of the roller was thus necessarily horizontal, we may suppose that a float in the open limb of a siphon barometer was attached to a cord which passed over two pulleys to a counterweight. A stylus or "punch" was fixed to the horizontal part of this cord, which passed near the roller. Such a development of his "wheel" barometer[29] would naturally occur to Hooke. The thermometer would be very like the barometer, except that an air thermometer bulb would be substituted for the long branch of the barometer tube.

As to the hygroscope, it is difficult even to guess. The miniature one made of an oat-beard and described by Hooke in the *Micrographia*

[27] *Musaeum Regalis Societatis or a Catalogue and Description of the Natural and Artificial Rareties Belonging to The Royal Society* [etc.] (London 1681), pp. 357–58.

[28] *Météorographie ou Art d'observer d'une manière commode et utile les Phénomènes de l'Atmosphère. Contenant la description de deux Barométrographes ou Baromètres qui tiennent note par des traces sensibles de leurs variations et des tems précis où elles arrivent; avec l'idée de plusieurs autres Instrumens Météorologiques* [etc.] (Paris, 1781).

[29] Hooke, *Micrographia* (London, 1665), preface. See p. 26 above.

was obviously unsuitable, and among the others which had been suggested by 1680 the most likely are the weighing hygrometer of Lana (1670) and an instrument, described anonymously in the *Philosophical Transactions* of 1676, which depended on the swelling of wood.[30]

The rain gauge was clearly a tipping bucket on a balance arm. The position of the bucket was indicated by one punch, and the other punch, which recorded the emptying of the bucket, must have been actuated by the bucket as it tipped, or perhaps by the energy of the falling water. It would seem that Hooke's choice of a horizontal axis for his roller must have greatly intensified his mechanical problems.

The anemometer was certainly a windmill ("Vane-Mill") and so must have been faced into the wind by a windvane. There must have been a mechanism with three stepped cams on shafts geared to the windmill in ratios of 1:100, 1:1,000, and 1:10,000. The first of these also operated one of the direction punches. The description is far from clear, but I incline to the view that the record showed merely whether the wind was from the north, east, south or west quarter, by the number of holes, from one to four, punched in the paper.

To return to the *Journal Books:* on January 9, 1678/9, before the Society actually began its meeting, "The President with Sr John Lowther Mr Henshaw Mr Thin [Thynne] & severall others went up into Mr Hooks Turret to see the further progress that had been made in the Clock contrived by Mr Hook for keeping an account of the Severall Variations of the Weather the which was very well approved by them. The President took the chair." This was only five weeks after they had desired him to proceed with it, so that it must have been causing a good deal of interest. However, the next entry that refers to it seems to be one on May 22, 1679: "A Stranger being present the Weather clock now finished by Mr Hook was omitted to be seen till the next day when it was to be carryed to Mr Hunts[31] lodgings that he might attend it and supply it with fresh papers." "The next day" evidently means the day of the next meeting, May 29, when the Society went "to take a View of the new weather clock which was set up in Mr Hunts lodgings, made to keep an Account of the Quantity and kind of all the Changes that happen in the Air as to its heat and cold, its dryness and moisture, Its gravity and Levity, Its motion in what Quarter and with what strength and Velocity, As also of the kinds and Quantity of the Rain Snow and hail that falls all which it sets down in Paper, so as to be very legible and certain."

This remarkable and complicated instrument did not endure. Hoff and Geddes have given good reasons for believing that, in common with most of The Royal Society's other instruments, it was derelict by

[30] Anon, *Phil. Trans.*, Vol. 11 (1676), pp. 647–53.
[31] Harry Hunt, "Operator" to the Society from 1676 to 1696. See E. G. R. Taylor, *The Mathematical Practitioners*, pp. 266–67.

1710 at the latest.[32] It was in trouble even in 1684, for on April 2 of that year we read that the Council of The Royal Society ". . . ordered, that Mr Hooke be desired to put into writing a description of the weather-clock and all its parts; and that it be delivered to the secretary to be entered in the register book: That Mr Hooke give his directions and assistance to Mr Hunt, to reduce into writing some of the first papers marked by the weather-clock, that thereby the Society might have a specimen of the weather-clock's performances before they proceed to the repairing it."[33]

The last sentence of this extract shows very clearly why instruments of this sort went out of fashion. If an able artisan like Harry Hunt had to be helped by the inventor to interpret the records of the weather clock, it is unlikely that many other people could succeed. In any event, counting large numbers of tiny holes in a sheet of paper is not a congenial task, and observing the meteorological elements in the ordinary way would not necessarily be much more arduous. The seventeenth-century interest in elaborate clocks, which Hooke undoubtedly shared, was the real source of this development, as has been suggested by Hoff and Geddes;[34] but I find it difficult to agree with these authors that the influence of Hooke's weather-clock is to be found in the anemograph of D'Ons-en-Bray,[35] much less the barographs of Cumming[36] and of Changeux,[37] although these barographs were indeed derived from Hooke's wheel barometer, which had become a very common instrument.

Another reason for the unpopularity of meteorographs in the eighteenth century is the growing appreciation of the importance of exposure. The thermometer and the hygrometer had to be exposed out of doors,[38] but the barometer would last longer if it were not. The rain gauge and the windvane were not obvious neighbors, especially after the dependence of the catch of a rain gauge on its height above ground had been discovered.[39] A composite recording instrument made an esthetic appeal to the instrumentmaker, not a practical one to the meteorologist.

Nevertheless there were a few essays of this sort towards the end of the eighteenth century, mainly in Italy. Antonio Maria Vassalli-Eandi, professor of physics and later director of the astronomical observatory at Turin, described a meteorograph in 1799,[40] and another

[32] *Isis*, Vol. 53 (1962), pp. 303–04.
[33] Birch, *History*, Vol. IV, p. 277.
[34] *Isis*, Vol. 53 (1962), p. 304.
[35] See above, p. 210.
[36] See H. Alan Lloyd, *J. Suisse Horolog.*, English ed., Vol. 78 (1953), pp. 371–81; Vol. 79 (1954), pp. 46–56.
[37] See note 28, this chapter.
[38] For the history of the exposure of thermometers, see my *A History of the Thermometer* (Baltimore, 1966), Chap. X.
[39] See Middleton, *A History of the Theories of Rain* (London, 1965, New York, 1966), pp. 98–99.
[40] Vassalli-Eandi, *Mem. Soc. Ital.*, Vol. 8 (1799), pp. 516–20.

in 1802.[41] In the earlier one, there was a siphon barometer and an astonishing thermometer consisting of a flask holding 6 lb. of mercury and with a neck 1 cm in diameter—an early use of the new metric system—each having a float operating a magnifying lever. The levers carried pens that recorded on a cylinder driven by a clock. Vassalli-Eandi suggested additional pens for a rain gauge, an atmometer, and a hygrometer. A recording windvane and anemometer would have needed a second drum, but he gave no details of all these additions. The later instrument was meant to record wind direction, wind force (with a pressure plate), temperature, and barometric pressure.

Instrumentmakers, of course, occasionally designed and built such instruments. A famous one was that described briefly in 1846 to the British Association by George Dollond of the famous London optical firm.[42] This "atmospheric recorder" included a barometer, thermometer, hygrometer, electrometer, rain gauge, atmometer, pressure-plate anemometer, and windvane. "I have found it to answer the purpose for which I intended it, in every way satisfactorily," boasted Dollond.

We know a great deal about this meteorograph because he set it up at the Great Exhibition of 1851,[43] when he was seventy-seven. The unusually long and detailed description in the official catalogue is prefaced by a not unusually sententious paragraph by James Glaisher: "Self-registering instruments, which move equally by clock-work or otherwise, and are made subservient to the registration of natural phenomena, are of the highest importance, and particularly so in meteorological investigations, where the changes of every element of research are perpetual, and those which accrue during the night are of equal importance to those happening during the day.—J.G." I shall rely on Figure 7.4 to give the reader a general picture of this apparatus, which at least avoided the Gothic style, and confine myself to a few notes on the meteorological instruments that it contained.

The barometer was simply a large siphon barometer with an almost counterpoised float and a lever system giving a magnification of three. The temperature was recorded by a system of ten similar mercury-in-glass thermometers side by side in a frame. The frame formed the beam of a balance, and the expansion of the mercury changed the position of the center of gravity, causing this to tilt. The idea of using ten thermometers was to multiply the turning moment without a corresponding decrease in the speed of response. The hygrometer was a strip of mahogany cut across the grain—an idea that, as we have seen, dates from the seventeenth century. The indexes of these three instruments terminated in sharp points, which were struck into the

[41] *Mem. R. Accad. Sci. Torino*, Vol. 12 (1802–3), pp. 426–44.
[42] Dollond, B.A.A.S., Southampton, 1846, *Sections*, p. 17.
[43] London, Great Exhibition of 1851, *Official Descriptive, and Illustrated Catalogue* [etc.] (London, 1851), part 2, pp. 414–16.

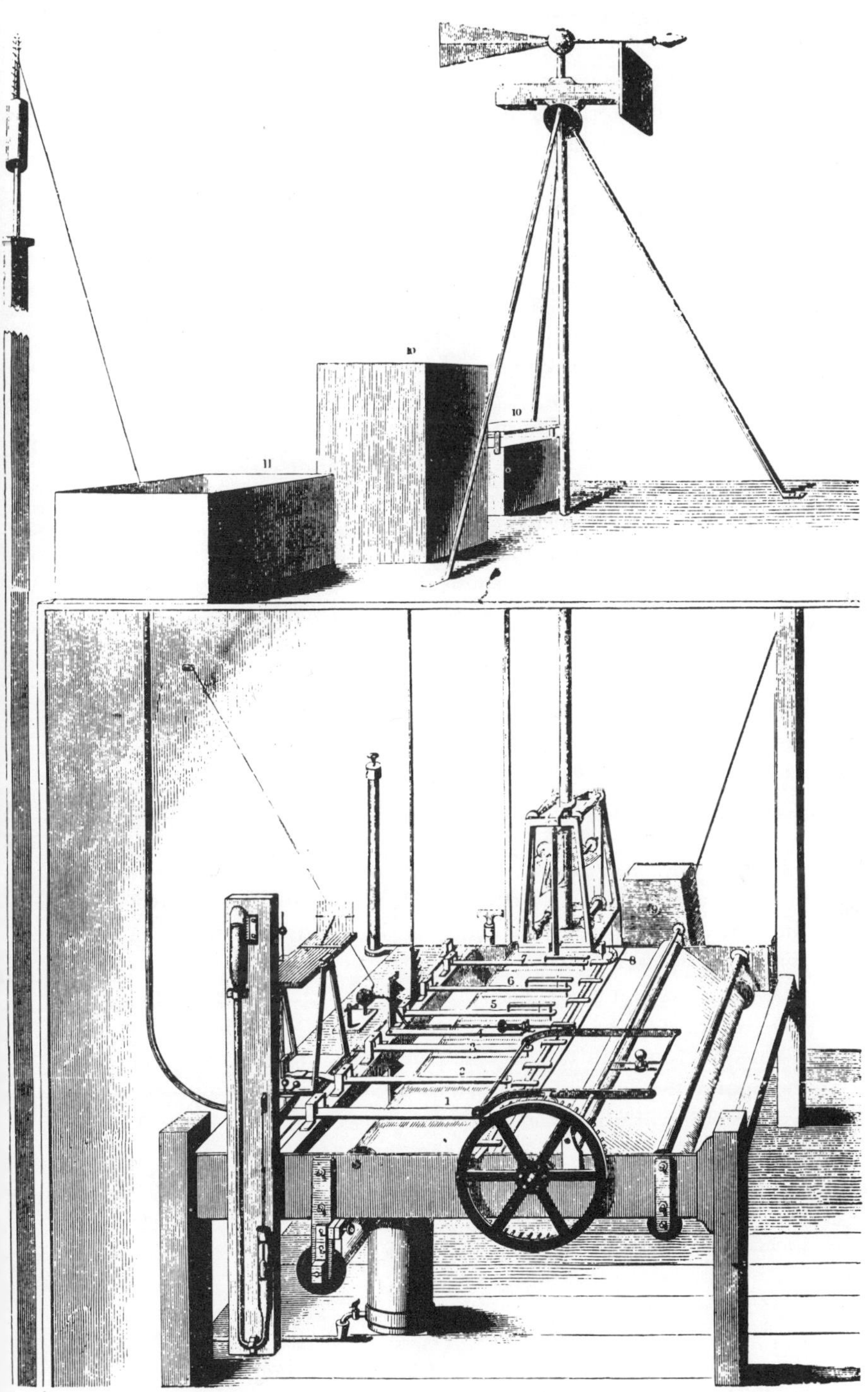

Fig. 7.4 Dollond's "atmospheric recorder," 1846.

paper every half hour by a falling bar. The bar also made time marks at each edge of the paper. The remaining instruments had pencils.

Of these the first was an electrometer for registering the electric field below a thunderstorm. I must admit that I find the description of this obscure. The rain gauge, with a receiver one foot square on the roof, has already been described.[44] The evaporation gauge made a record in some incomprehensible manner by means of a float.

The anemometer had a pressure plate one foot square. As it was moved back by the pressure of the wind, it raised a succession of weights. The vane that faced this pressure plate into the wind recorded its direction, but we are not told how—or whether—the problem presented by the vane making several complete revolutions was dealt with.

George Dollond's pride in this unusual instrument is evident, and we may excuse it if we reflect that he can scarcely have hoped to profit financially. If he did, he had not time to regret his mistake, for he died in the following year.

The next important meteorograph, much better than that of Dollond, was designed at Rome by the noted astronomer Angelo Secchi, Director of the Observatory of the Collegio Romano. There were two versions of this, the first being described in 1857.[45] The later and more highly developed one, which Secchi was proud to send to the Paris Exhibition of 1867,[46] will be dealt with here. This great instrument, which in 1964 was still preserved in the Museo Copernicano on the Monte Mario in Rome, bears a brass plate bearing the name "E. Brassart, Roma." Two views of it, taken from the work by Lacroix, are shown in Figure 7.5. It should be noted that this includes only the upper part of the instrument, the sub-base actually being 60 cm high, and containing a cupboard for the Daniell cells that supplied the current for the electromagnets. The entire instrument is more than three meters high.

What is not quite obvious from the figure is that the meteorograph has two entirely separate rectangular frames carrying charts, each sliding in grooves in two of four columns. The front one, shown at the right of the figure, is let down by the clock in 2½ days, and carries records of the pressure, the dry-bulb and wet-bulb thermometers, and the time of occurrence (but not the amount) of rain. The accumulated rainfall, received from a large funnel in a float-chamber R, is recorded on a paper disk p by a pencil moved radially at a constant speed.

[44] See p. 147 above.

[45] Secchi, *Nuovo Cimento*, Vol. 5 (1857), pp. 14–17; also *Atti dei nuovi Lincei*, Vol. 10 (1857), pp. 137–45.

[46] See E. Lacroix, *Etudes sur l'exposition de 1867* (Paris, 1867), Vol. 2, pp. 313 ff. Secchi also published a pamphlet about it: *Notice sur le météorographe du P. Secchi, S.J.* (Paris, n.d. [1867]), 7 pp.; and two descriptions in Italian: *Boll. meteorol. Coll. Romano*, Vol. 5, no. 4 (April 30, 1866), pp. 25–44; *Descrizione del meteorografo dell'Osservatorio del Collegio Romano* (Rome, 1870).

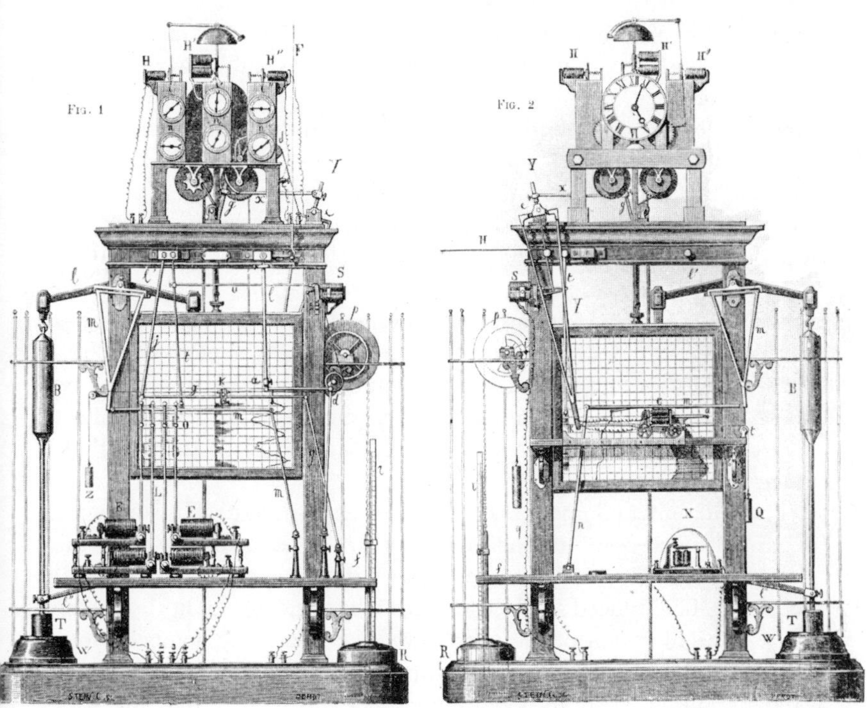

Fig. 7.5 Secchi's meteorograph, 1867, front and back views.

The chart at the back of the instrument carries records of pressure, wind speed and direction, and the temperature measured by the relative dilatation of a long brass wire and a pine beam. This chart descends in ten days.

Returning to the front of the instrument, let us first consider the barograph. This is a balance barometer with a very large tube made in a gun factory out of a single forging; the inner diameter of the chamber *B* is 5 centimeters. The counterpoise is out of sight between the two charts. The barograph is the only component of the meteorograph, with the exception of the wire thermometer, that does not demand any electrical apparatus. A Watt parallel motion *mmm* carries the pen.

The wet-bulb and dry-bulb thermometers are not shown in the drawing; they were installed in a more appropriate position out of doors. They are mercury thermometers open at the top, with platinum wires sealed into their bulbs. Two vertical platinum wires mounted on a chassis that can move up and down under the control of a wire *H* can dip into the mercury of the thermometers. Let us see how this apparatus functions. Every fifteen minutes the wheel *g* is turned by the striking train of the clock, pulls the rod *x* from left to right, and then lets it go back again. This turns the lever *Y* first clockwise and then counterclockwise, slacking off the wire *H* and then pulling it back, so that the two platinum wires dip slowly into the mercury of the two thermometers,[47] and then emerge. At the same time, the lever *Y* is pulling a little carriage *c* to the left and then back again. This carries an electromagnet which is energized whenever only the platinum wire in the dry-bulb thermometer is making contact, but not, thanks to the back-contact relay *X*, when the wire in the wet bulb thermometer is doing so, and of course not when neither contact is being made. Thus as the carriage moves to the left, the pencil will be pressed against the paper as soon as the dry-bulb contact is made, and leave it upon the completion of the circuit through the wet-bulb thermometer. On the way back, another mark will be made between the same limits. The envelopes of the series of short horizontal lines made every quarter of an hour will be the two thermograms.

The remaining components require less explanation. The duration of rain is marked by a pencil deflected by an electromagnet in the circuit of a battery and a switch operated intermittently by a small overshot wheel placed "under a rain-spout somewhere on the building." The direction of the wind is recorded to eight points by four electromagnets *E*, operated singly or in pairs by contacts on a windvane.[48] The wind speed is registered very ingeniously, starting with a cup anemometer that makes an electrical contact after a certain number of revolutions. The resulting pulses operate a counter H'', the last wheel of which winds up a chain *dd* that carries a pencil at *K*. The pencil thus moves to the right a certain distance for every kilometer of wind, until at the end of an hour a mechanism operated by the striking train lets the weight *Z* pull the pencil *K* back to zero again. This produces a line on the chart, of a length proportional to the total wind during the hour just elapsed. I am unable to understand why this eminently readable arrangement has not been more widely used in connection with cupwheels.

Such, in outline, is the great meteorograph of Angelo Secchi. The philosophy behind the instrument is made clear in the penultimate paragraph of his booklet, which I shall translate in full. "All scholars are agreed on this point, that meteorology can only advance by having machines that record all the phenomena automatically. The construction of a complete apparatus like the one that has just been described is rather an expensive matter, but its essential parts could be procured at a relatively moderate price, if a great number of them had to be made. The problem of the progress of storms over the continent will soon be resolved by these methods."[49]

At about the time that Secchi was building his instrument, A. G.

[47] This idea is due to Charles Wheatstone; see B.A.A.S., Cork, 1843, *Report*, pp, xl–xlii. See also p. 317 below.
[48] See p. 219 above.
[49] Secchi, *Notice*, p. 7.

Theorell was busy with a complicated meteorograph for the Academy of Sciences at Stockholm, another example of which was built for the Copenhagen Academy, and a third, improved one for the Observatory of Uppsala.[50] This meteorograph recorded only pressure, dry-bulb temperature, and wet-bulb temperature, all three by a modification of Wheatstone's method; the modification was that the contact wires were caused to touch the mercury surfaces and immediately retreat without dipping into the mercury.

The general appearance of Theorell's meteorograph is shown in Figure 7.6. The barometer *F* is a siphon barometer; the thermom-

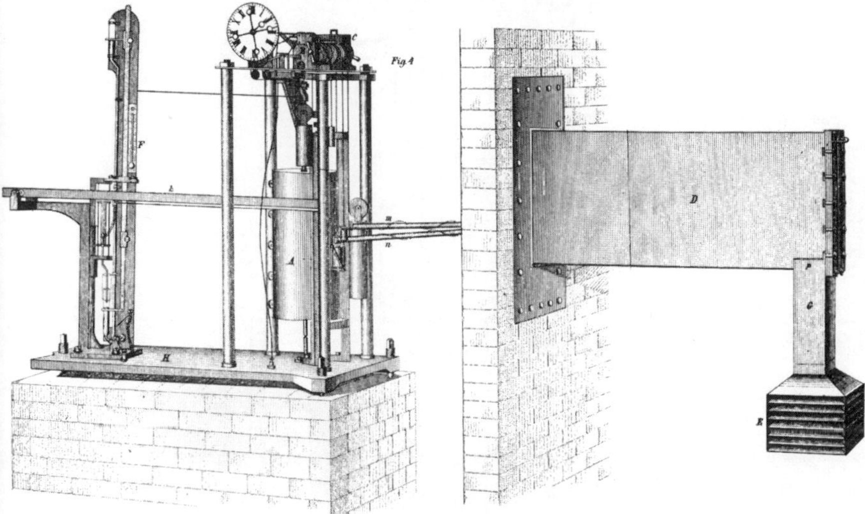

Fig. 7.6 Theorell's meteorograph, 1869.

eters, which are in the screen *E*,*G* at the other end of the instrument, are mercury thermometers open at the top, with steel rods dipping into their tubes. These rods are moved by the levers *m*,*n* with fulcrums at the left-hand end of the chamber *D*, which is hermetically sealed and provided with a drying agent, a feature that Theorell found essential to the continued functioning of the instrument. These levers have equal arms. The lever *k* moves a probe into and out of the mercury in the open limb of the barometer.

The three levers are attached to electromagnets *1*, *2*, and *3* (Fig. 7.7), each of which can move vertically on pairs of steel wires stretched between the ends of the frame *R*. Once every fifteen minutes in the Copenhagen instrument, which is illustrated, or once every ten minutes in the one at Uppsala, magnet *1* is let down, and magnets *2* and *3*

[50] Theorell, *Repert. phys. Techn.*, Vol. 5 (1869), pp. 121–33; also *Z. österr. Ges. Meteorol.*, Vol. 4 (1869), pp. 497–504; 522–28.

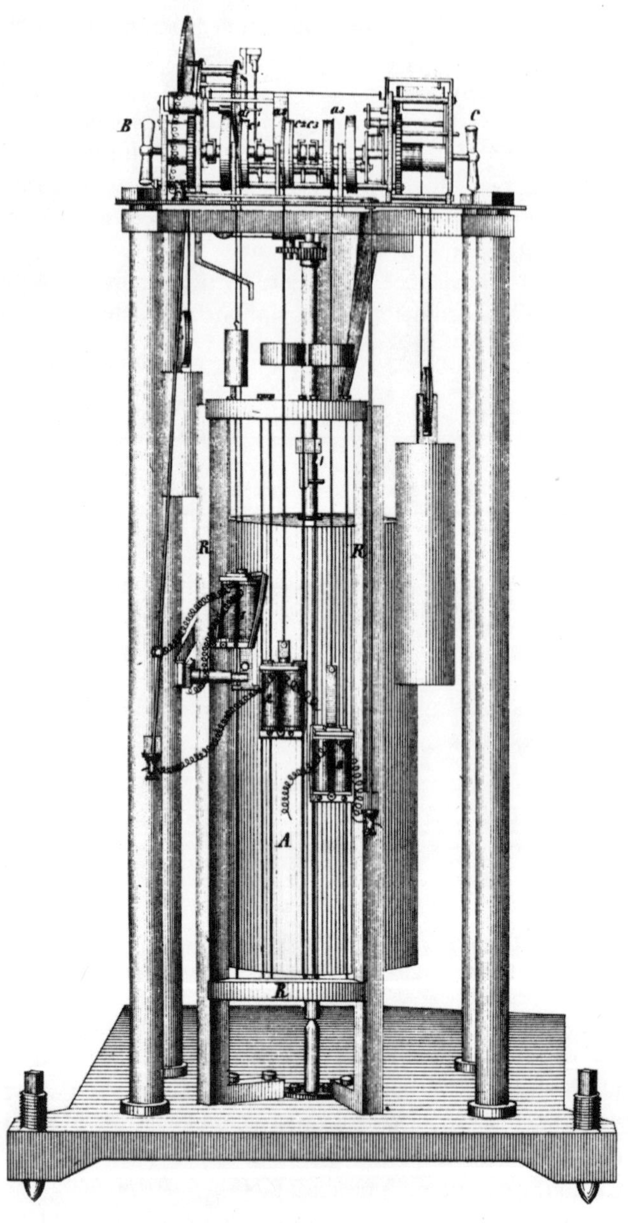

Fig. 7.7 Theorell's meteorograph, recording part.

are raised, until the probes at the far ends of the levers touch the corresponding mercury surfaces. Whenever this happens, an electromagnet causes a dot to be made on the paper on the drum *A*, and the motion is immediately reversed by the extremely ingenious (and complicated) system of clutches at the top of the instrument. I do not feel constrained to describe this.

A few years later Theorell fitted his meteorograph at Uppsala with a printing mechanism that recorded the results hourly in figures.[51] This instrument, installed as it was originally, in a small brick building, with the thermometer to the north, was still at the Meteorological Institute of the University when I visited Uppsala in 1965. It was not working, and not quite in order, but it was very well preserved.

Several more meteorographs were built in various places during the period between 1860 and 1880. G. W. Hough of the Dudley Observatory built one for the United States Army Signal Corps,[52] on the same principles as Theorell's, but less neatly designed and made. Of quite a different order was that built by Peter Stevenson, an instrument-maker of Edinburgh, for the eighth Marquis of Tweeddale, who died in 1876.[53] This recorded pressure, air temperature, wet-bulb temperature, and rainfall, all very simply and entirely mechanically, the whole thing being contained in a cylindrical cast-iron box, with the funnel of the rain gauge set into the top. It had apparently been kept working for several years.

The truth is that by this time large meteorographs were out of date, and there were two reasons for this. In the first place it was just beginning to dawn on a few meteorologists that the rapid spatial and temporal fluctuations on the meteorological elements put a limit on the desirable or indeed attainable precision of measurement. The second was the introduction in the 1880's of a series of small, light, and inexpensive recording instruments by the Paris firm of Richard Frères. The great meteorographs, like the huge reptiles of the Jurassic era, simply could not compete.

[51] Theorell, *Kongl. Svenska Vet.-Akad., Handl.*, Vol. 10 (1871), pp. 3–10.
[52] Hough, *Annals of the Dudley Observatory, Albany, N.Y.*, Vol. 2 (1871), pp. xxxvi–xlii.
[53] Stevenson, *Trans. R. Scottish Soc. Arts*, Vol. 11 (1887), pp. 139–41 (read Feb. 25, 1884).

Fig. 8.1 The height of clouds by triangulation.

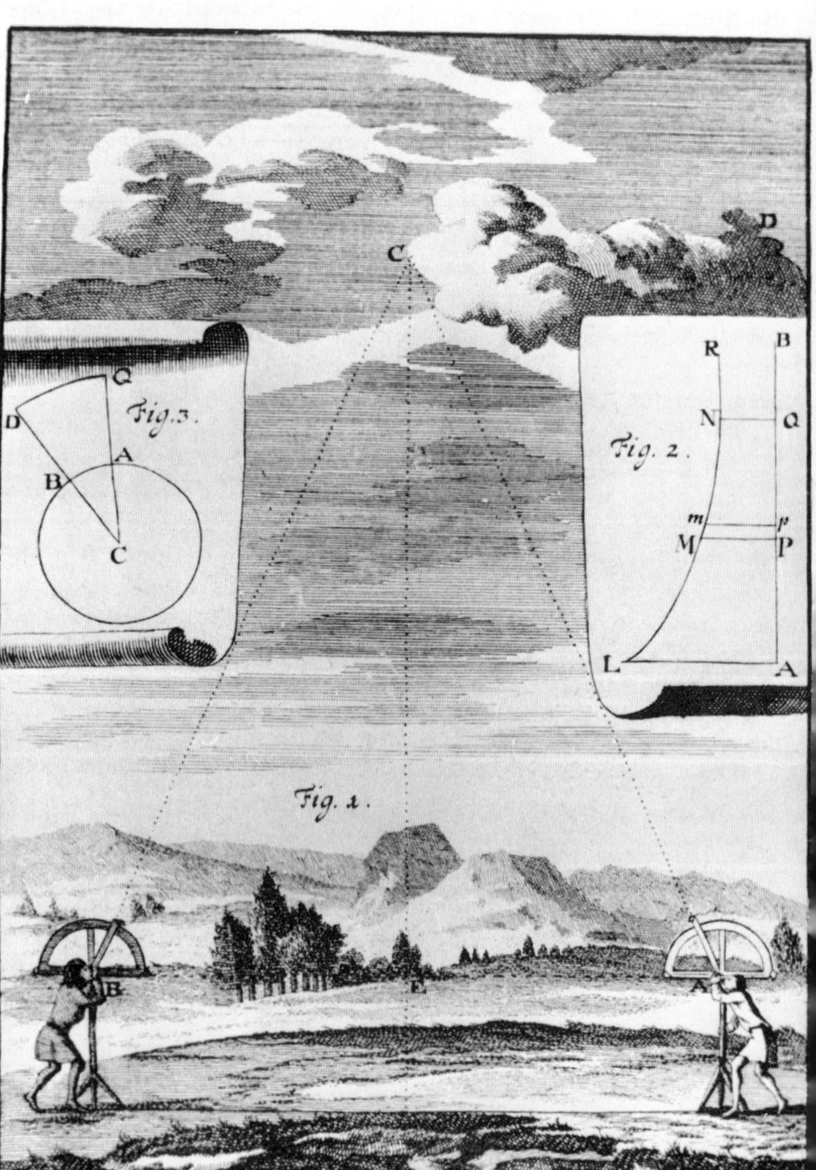

VIII

Upper winds and the height and motion of clouds

---

1. Seventeenth- and eighteenth-century suggestions. The great Johannes Kepler was not above considering how to measure the height and motion of clouds.[1] He knew very well that if you have two observers with instruments at the ends of a measured baseline, you can determine the height of a cloud. This idea seems to pervade the early literature on the subject, and I reproduce in Figure 8.1 the frontispiece of a little book published about a century later.[2] Kepler knew that it is uncommon to be so well provided with observers, and went on to show how one observer can determine the height of a cloud if the sun is shining and he can see where the shadow of the cloud is. In Figure 8.2 C is the cloud, S its shadow, A the observer, and E a point on the ground below the cloud. If AS is known, the observer has only to measure the zenith angles, SCE of the sun, and ACE of the cloud, and solve triangles. A hole in a layer of cloud will serve just as well. Kepler treated only the plane problem, with the cloud and the sun in the same azimuth.

In 1651 G. B. Riccioli of Ferrara, after citing Kepler's method, gave a second one which he credited to Simon Stevin. The shadow of a cloud is measured on the ground, and at the same time an observer in the shadow determines the angular subtense of the cloud.[3] This would be easier if the cloud remained at rest. In 1688 Jacob Bernouilli gave a "new method of measuring the height of a cloud" after sunset or before sunrise by observing the shadow of the earth on it, working out formulas for any position of the cloud.[4]

In 1766 Alexander Brice of Kirknewton in Scotland measured the speed of the wind aloft by timing the motion of the shadows of detached clouds,[5] a direct and simple idea usually credited to J. H.

[1] Kepler, *Epitome astronomiae Copernicanae* [etc.] (Linz, 1618), pp. 70–71.
[2] Johann Frederic Glöckner, *Specimen novum nephelemetriae seu dissertatio . . . de pondere nubium* (Dissertation, Halle, 1722).
[3] Riccioli, *Almagestum novum, astronomiam veterem novamque complectens* [etc.] (Bologna, 1651), p. 82.
[4] Bernouilli, *Acta Eruditorum* (1688), pp. 98–103.
[5] Brice, *Phil. Trans.*, Vol. 56 (1766), pp. 224–29.

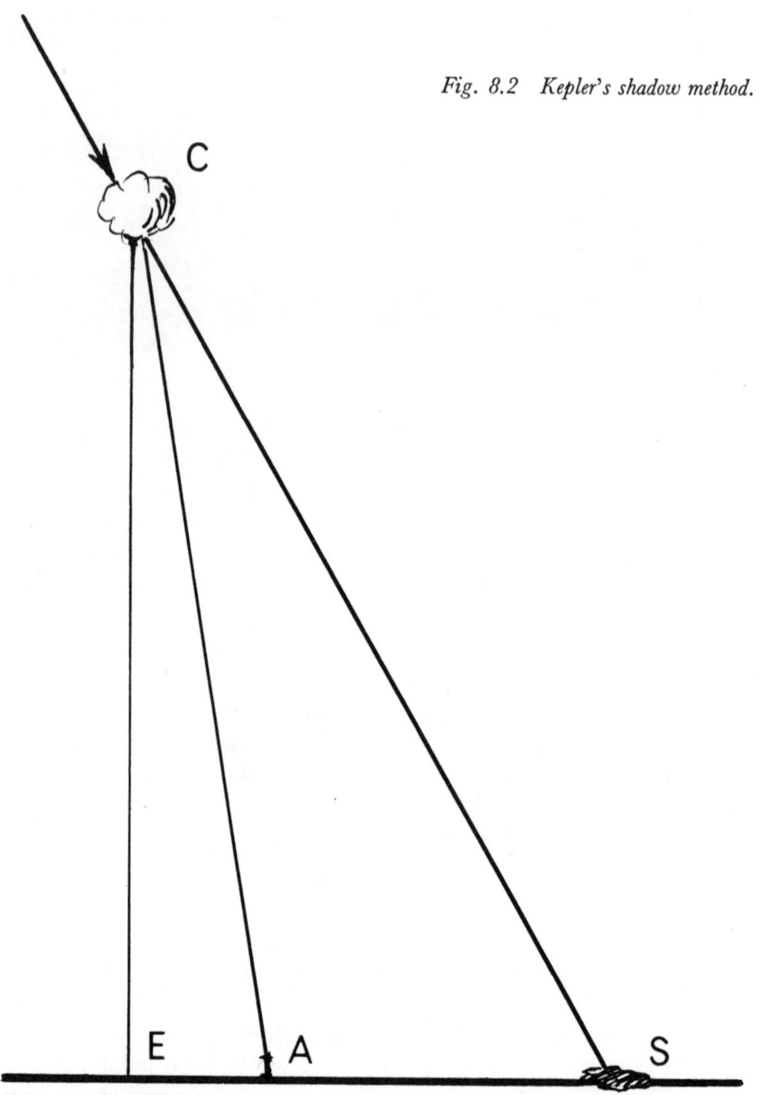

Fig. 8.2 Kepler's shadow method.

Lambert,[6] who also used a camera obscura to measure the angular height and azimuth of the cloud itself. Two measurements of the latter quantities at the moments when the shadow passed two features of the landscape that were recognizable on a map would yield all the data necessary for the computation of the altitude of the cloud and the speed and direction of its motion. A hill rising above a city, as Lambert pointed out, is an ideal place for such measurements.

None of these seventeenth- and eighteenth-century methods made use of any specially devised instruments. These began to appear in

[6] Lambert, *Nouv. Mém. Acad. Berlin* (1773), pp. 42–46.

the nineteenth century, and fell into several classes that will be dealt with separately.[7]

2. The use of a horizontal reflector. A method very attractive in principle was invented at about the same time by A. Bravais[8] at Lyons, E. Wartmann[9] at Geneva, and William Whewell[10] at Cambridge, although only Bravais seems to have used it extensively and developed it. In Figure 8.3, an observer at O has a surface of still water below

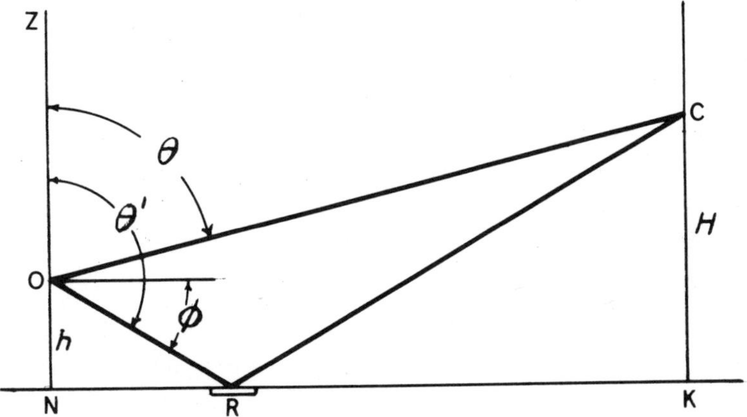

Fig. 8.3 The use of a horizontal reflector.

him at R, the vertical difference of level being $ON = h$. Some recognizable feature of a cloud at C is observed both directly and by reflection in the water at R. The zenith angles $ZOC = \theta$ and $ZOR = \theta'$ are measured, as is their common azimuth. It is easy to show that the height of the cloud CK ($= H$) is

$$H = \frac{h \sin (\theta' - \theta)}{-\sin (\theta' + \theta)}$$

and that the horizontal distance NK $(= X)$ is

$$X = \frac{2h \sin \theta \sin \theta'}{-\sin (\theta' + \theta)}$$

Bravais' mirror at R was a shallow tank of water darkened with ink, and h was 21.8 meters.

[7] Readers particularly interested in this subject should refer to Cleveland Abbe's *Treatise on Meteorological Apparatus and Methods* (Annual Report of the Chief Signal Officer for 1887, Washington, D.C., 1888, Appendix 46), pp. 312–36, where it is dealt with in great detail, with the necessary theory.
[8] Bravais, *Ann. Chim. Phys.*, Ser. 3, Vol. 24 (1848), pp. 497–501.
[9] Wartmann, *Bull. Soc. Vaudoise Sci. Nat.*, Vol. 1 (1842), pp. 21–25.
[10] Whewell, B.A.A.S., Southampton, 1846, *Sections*, pp. 15–16.

He developed a special instrument for the method. A plane mirror, part of which was unsilvered, was attached to a vertical circle like that of a transit instrument. After determining the angle of dip $\phi = \theta' - 90°$ of the image of the cloud in the water, he turned the plane through the small angle ω necessary to make the two images seen in it, the one of the cloud in the silvered part and the one of its reflection from the pool in the unsilvered part, coincide. This procedure made it easy to be sure that the same part of the cloud was being used. Then

$$H = \frac{h \sin (2\theta - 2\omega)}{\sin 2\omega}$$

In 1880 Francis Galton suggested a modification of this method, using two strictly horizontal surfaces, a pool of water below and a small mercury surface near the observer.[11] But such methods are so greatly restricted, by wind as well as by the small proportion of the sky that can be studied, that they have been little used except by Bravais.

3. The mirror nephoscope. Bravais, who started using these methods about 1842 at Lyons, made a further use of his reflecting pool, observing the time taken for the image of a detail of the cloud to describe a path of about one meter on it, and also noting the azimuth of the motion. As he already knew the height of the cloud, this gave him all the data needed to find its absolute speed and direction.

A horizontal mirror near the observer can be used to find the direction and relative speed of the motion of the cloud, but not the absolute speed, unless the height is otherwise determined. A special instrument of this sort was first constructed by G. Aimé in 1846.[12] He called it "l'anémomètre par réflexion"; instruments of this type are now called mirror nephoscopes.[13]

Aimé's instrument consisted of a mirror and a compass mounted on the same rotatable base. The mirror was ruled in squares. There were some vertical brass rods of adjustable length around the mirror, terminating in points. The eye was held so that the image of one of these points coincided with one of the corners of the squares, and the image of a cloud was watched to see if it moved along one of the lines. If it did not, the whole instrument was rotated until it did, the direction of the motion read from the compass, and the image timed across one or more squares. If the height of the pointed rod was measured, the distance that the cloud had moved in relation to its height could be found by simple proportion.

[11] Galton, B.A.A.S., Swansea, 1880, *Sections*, pp. 459–61.
[12] Aimé, *Ann. Chim. Phys.*, Ser. 3, Vol. 17 (1846), pp. 498–501.
[13] The word nephoscope seems to have been coined by C. Braun, *Z. österr. Ges. Meteorol.*, Vol. 2 (1867), pp. 337–52, who designed such an instrument.

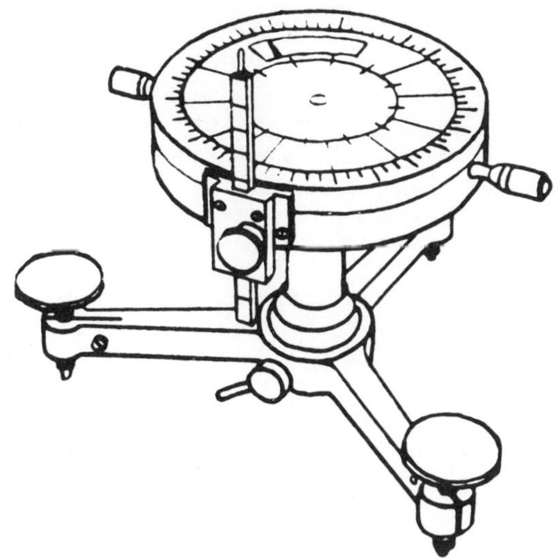

Fig. 8.4 Fineman's nephoscope.

This first instrument had all the essential features of later mirror nephoscopes such as those of Braun, Fineman,[14] and Marvin.[15] Improvements were generally in the matter of convenience, the design of Fineman (Fig. 8.4) being outstanding. This consists of a disk of black glass mounted on a tripod stand fitted with leveling screws. A compass needle is built in for purposes of orientation. A vertical pointer is arranged so that it can be rotated around the center of the disk, and set at various known distances above the reflecting surface. The disk is engraved with concentric circles at a known distance apart.

In use the pointer is placed in such a position that the image of a cloud appears in line with the pointer and the center of the disk. The image is then watched, being kept in line with the pointer, and the direction of motion of the cloud is read from the graduations on the edge of the disk. It need scarcely be said that if the instrument is set up with the 180° mark pointing north, the direction from which the cloud is moving will correspond to the graduation toward which its image moves.

The observation of cloud shadows as a complement to measurements with the nephoscope was recommended by Aimé.[16] C. Braun, in a second paper,[17] described a special fitting attached to his nephoscope for facilitating these measurements. But before this, Professor Temple Chevallier had demonstrated a very simple device for measuring the height of any detached cloud if the cloud and its shadow are

[14] C. G. Fineman, *Zeits. Instrum.*, Vol. 6 (1886), pp. 206–08.
[15] C. F. Marvin, *Monthly Weather Rev.*, Vol. 24 (1896), pp. 9–13.
[16] Aimé, *Ann. Chim. Phys.*, Vol. 17 (1846), p. 500.
[17] Braun, *Z. österr. Ges. Meteorol.*, Vol. 9 (1874), pp. 257–61.

both identifiable.[18] A hinged scale like a carpenter's rule, graduated from the center of the joint (Fig. 8.5), is laid on a horizontal table, and a slider is provided, carrying a vertical pointed rod of known length, say one unit. The part of the scale carrying the slider is directed toward a point perpendicularly below the cloud, and the other branch toward the shadow, which must be located on a map.

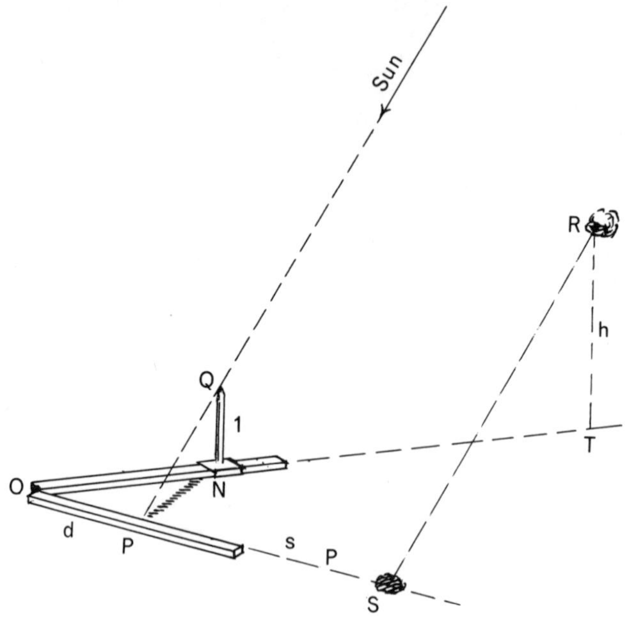

Fig. 8.5 Illustrating Chevallier's method.

The slider is then moved so that the shadow of the point Q just touches the inside top edge of the scale at P. Referring to Figure 8.5, since PQ and SR are necessarily parallel, the height h of the cloud can be found from the equation

$$h/1 = s/d$$

where $d = OP$ and $s = OS$. The latter can be read from the map, and there is no need to know the zenith angle of the sun. Similarly, if ON is read off, the distance OT can be found.

4. *The direct-vision nephoscope.* The function of the mirror nephoscope is to superimpose a coordinate system on a reflected image of the sky. The same result can be obtained by building some sort of coordinate system over the observer's head. This was apparently first done by a Dr. [J. T. ?] Bagot of Dublin, who exhibited, at the International Exhibition of 1862, a disk marked with a compass rose, and mounted

[18] Chevallier, B.A.A.S., Newcastle, 1863, *Sections*, p. 21.

on tall legs so that an observer could sit beneath it.[19] To be practical this would have had to be large. In 1897 L. Besson introduced his *herse néphoscopique*,[20] literally a nephoscopic harrow. In other countries, this apparently was suggestive of a rake (*Wolkenrechen*) or a comb (comb nephoscope). It consists of a mast on which is supported a rotatable horizontal bar, four meters above eye-level. This bar carries seven equidistant points or spikes 0.4 meters apart. By means of cords, the observer can rotate the bar until it is parallel to the motion of the clouds. An improvement was introduced in 1928 in Norway by replacing the "comb" by a rectangle divided into ten squares (Fig. 8.6).[21]

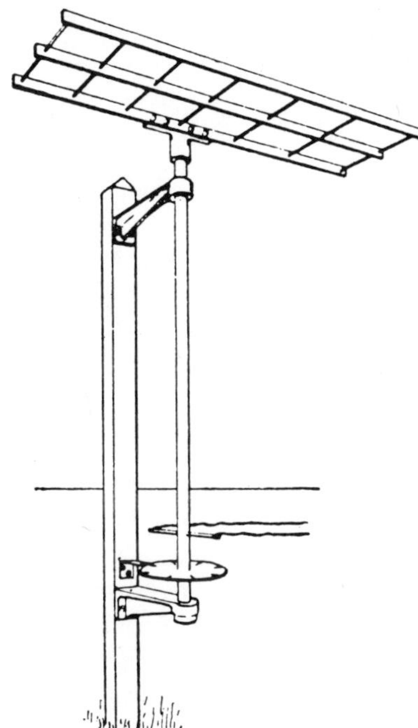

Fig. 8.6 Devik's nephoscope.

5. *Triangulation from two stations.* The introduction of the telephone in the 1870's made the determination of the position of a cloud by simultaneous observations from two stations a practical possibility, since the observers could discuss the appearance of the cloud and agree on the feature that was to be selected for observation. N. Ekholm and K. L. Hagström used this method with success during the Swedish expedition of 1882 to Spitzbergen.[22] A telescope is a hin-

[19] London, International Exhibition of 1862, *Reports by the Juries* [etc.] (London, 1863), pp. XIII–39.
[20] Besson, *Ann. Soc. Météorol. France*, Vol. 45 (1897), p. 172.
[21] Olaf Devik, *Zeits. Instrum.*, Vol. 48 (1928), pp. 224–25.
[22] N. Ekholm and K. L. Hagström, *Nova Acta. R. Soc. Sci. Upsal.*, Ser. 3, Vol. 12, no. XV (1885).

Fig. 8.7 Ekholm and Hagström's altazimuth instrument.

drance rather than a help in observing clouds, and Ekholm and Hagström used two special altazimuth instruments (Fig. 8.7) in which cross-wires were used in place of a telescope objective and a 3-mm hole instead of an eyepiece. The cross-wires were on a ring that could be rotated and had a scale of degrees.

Another nineteenth-century invention that was exploited for the purpose of triangulation is photography. As early as 1855 C. S. M. Pouillet proposed the use of two cameras of about 60-cm focal length, pointed toward the zenith and separated by 100, 300, or 600 meters for the observation of lower, middle, or high cloud, respectively.[23] Exposures would be made simultaneously. He noted that if this method was to be extended to other parts of the sky the apparatus would become more complicated, because the optical axes of the two cameras had to be parallel.

In 1871, Abbe informs us,[24] "apparatus was purchased for photographic work by the [United States] Signal Office," but no details are given. In the summer of 1876 Arnulf Malloch of Witham in England made simultaneous cloud photographs with two cameras, with appar-

[23] Pouillet, *Compt. Rend.*, Vol. 40 (1855), pp. 1157–64.
[24] Abbe, *Treatise*, p. 312.

ent success, but he gave no details in his short letter to *Nature*.[25] In 1879 W. de W. Abney, who has an important place in the history of photography, was asked to make recommendations about the matter, and reported to the Meteorological Council.[26] He called for two cameras mounted on altazimuth mountings (what we would now call phototheodolites) using the new dry-plate process rather than collodion emulsions. Measurements should be made on the negatives rather than on the prints. Abney suggested that a second later exposure at one station would give the speed and direction of motion, and in this connection Abbe[27] was quite prepared to make two exposures on the same plate "at an interval of a few seconds."

In 1881 Abney reported again.[28] He had had two cameras made specially by the Cambridge Scientific Instrument Company, with electrical shutter releases. It is noteworthy that he found it best to prepare his own dry plates. Observations were started at Kew, but in succeeding reports there seems to have been less and less enthusiasm for the project, although Sir George Stokes devised a quick optical method of projecting the cloud points obtained in this way from the plane of the photographic plates onto a horizontal plane.[29]

Meanwhile, John Harmer had suggested mounting two "small" cameras on a wooden truss about fifty feet long that could be revolved around a vertical axis, with accurate adjustments for vertical angle, and fiducial wires or points in front of the plates.[30] The resulting plates would be placed in a stereoscope, and calibration would be performed by photographing terrestrial objects at known distances. I do not know whether this project was ever carried out.

In Germany Zenker[31] used a method similar to that of Abney, but with the shutters synchronized electrically. Nothing very striking seems to have occurred in this field after that until in 1898 A. Sprung described the automatic zenith cameras that he had had built for the Potsdam observatory by R. Fuess.[32] When a switch was closed, either manually or by a clock, an electromagnet first removed a rain shield, and then the camera transferred a dry plate from the top of a pile to the focal plane, exposed it, and put it on the top of a second pile. The rain shield then returned to its original position. It seems that this apparatus was kept in operation at Potsdam for 25 years.[33]

[25] Malloch, *Nature*, Vol. 15 (1877), p. 313.
[26] *Rep. Meteorol. Council for 1879* (London, 1880), pp. 25–26.
[27] *Treatise*, p. 313.
[28] *Rep. Meteorol. Council for 1881* (London, 1882), pp. 33–34.
[29] Stokes, *Rep. Meteorol. Council for 1886* (London, 1887), pp. 22–23.
[30] Harmer, *Photographic News*, Vol. 25 (1881), pp. 126–27.
[31] W. Zenker, *Meteorol. Zeits.*, Vol. 1 (1884), pp. 4–11.
[32] Sprung, *Zeits. Instrum.*, Vol. 19 (1899), pp. 111–18; 129–37.
[33] R. Süring, in: E. Kleinschmidt, ed., *Handbuch der meteorologischen Instrumente* (Berlin, 1935), p. 597.

6. The cloud searchlight or "ceiling projector." None of the methods so far referred to is of any use in measuring the height of a uniform overcast,[34] and none can be used on a dark night. Methods using artificial light are clearly required for night use. In 1871 Paul La Cour in Denmark suggested that at night, the diffuse spot of light from a town can be used to estimate the height of an overcast if the observer is at a suitable distance.[35] He also stated that on winter days the height of an overcast could be measured if one were at some distance from the boundary between large areas of snow-covered ground and of forest or sea. In discussing the nocturnal method, however, he assumed that the illuminated spot on the clouds would be half-way between the observer and the source of light. This is not a safe assumption, for a cloud does not reflect light as a mirror does, but diffuses it in an extremely complicated way.[36] Plainly, a directed beam of light is necessary.

Cleveland Abbe claimed to have been the first to suggest the use of a searchlight beam for the measurement of the height of cloud at night.[37] This was in 1872. In 1882, O. Jesse of Berlin published such a proposal.[38] His first idea was to have a searchlight directed at a cloud, and two observers on a four-kilometer base with measuring instruments, or, more economically, the direction of the beam could be measured, and also that of the spot seen from some distance. By 1896 a vertical beam and a theodolite were being used in a routine manner at Blue Hill Observatory in Massachusetts.[39] This technique has persisted, except that a special altitude instrument or "clinometer" on a permanent stand became the usual measuring equipment at every airport for measuring the angular elevation of the spot of light.

In 1939 I suggested that a sufficiently powerful *modulated* light beam might make it possible to employ this principle in the daytime, and that a tuned amplifier could be used with a phototube to distinguish the reflected modulated light from the nearly steady background. I also made theoretical estimates of the signal to be expected.[40] Making use of the newly invented high-intensity mercury arc, Maurice Laufer and Laurence Foskett of the National Bureau of Standards developed a satisfactory apparatus at the request of the United States Weather Bureau.[41] With the most advanced electronic techniques then avail-

[34] In the United States the height of the base of cloud has been called the "ceiling" ever since the beginnings of commercial aviation.

[35] La Cour, *Dansk. Vid. Selsk. Oversigt* (1871), pp. 75–88 (in Danish; French summary on pp. 27–30).

[36] See Jean Bricard, *Physique des nuages* (Paris, 1953), Ch. VIII.

[37] Abbe, *Treatise*, p. 323.

[38] Jesse, *Z. österr. Ges. Meteorol.*, Vol. 17 (1882), pp. 181–86.

[39] H. Helm Clayton, *Meteorol. Zeits.*, Vol. 13 (1896), p. 140.

[40] Middleton, *J. Opt. Soc. Amer.*, Vol. 29 (1939), pp. 340–49.

[41] Laufer and Foskett, *J. Aeron. Sci.*, Vol. 8 (1941), pp. 183–87.

able, they designed a photoelectric telescope that scanned the modulated vertical beam of a searchlight automatically from the horizon to the zenith; the base of cloud was indicated by a sudden peak in the response. I shall not take the story any further, except to note that the advent of the laser has recently made this sort of measurement much easier.

7. The use of range finders. A range finder consists fundamentally of two telescopes separated by some fixed distance, with a means of turning the axis of one of them through a small measurable angle until the two axes intersect at the object whose range is to be obtained. There are two fundamentally different kinds of range finder: the coincidence range finder, in which two superimposed or adjacent images are made to coincide; and the stereoscopic range finder, in which by optical means an artificial mark is made to seem to advance or recede in the field of view until it appears to be at the same distance as the target. Not everyone can be trained to use the stereoscopic range finder successfully.

With a baseline of any manageable dimensions, the angles to be measured are very small, and although Krecke described an elementary instrument in 1849 and suggested its use with clouds,[42] it was not until the twentieth century that range finders, designed for military purposes, became sufficiently accurate and stable. In 1920 W. Hartmann reported on the use of a coincidence range finder on clouds,[43] and in 1931 Vilho Väisälä used two stereoscopic instruments,[44] which would probably be the better kind for the purpose, but range finders have seldom been employed for such measurements.

8. The use of pilot balloons. None of the techniques so far described permits the measurement of the upper winds in the absence of cloud, or at night. With the invention of the balloon in 1783[45] an urgent requirement for such information immediately developed, as is indeed shown by the term "pilot balloon," signifying a balloon sent up shortly before the ascent of a manned balloon, so that the aeronauts might have some idea of the direction and speed with which they were likely to travel. It is therefore very interesting to learn that on December 1, 1783, just before the ascent of the very first manned hydrogen balloon

[42] F. W. C. Krecke, *Description de l'observatoire météorologique et magnétique à Utrecht* (Utrecht, 1849).
[43] Hartmann, *Das Wetter*, Vol. 37 (1920), pp. 165–70.
[44] Väisälä, *Beitr. Phys. freien Atm.*, Vol. 17 (1931), pp. 220–26.
[45] See also p. 287 below.

from the Tuileries in Paris, a pilot balloon six feet in diameter was sent up.[46]

On January 7, 1785, Jeffries and Blanchard flew across the English Channel, and before they left they "raised a paper kite" and launched "a paper Montgolfier [a hot-air balloon], and a small gaz balloon," and observed that they all went in the desired direction.[47]

The first pilot-balloon observations not directly inspired by the necessities of aeronautics were made, as far as I know, in 1809. On October 25 of that year "a gentleman, named Wallis, sent up a small inflammable air [hydrogen] balloon from Clapton, in Hackney."[48] It was observed with a telescope, and found to have a very complicated path, indicating four different currents, one above the other. Mr. Wallis later sent up more than thirty ballons, and most of them indicated several currents of air.

I was surprised to find nothing whatever about pilot ballooning—except, of course, as a direct aid to aeronauts—between 1815 and 1872, and I should have been convinced that I lacked diligence except for the fact that Cleveland Abbe, in his immensely erudite *Treatise on Meteorological Apparatus and Methods* of 1888, refers to nothing earlier than 1872, and that reference is to a suggestion that he himself made to the Signal Corps in that year:

This method, which is not expensive, is worthy of being made a special feature of all first-class meteorological stations and national services. The flight of such a balloon during five or ten minutes would be an item more important for the daily weather map than the usual local record of wind, and, as recommended by me in 1872, should be added to the ordinary weather telegram. To this end the small Montgolfier may be used in dry-weather or at night time, and the small gas balloon in rainy or foggy weather and in the day time. Such balloons should be numbered consecutively and carry a corresponding numbered postal card, requesting the finder to inscribe the exact time and place of finding, and return it to the central office.[49]

We may suppose from Abbe's words that his suggestion had hung fire. He continues with a practical method of determining the position of the balloon in space at a given moment:

The balloon should carry a suspended light thread from 50 to 500 feet long, at the bottom of which hangs suspended a light object. The observer can at any time ascertain the linear distance and altitude of the balloon by observing the apparent angular altitude of the upper and lower end of the vertical line thus carried by the balloon . . .[50]

[46] C. H. Turnor, *Astra castra: experiments and adventures in the atmosphere* (London, 1865), p. 48.

[47] *A Narrative of the two Aerial Voyages of Doctor Jeffries with Mons. Blanchard; with Meteorological Observations and Remarks* [etc.] (London, 1786), p. 41. I am indebted to Mr. F. R. Maddison for the reference.

[48] Thomas Forster, *Researches about Atmospheric Phaenomena* (2nd ed., London, 1815), p. 263.

[49] Abbe, *Treatise*, pp. 310–11.

[50] *Ibid.*, p. 311.

and he gives the necessary theory. We shall see later how this idea was developed by others.

In 1886, Paul Schreiber, who was notoriously fond of analytical exercises, gave a theory for calculating upper winds from observations of a balloon at two stations,[51] and recommended that rubber balloons should be used, instead of the paper ones generally employed at the time. Not much attention seems to have been paid to this article, and in 1893 V. Kremser[52] proposed the determination of the paths of pilot balloons by means of one theodolite fitted with a micrometer eyepiece so that the subtense of the balloon could be measured. This was presumably to be a paper balloon, the diameter of which would remain constant, and a capacity of about one cubic meter was envisaged. The closed rubber balloon first came into meteorological use as a sounding balloon a decade later.[53]

About the time that Kremser was writing his paper, Gustave Hermite and his friend the aeronaut Besançon were sending up, from their Paris apartment in the Boulevard de Sébastopol, balloons furnished with questionnaire cards, about half of which were returned, from places as far as 150 km away.[54] These balloons were not followed optically, but on August 5, 1896, Hermite and Besançon brought their *dromographe* into use.[55] This was a recording theodolite in which two pens, using red ink for altitude and black for azimuth, were moved axially along a vertical drum by gearing. The drum was rotated by clockwork. I reproduce Figure 8.8 in case it may make this any clearer.

Meanwhile, the technique of the sounding balloon was being developed at Trappes by Leon Teisserenc de Bort,[56] and at the Paris Exhibition in 1900 he showed a number of sounding-balloon trajectories, some of them reaching a height of 13 km., determined with two theodolites several years before, but apparently not published.[57] These had inspired A. de Quervain, who at Strasbourg on July 2, 1903 followed a balloon to an elevation of 15.34 km., using a theodolite of his own make. Two extra observers read altitude and azimuth. In a footnote De Quervain tells us that a recording theodolite, not specially for work with balloons, was made at that time by T. Rosenberg of Berlin, and cost 450 marks. It apparently recorded on a disk that could be measured later with an auxiliary apparatus.

In the following year De Quervain described his own special theodolite for balloon observations,[58] which is the ancestor of many later

[51] Schreiber, *Meteorol. Zeits.*, Vol. 3 (1886), pp. 341–45.
[52] Kremser, *Meteorol. Zeits.*, Vol. 10 (1893), pp. 198–99.
[53] See below, p. 303.
[54] W. de Fonvielle, *Les ballons-sondes de MM. Hermite et Besançon* (Paris, 1898), pp. 10–11.
[55] *Ibid.*, pp. 80–81.
[56] See below, p. 301.
[57] A. de Quervain, *Beitr. Phys. freien Atm.*, Vol. 1 (1904–5), pp. 47–54.
[58] De Quervain, *Zeits. Instrum.*, Vol. 25 (1905), pp. 135–37.

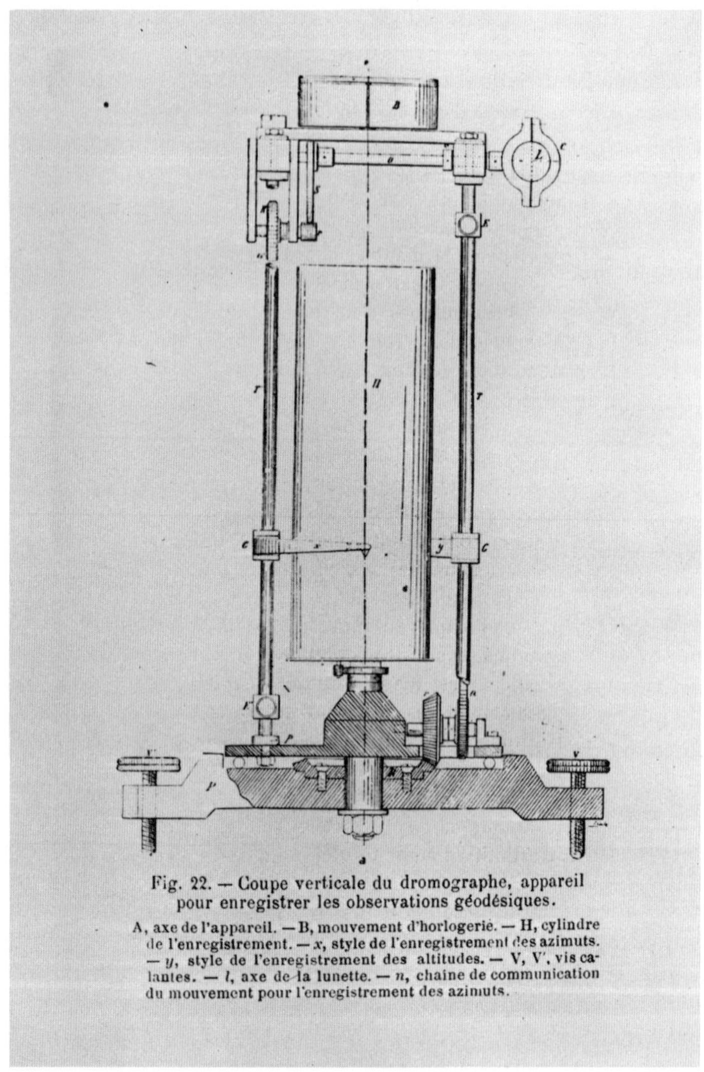

Fig. 22. — Coupe verticale du dromographe, appareil pour enregistrer les observations géodésiques.

A, axe de l'appareil. — B, mouvement d'horlogerie. — H, cylindre de l'enregistrement. — *x*, style de l'enregistrement des azimuts. — *y*, style de l'enregistrement des altitudes. — V, V', vis calantes. — *l*, axe de la lunette. — *n*, chaine de communication du mouvement pour l'enregistrement des azimuts.

Fig. 8.8 The "dromograph" used by Hermite and Besançon.

designs. Its essential feature, as shown in Figure 8.9, is that the optical axis is bent by means of a prism, so that the observer always looks horizontally. He can also read the vertical circle. His assistant keeps time and reads the horizontal circle, which is the reason that the vernier is at the back, almost hidden by the pillar.

These theodolites, made by Bosch, were an instant success, and were imitated elsewhere with small variations. In the course of time more elegant and convenient pilot-balloon theodolites appeared on the market, but for a description of these the reader must be referred to

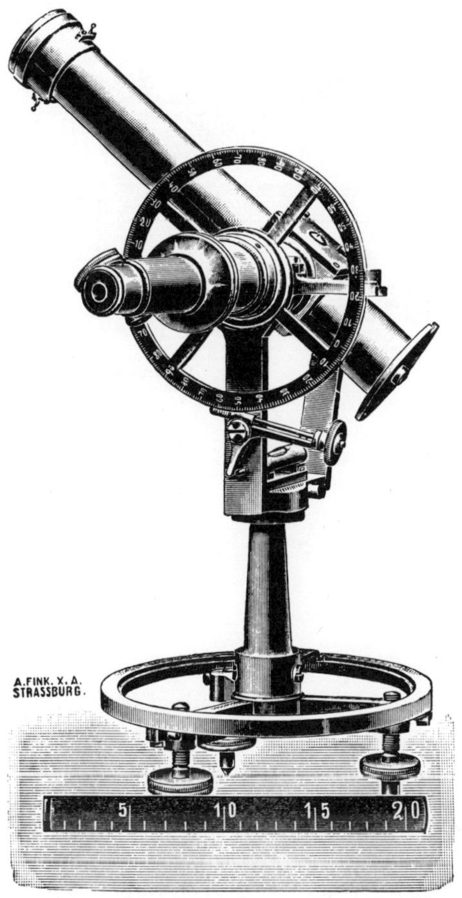

Fig. 8.9 De Quervain's balloon theodolite, 1905.

the instruction manuals issued by the various meteorological services. Keeping within the scope of this book, I shall here notice only some early recording theodolites.

Such instruments are of two classes. In one, represented by the "dromograph," azimuth and altitude are recorded separately as a function of time. In the other, means are provided for making a horizontal projection of the path of the balloon automatically. In the first class the outstanding instrument is the one invented by J. S. Dines about 1910.[59] This is shown in Figure 8.10, in which it will be seen that the recording pens move parallel to the axis of a horizontal drum. The pens are attached to chains moved very simply by being coiled around the edges of the divided circles of the theodolite. The

[59] J. S. Dines, *Rep. Memor. Adv. Comm. Aero., London,* No. 28, 1910.

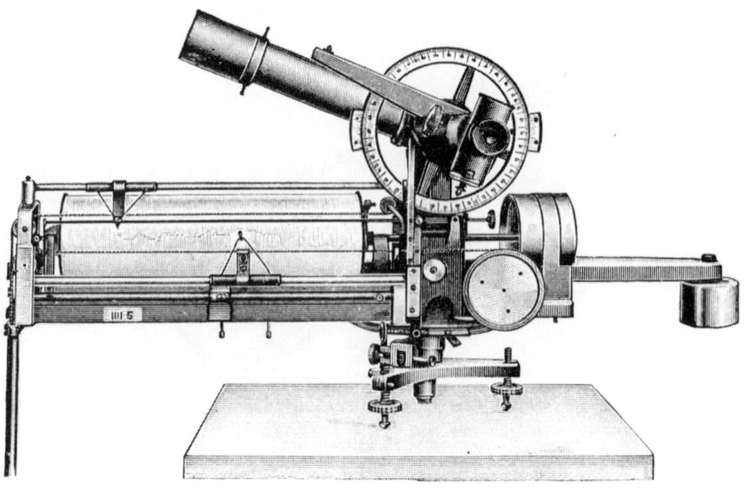

Fig. 8.10 The recording theodolite of J. S. Dines. 1910 (courtesy of Macmillan & Co.).

chains are kept taut by hanging weights which, with the drum, move around the vertical axis as the azimuth of the balloon changes. The original prototype instrument is in the collection of the Meteorological Office, Bracknell, England.[60] A theodolite of this class, differently arranged, was described two years later by J. Pircher.[61]

A theodolite of the second class was developed in the Netherlands Weather Service, beginning about 1910, by C. Schoute.[62] Figure 8.11 shows the first model, in which the recording device was simply built onto an ordinary balloon theodolite of the period. The principle is demonstrated in Figure 8.12. Suppose the balloon to have been launched from near the theodolite at *P*, and after a given time to have reached *B*, at a height *H* above the plane *PF* of the theodolite axis. Let *AA* be the disk on which the path of the balloon is to be recorded, and let some means be provided of lowering this through a distance *h*, some known fraction of *H*. Arrange to make a mark at *S*, where *PS* is a prolongation of *BP*. Then, because the triangles *PBF* and PSP_1 are similar, the point *S* will be a horizontal projection of *B* on the scale h/H.

In the original Schoute instrument, the plotting disk was geared to the horizontal circle through a long pinion, so that it would rotate in synchronism but would be free to move vertically. The vertical motion was made by hand in equal steps. The stylus was at the end of a spring-loaded sliding rod fastened to the vertical circle and inter-

60 Inv. no. 1502B.

61 Pircher, *Int. Komm. wiss. Luftschiffahrt*, *7te Sitzung* (Vienna, 1912), pp. 170–71.

62 Schoute, *Kon. Ned. meteorol. Inst., Meded. Verhandl.*, No. 26 (Utrecht, 1921).

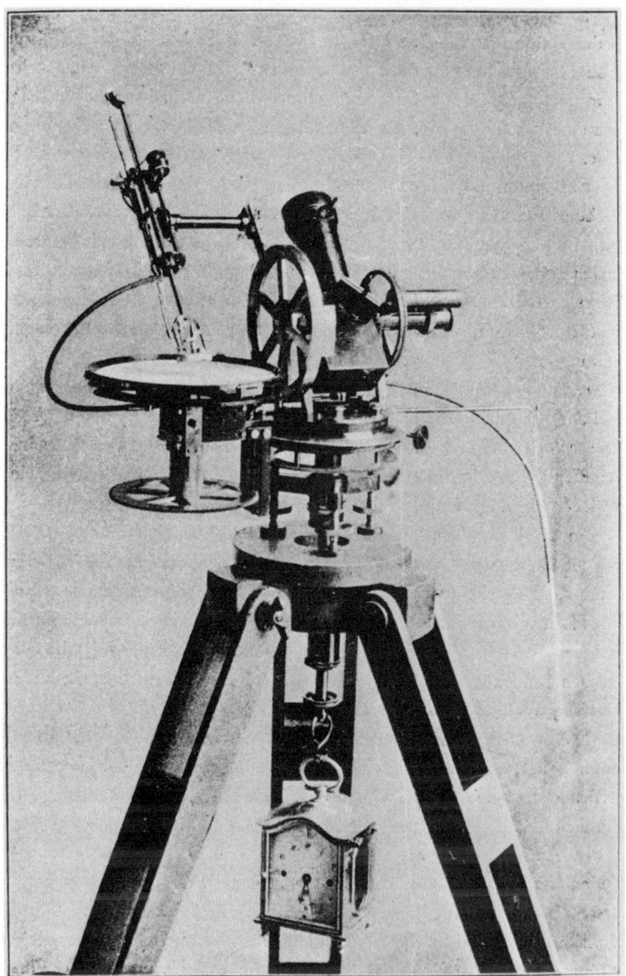

Fig. 8.11 The recording theodolite of C. Schoute, first model.

secting the prolongation of the horizontal axis of the telescope. This rod was moved down by a cable-operated actuator and then the stylus, when it had touched the paper, was pressed vertically into it.

I have illustrated the original model because it shows the principle of the device so well, and also because it was the first one. Schoute realized at once that it was not durable enough, and much more sophisticated models followed, with the co-operation of the firm of Goerz.

If the height of the balloon is to be represented by equal steps corresponding to equal intervals of time, it follows that the rate of ascent of the balloon must be constant and must be known. There is a large literature about the rate of ascent of small rubber balloons and its relation to the free lift, that is to say, the excess of the buoyancy of the

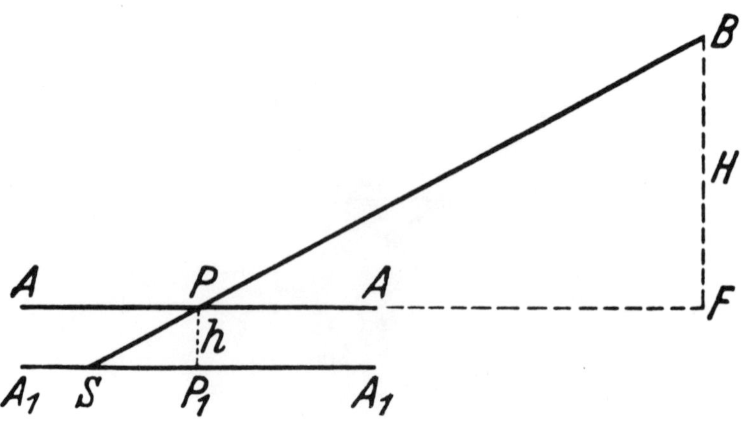

Fig. 8.12 Principle of the Schoute theodolite.

balloon over its weight plus that of any attachments.[63] It did turn out that the rate of ascent varies little with height. It is inversely proportional to the sixth root of the atmospheric pressure, as was predicted on theoretical grounds by H. Hergesell in 1903.[64] The rate of ascent V turns out to be related to the free lift L and the weight W of the balloon (including the gas and any attachments) by the formula

$$V = aL^{\frac{1}{2}}/(L + W)^{\frac{1}{3}}$$

The measurement of the constant a gave a great deal of trouble.

There are two possibilities: to release balloons in very large buildings,[65] and time their ascent through a known distance, or to make large numbers of two-theodolite ascents in the open.[66] It was found that in the open the balloons went up a good deal faster, and Hergesell ascribed this to ascending currents, but in 1917 R. Wenger, noticing that Hesselberg and Birkeland, working in a large stone church, had found lower values of a than Hergesell, who worked in a building with a glass cupola, thought that turbulence must increase the rate of ascent,[67] and adduced experiments on the resistance of spheres. This view prevailed.

Because routine two-theodolite observations are expensive, the "tail method" suggested by Abbe in 1887[68] was given much attention. In 1908 C. H. Ley, in Ireland, fitted a large theodolite with a forty-five-

[63] This is summarized by J. Reger in: E. Kleinschmidt, ed., *Handbuch der meteorologischen Instrumente* (Berlin, 1935), pp. 446–52.
[64] Hergesell, *Illustrierte äeronaut. Mitt.*, Vol. 7 (1903), pp. 163–68.
[65] H. Hergesell, *Comm. int. pour l'aérostation scientifique, 6e réunion à Monaco, 1909* (Strassburg, 1910), Annexe 10, pp. 86–103; T. Hesselberg and B. J. Birkeland, *Beitr. Phys. freien Atm.*, Vol. 4 (1912), pp. 196–216.
[66] H. Hergesell, *ibid.*
[67] Wenger, *Ann. Hydr. marit. Meteorol.*, Vol. 45 (1917), pp. 121–37.
[68] See p. 276 above.

power eyepiece and two cross-hairs slightly converging, with a number of vertical hairs at a known spacing.[69] This was used to estimate the apparent diameter of the balloon itself. In the following year he introduced his "subtense method,"[70] hanging a balloon inflated with air thirty-two feet below the one filled with hydrogen, and using a micrometer eyepiece. Hergesell[71] filled the second balloon with a mixture of gas and air to give a slight downward force, and found the swinging of the "tail" to be greatly reduced. Two years later he reported on the use of various paper shapes instead of the second balloon.[72] It is interesting that he thought he could measure vertical currents in this way, by noting the changes in upward speed.

As pilot-balloon observations became a routine matter, special devices were invented for plotting the horizontal projection of the path of the balloon. The first two of these appeared in the literature in 1912.[73, 74] Each depends on the provision of extensive tables of the quantity $V.t.cot\ \alpha$, where V is the assumed rate of ascent, α the angle of elevation, and t the number of minutes since the start of the flight. To each value of α and t will correspond a radial distance, expressed in terms of some adopted scale.

The plotting board can be thought of as a map, with the theodolite in the center. That of Jonas (Fig. 8.13) consists of a circular disk carrying a coordinate network, and around its edge a reversed compass scale. The disk is mounted on a baseboard and is rotatable. Tracing paper is stretched over it, leaving a little of the circumference exposed so that the disk can be turned, and marks are made on the tracing paper to represent north and south. Each point is plotted by turning the board until the north point coincides with the azimuth reading, and the appropriate distance is plotted over the resulting position of the 0°-line. When two or more points have been plotted, wind directions are found by turning the board until the 0° to 180° diameter is parallel to the line joining two adjacent points, and reading the graduation under the *south* mark, because the wind direction is that *from which* the wind is blowing. The wind speed corresponds to the distance between the two points.

Hesselberg's board was a large compass rose. Dividers were used for plotting the points on tracing paper stretched over it, and a scale and a protractor must have been used for reading off the results. Most subsequent plotting boards have been variations on that of Jonas, sometimes rather well disguised.

[69] Ley, *Quart. J. Roy. Meteorol. Soc.*, Vol. 34 (1908), pp. 27–45.
[70] *Ibid.*, Vol. 35 (1909), pp. 15–29.
[71] Hergesell, *Int. Komm. wiss. Luftschiffahrt, 7te Sitzung, Wien, 1912*, Anhang 25, pp. 140–42.
[72] *Beitr. Phys. freien Atm.*, Vol. 6 (1914), pp. 187–224.
[73] T. Hesselberg, *Beitr. Phys. freien Atm.*, Vol. 4 (1912), pp. 217–21.
[74] G. Jonas, *ibid.*, pp. 1–12.

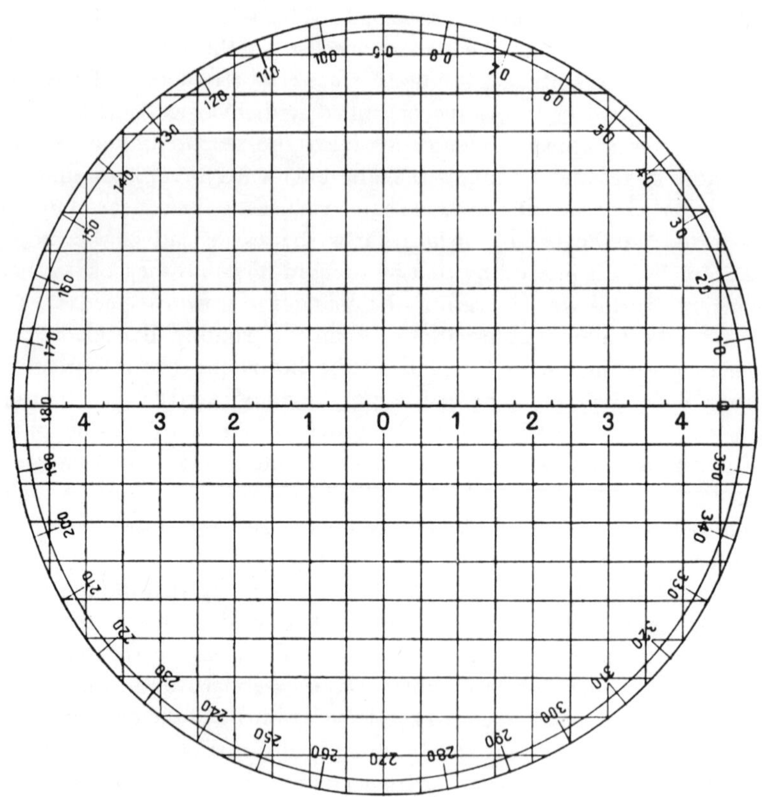

Fig. 8.13 The plotting board of Jonas.

9. *Tracking of balloons by radio.* An obvious limitation of pilot balloons is the impossibility of seeing them through clouds. I shall end this chapter by an account of the early history of the use of radio direction-finding to track balloons, confining myself to the methods that were developed, although not all described, before the war of 1939–45.

This story began in 1923 and 1924, when at the United States Signal Corps laboratory at McCook field, a balloon carrying "a simple buzzer transmitter weighing just under a pound" was tracked for 20 minutes, or to a height of 4,000 yards.[75] In 1928 a vacuum-tube transmitter using one tube, a 4.5-volt battery, and a buzzer transformer was employed, in the center of a dipole 80 feet long, and the signals were received by a rotatable loop antenna and a receiver that was very advanced for the time. Comparisons between radio and optical bearings showed a general agreement to within one degree of

[75] W. R. Blair and H. M. Lewis, *Proc. Inst. Radio Engrs.*, New York, Vol. 19 (1931), pp. 1531–60.

arc, with occasional larger discrepancies ascribed to the swinging of the antenna attached to the balloon. Angles of elevation were not measured, as indeed they could not have been with the relatively low frequency employed, about 2.4 Mc/sec. Stimulated by the paper of Blair and Lewis, Joseph Kolzer and Kurt F. Möller made experiments in Germany, partly reported in 1933.[76] Their transmitter, weighing 740 gm., emitted continuous waves at 1.50 Mc/sec.

The first operational system seems to have been that of the Office National Météorologique in France; a general description was published in 1935.[77] In this the balloon carried a transmitter sending out waves at about 2.5 Mc/sec, modulated "at a musical frequency." Two or three radio-goniometers were installed on the ground, with a baseline or baselines of several tens of kilometers. The superheterodyne receiver at each of the stations recorded a point at the minimum of the signal from the loop antenna; this was done by an ingenious electromechanical arrangement for bisecting the angle between equal small signals. Still, of course, only the bearing could be measured, which meant that the rate of ascent had to be known, or else the altitude had to be deduced from pressure and temperature signals from the balloon, which in that case carried a radiosonde.

The techniques of producing and dealing with shorter and shorter radio waves developed rapidly in the 1930's. Just before World War II, H. A. Thomas of the National Physical Laboratory of Great Britain patented[78] a system of direction-finding in which the antenna array consists of two vertical dipoles spaced more than one wave length apart, connected by transmission lines in such a way that the whole mechanism can be rotated about a vertical axis. The signals from the two dipoles can be added or subtracted, and this procedure gives six minima from which both azimuth and elevation can be deduced. In 1939 this antenna array was built into a direction-finding system for the Meteorological Office,[79] but the possibility of measuring elevation was not used, as the information provided by a radiosonde was considered essential in any event. At each installation three direction-finders were set up in a triangle with sides from of 30 to 75 km. The transmitted frequency was about 28 Mc/sec.

It was found that most of the errors in bearing were due to obstacles on the ground very near the antennas, and a "ground calibration" had to be made. Nevertheless, the system was used with great success for wind-finding out to long ranges, an important operation in wartime; but it is now quite obsolete, having been superseded by more sophisticated methods using much higher radio frequencies.

[76] Kolzer and Möller, *Meteorol. Zeits.*, Vol. 50 (1933), pp. 297–300.
[77] Corriez and Perlat, *La Météorologie* (1935), pp. 368–84.
[78] British Patent 504,293, April 24, 1939.
[79] R. L. Smith-Rose and H. G. Hopkins, *Proc. Phys. Soc., London*, Vol. 58 (1946), pp. 184–200. This paper was written in the spring of 1939. See also D. N. Harrison, *Meteorol. Mag.*, Vol. 76 (1947), pp. 217–25.

IX

Upper-air soundings without telemetry

1. Introduction. Any student of the history of meteorology in the seventeenth, eighteenth, and indeed the nineteenth century must be impressed by the tremendous difficulty of the subject in the absence of observations made more than a few meters above the surface of the ground. Whatever observations could be made on mountains were of course eagerly studied, and it is no accident that the two most active meteorological theorists of the eighteenth century, J. A. Deluc and H. B. de Saussure, were natives of Geneva and enthusiastic mountaineers. The invention of the balloon in 1783 provided at least the possibility of making observations aloft, and the realization of this possibility, first with manned balloons, later with captive balloons and kites, and finally with free, unmanned balloons, depended upon the development of the instruments that form the subject of this chapter. It seems desirable to treat first the instruments carried on flights of manned balloons, then those devised for kites and captive balloons, next the remarkably rapid development, at the end of the nineteenth century and the beginning of the twentieth, of the technique of the free, unmanned balloon or *ballon-sonde*, and finally the meteorological instruments for use on aircraft.

2. Instruments for manned balloons. The year 1783 marks the first manned free flight of the hot-air balloon of the brothers Montgolfier, and also of a manned balloon filled with hydrogen. On December 1 of that year the physicist J. A. C. Charles, accompanied by one Mr. Robert, ascended with a hydrogen balloon from the garden of the Tuileries in Paris,[1] and it is interesting to note that he took with him, and observed, a barometer and a thermometer. Only a fortnight later Joseph Priestley, having heard about the new balloons from Sir Joseph Banks,[2]

[1] For an entertaining, though rather eccentric, account of early ballooning, see C. H. Turnor, *Astra castra: experiments and adventures in the atmosphere* (London, 1865).

[2] This letter is not in Warren R. Dawson, ed., *The Banks letters. A calendar of the manuscript correspondence* [etc.] (London 1958).

wrote a reply[3] containing the sentence: "I thank you also for your account of the air balloons which though at present they only amuse the idle, may in time answer some important purposes in philosophy, enabling us to explore the upper regions of the atmosphere."

Apart from the many barnstorming balloonists who quickly learned how to "amuse the idle" for whatever money they could collect, aeronauts generally took up a barometer and often a thermometer, if only in order to estimate the height that they had attained. In 1786 Dr. John Jeffries went up from London with the professional aeronaut Jean Pierre Blanchard, armed with a thermometer, a barometer, an electrometer, a "hydrometer," which turned out to be a hygrometer of some kind, a watch, and a mariner's compass.[4] The meteorological observations that he made seem to have been fairly frequent but entirely unsystematic, and their historical importance is that they show clearly that the earliest balloonists had a sense of the scientific possibilities of the new invention.

Apart from a rather badly conducted ascent in January, 1804, at St. Petersburg, the first balloon ascent made entirely for scientific purposes was probably that of Gay-Lussac and Biot, on 9 fructidor An 12 (August 27, 1804).[5] Their main purpose was to observe the magnetic field of the earth, but they also carried barometers, thermometers, electrometers, and hygrometers. Naturally, these were not specialized instruments, but it is worth noting that the two men realized the importance of shielding the thermometers from the sun. Three weeks later Gay-Lussac risked his life by ascending alone to an estimated height of 6,636 meters, also carrying evacuated flasks for obtaining air samples.[6] His observations of the temperature gradient were accepted as being representative for many years.

On July 27, 1850, the aeronauts Barral and Bixio made an ascent with a set of instruments specially made by Fastré, under the direction of Regnault.[7] A certain lack of trust is shown by the fact that Regnault did not give the balloonists the calibration tables for the instruments, which had arbitrary scales. Much thought was given to the thermometers. Following a scheme suggested by J. B. J. Fourier,[8] three thermometers with cylindrical bulbs were exposed in sunshine in front of a polished silver plate, one with a plain bulb, one with its bulb blackened, and one with a silvered bulb. There was also a

[3] Priestley to Banks, December 14, 1783, quoted by Sidney M. Edelstein, *Chymia*, Vol. 1 (1948), p. 123.

[4] *A narrative of the two aerial voyages of Doctor Jeffries with Mons. Blanchard; with meteorological observations and remarks* [etc.]. (London, 1786), p. 13. The barometer is now in the U.S. National Air Museum, inventory no. 1963–27.

[5] L. J. Gay-Lussac and J. B. Biot, *Obs. sur la Phys.*, Vol. 59 (1804), pp. 314–20.

[6] Gay-Lussac, *Obs. sur la Phys.*, Vol. 59 (1804), pp. 454–61.

[7] F. Arago and V. Regnault, *Compt. Rend.*, Vol. 31 (1850), pp. 122–31.

[8] Fourier, *Ann. Chim.*, Vol. 6 (1817), pp. 259–303. See my *A History of the Thermometer*, p. 213, for more details.

thermometer with its cylindrical bulb placed on the axis of several concentric polished tin tubes.

The next special instrumentation for a balloon flight was made by Patrick Adie for John Welsh's four balloon ascents in 1862; it included a siphon barometer calibrated so that it could be read at one end only.[9] There were dry-bulb and wet-bulb thermometers, ventilated by a "concertina" bellows gradually opened by a weight. Air was first drawn over the dry bulb and then down past the wet one, at a speed of about four meters per second. A polished conical shade was provided around each bulb, and there was also a second pair of thermometers, similar to the first but without the aspirator. It is evident from the results that the aspirator was essential. We may suppose that the conical shades were, too, in view of the results obtained in 1862–66 by James Glaisher who, with the assistance of Coxwell, made twenty-eight ascents in various parts of England under the auspices of the British Association for the Advancement of Science.[10] If the illustration of his instrumental armament (Fig. 9.1) is accurate, it would seem that the thermometers had no shield. Great doubt was later cast[11] on Glaisher's results for the vertical gradient of temperature and humidity, the determination of which had been the main object

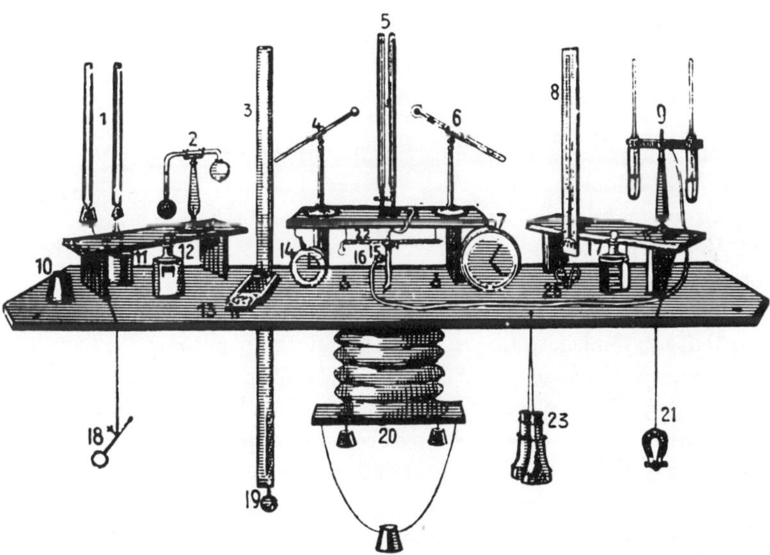

Fig. 9.1 Glaisher's balloon instruments.

[9] Welsh, *Phil. Trans.*, Vol. 143 (1853), pp. 311–46.
[10] Glaisher, B.A.A.S., Cambridge, 1862, *Report*, pp. 376–503, and further papers in subsequent *Reports*, 1863–1866.
[11] R. Assmann, *Meteorol. Zeits.*, Vol. 12 (1895), pp. 336–39; also *Wissenschaftliche Luftfahrten* (Braunschweig), Vol. 1 (1899), pp. 47 ff. Assmann said that it was "with great regret" that he made this criticism. Glaisher was then in his ninetieth year.

of the ascents. There had been other purposes: to compare Daniell's hygrometer, Regnault's hygrometer, and the psychrometer; to compare the aneroid barometer with the mercury barometer; and to make other electrical, magnetic, acoustic, and chemical observations, not to mention those on clouds and weather. Glaisher compared an aneroid with a siphon barometer at pressures down to seven inches of mercury and considered the aneroid to read "correctly," but he did not seem concerned about differences of a 0.1 inch or so.[12] Mercury barometers for manned balloons were generally of the siphon type, but toward the end of the nineteenth century we find a fixed-cistern barometer with elaborate baffles in the cistern, as shown in Figure 9.2.[13]

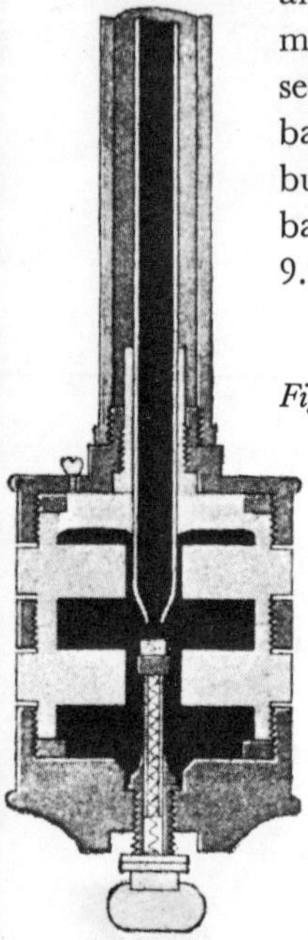

Fig. 9.2 Assmann's design for a balloon barometer cistern.

About thirty years after Glaisher's ascents there was an upsurge of interest in ballooning in Germany, where a society was formed with the patronage, financial as well as honorific, of the Emperor. In 1893 and 1894, forty-seven ascents were carried out with scientific help from Richard Assmann, who summarized the results in 1895.[14] Finding the lapse-rate of temperature to be much greater than that reported by Glaisher, he ascribed the difference to the effects of radiation. As the inventor of the well-known aspirated psychrometer, he commanded attention; the lengths to which he was willing to go are shown by Figure 9.3, a photograph of an aspirated meteorograph for manned balloons.[15] This elaborate instrument, first used on December 15, 1893, was hung on a cord below the basket of the balloon, the fan being driven by the weight of the meteorograph (8.5 kg) as it slowly sank along the cord. There were separate drums for pressure, temperature, and humidity. The barometer had four aneroid chambers; there was a hair hygrograph; and the thermometer was a Bourdon tube filled with spirit.

The manned balloon was too expensive and sporadic a means of

[12] B.A.A.S., Cambridge, 1862, *Report*, pp. 481–82.
[13] Assmann, *Wiss. Luftfahrten* (Braunschweig), Vol. 1 (1899), p. 166.
[14] Assmann, *Meteorol. Zeits.*, Vol. 12 (1895), pp. 334–44.
[15] *Wiss. Luftfahrten* (Braunschweig), Vol. 2 (1900), p. 630.

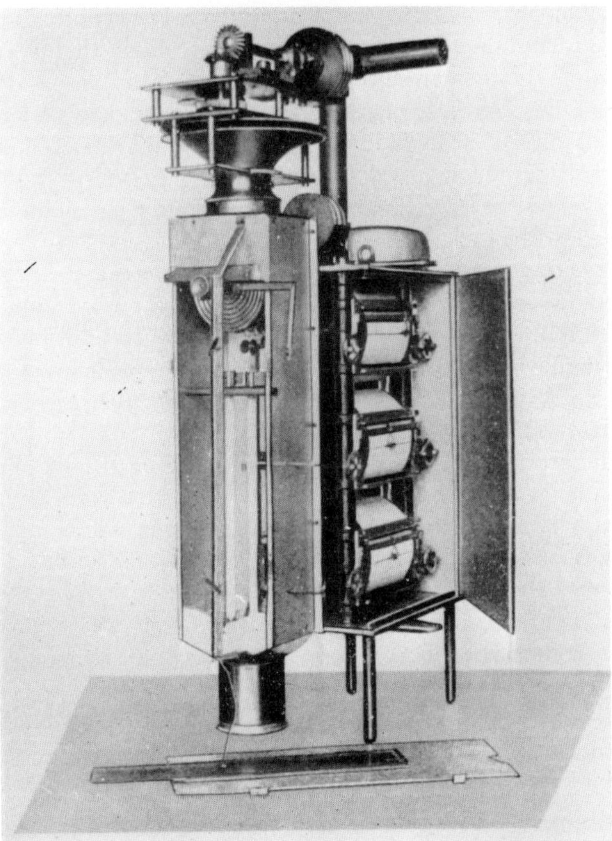

Fig. 9.3 Assmann's aspirated meteorograph for manned balloons.

collecting data from the upper air, and various events contributed to the demise of this kind of scientific ballooning. In the 1890's there was a rapid development of the use of kites and captive balloons. The sounding balloon was invented in the same decade, and finally the mastery of heavier-than-air flight after 1903 shifted interest quite away from the manned balloon as anything but a piece of sporting equipment.

3. Instruments for kites and captive balloons. The first attempt to use kites for obtaining information about the temperature in the free air seems to have been made in 1749, three years before Benjamin Franklin's celebrated electrical experiment with the kite, by Alexander Wilson, professor of astronomy at Glasgow.[16] With the help of

[16] Patrick Wilson, *Trans. Roy. Soc. Edin.*, Vol. 10 (1825), pp. 279–97. This biographical memoir of his father had been read in 1789, but withdrawn for alterations, and was discovered among Patrick Wilson's papers after his death.

Thomas Melvill, he made "half a dozen large paper kites" and flew them in tandem on a long line. Then: "To obtain the information they wanted, they contrived that thermometers, properly secured, and having bushy tossels [*sic*] of paper tied to them, should be let fall at stated periods from some of the higher kites; which was accomplished by the gradual singeing of a match-line."[17] We are not told of the results. It must be remembered that no maximum or minimum thermometer was then available.

Before Wilson's paper was published, the Reverend George Fisher went with Captain W. E. Parry's expedition to the Arctic, and in the winter of 1822–23, investigating atmospheric refraction, he tried to determine the vertical gradient of temperature "by means of a paper kite, to which was attached an excellent register thermometer, in a horizontal position. Its height above the level of the frozen sea, upon which the experiment was made, was determined by two observers in the same vertical plane, taking its altitude at the same time above the distant horizon; and from thence its height was computed."[18] The "register thermometer" was most probably the spirit minimum thermometer invented by Rutherford.[19] Quite apart from the effect of radiation, the horizontality of the instrument must have been doubtful, and the prevalence of deep inversions in the Arctic winter must have complicated the matter still further.

Plainly, recording instruments were needed. These tend to be heavy, so that improved, or at any rate larger, kites were required. Larger kites had a stronger pull, and wire rather than cord was indicated, an improvement usually credited to E. Douglas Archibald in the eighties.[20] In 1837 James Swaim, investigating atmospheric electricity in the United States, had flown kites on "wire No. 30," using a rather elaborate reel with a glass axle,[21] and about 1842 some workmen at a wire mill in Bradford, England, were using iron wire to fly kites for amusement.[22]

The kites used were of the ordinary kind, almost flat and four- or six-sided, until in 1893 the Australian Lawrence Hargrave, in the course of a rather strange series of papers on model flying machines, announced the box or Hargrave kite.[23] The advantages of this construction, which can be made to fly more steadily, and at a higher angle, were immediately appreciated, and it may fairly be said to have revolutionized meteorological kite-flying,[24] to which a great deal of

[17] *Ibid.*, p. 286.
[18] G. Fisher, *Quarterly J. Sci. & Arts*, Vol. 21 (1826), p. 348.
[19] See p. 74.
[20] Archibald, *Meteorol. Zeits.*, Vol. 2 (1885), pp. 47–52.
[21] Swaim, *Amer. J. Sci.*, Vol. 32 (1837), pp. 304–07.
[22] Epistolary evidence was presented by S. P. Fergusson, *U.S. Monthly Weather Rev.*, Vol. 25 (1897), p. 135.
[23] Hargrave, *J. Roy. Soc. New South Wales*, Vol. 27 (1893), pp. 75–81. *Re* Hargrave see A. Inglis, *Rec. Australian Acad. Sci.*, Vol. 1 (1966), pp. 18–41.
[24] The development of kites did not stop there, as may be seen from the extrava-

attention was being paid at the time, especially in the United States; even the theory of flying kites was investigated.[25] Both at Blue Hill and at Washington, and later at Lindenberg and Hamburg in Germany and at Trappes in France, the auxiliary ground equipment was highly and (especially at Lindenberg) expensively developed.[26] The evolution of these observatories often makes interesting and sometimes exciting reading,[27] but I must concentrate on the meteorological instruments raised by the kites.

As far as I can determine, a recording instrument was first raised in this way at Blue Hill, near Milton, Massachusetts, on August 4, 1894. Fergusson[28] tells us that in July of that year William A. Eddy of Bayonne, New Jersey, who had flown kites to great altitudes, came to Blue Hill to experiment with raising recording instruments. This one, Fergusson says, was a thermograph, "an ordinary Richard Thermograph . . . altered for use in the experiments,"[29] so as to weigh only 1.1 kg. In August, 1895, a barothermograph was flown, and "a baro-thermo-hygrograph of aluminum, constructed by Richard of Paris, was used for the first time on April 8 [1896]."[30] This weighed only 1.3 kg. It seems hard to believe that Fergusson, the instrument man at Blue Hill, could be wrong about the sequence of events, but it must be recorded that Eddy published a slightly different version, namely that the barothermograph was flown in 1894. Four years later he wrote: "On August 4, 1894, I used several of my kites to raise a meteorograph at Blue Hill Observatory. A Richard thermograph, giving also barometric pressure, had been remodeled by Fergusson at the expense of the observatory, and was lightened by the use of aluminum.... This meteorograph, the first in the world ever lifted by kites, reached a height of 1400 feet, and brought back to earth, at times, an admirably clear record."[31]

Whether it was flown in 1894 or 1895, the barothermograph probably looked like the one in Figure 9.4,[32] with a Bourdon thermometer and two aneroid chambers to measure atmospheric pressure. Fergusson does not mention any alterations to the barothermograph.

The celebrated barothermohygrograph, made by Richard Frères

gantly full account in the *Handbuch der meteorologischen Instrumente*, E. Kleinschmit, ed. (Berlin, 1935), pp. 474–516.

[25] *Cf.* C. F. Marvin, *U.S. Monthly Weather Rev.*, Vol. 24 (1896), pp. 113–23, 156–66, 199–206, 238–55.

[26] See for example R. Assmann, *Das königlich preussiche aeronautische Observatorium Lindenberg* (Brunswick, 1915); also the annual *Arbeiten* of the Observatory from 1900 onwards.

[27] E.g., S. P. Fergusson, *Ann. Astron. Obs. Harvard College*, Vol. 42, part 1 (1897), pp. 41–128; W. Köppen, *Archiv. dtsch. Seewarte*, Vol. 24 (1901), pp. 1–104. The very length of these papers shows how seriously the subject was taken.

[28] *Ann. Astron. Obs. Harvard College*, Vol. 42, part 1 (1897), pp. 41–42.

[29] *Ibid.*, p. 41.

[30] *Ibid.*, p. 42.

[31] Eddy, *U.S. Monthly Weather Rev.*, Vol. 26 (1898), p. 451.

[32] From W. de Fonvielle, *Les Ballons-sondes* [etc.] (Paris, 1898), p. 28.

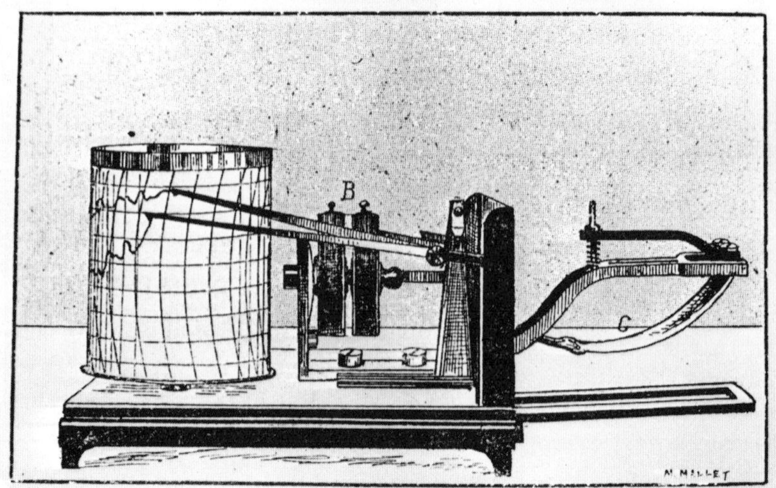

Fig. 9.4 Richard barothermograph, about 1893.

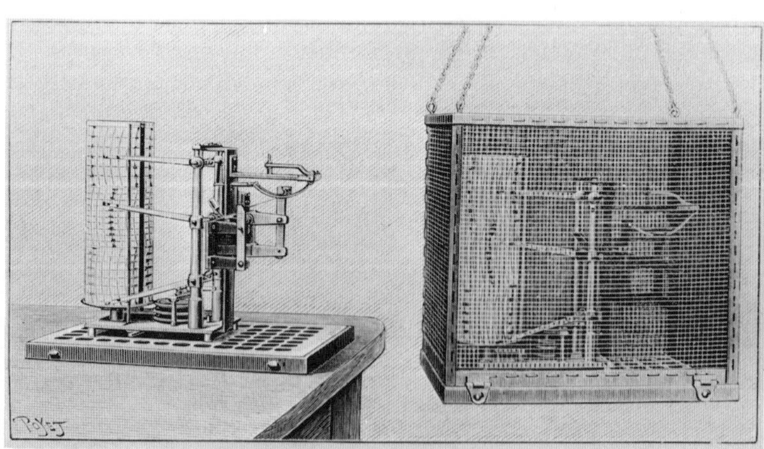

Fig. 9.5 Richard barothermohygrograph, 1896.

of Paris at the suggestion of the Director of Blue Hill Observatory, Professor A. Lawrence Rotch, is shown in Figure 9.5.[33] The Bourdon thermometer is seen at the top, and the three aneroid capsules are at the bottom. In the middle is the hair hygrometer, fitted with the Richard "correcting cams," which make the scale linear in relative humidity at the expense of more resistance for the hair to overcome. It will be noticed that the thermometer is not screened from radiation.

This meteorograph was the parent of dozens of later designs for use with kites, captive balloons, *ballons-sondes*, and aircraft.[34] Later on it was fitted with a cup anemometer by Assmann at Lindenberg.[35] The anemometer projected from the bottom of the instrument and jogged a pen by means of gearing and a cam. The thermometer was relocated in a horizontal duct.

To return to Blue Hill, it is remarkable that the original ascents were still made with the use of string, and only on January 27, 1896, was music wire substituted. Hargrave kites had been adopted in September, 1895.[36]

There was some exchange of ideas between Rotch and Fergusson at Blue Hill and C. F. Marvin—later Chief of the Weather Bureau and a notable authority on meteorological instruments—in Washington, D.C., where kite-flying was also being taken very seriously from 1895 onward. Indeed, it was taken so seriously that sixteen kite stations had been opened in various parts of the United States by 1898.[37] These were all equipped with the remarkable meteorograph designed by Marvin and made by Schneider Brothers in New York. This is shown without its light aluminum case in Figure 9.6. The figure is taken from the instruction book[38] written by Marvin for the kite stations, an engaging document, full of homely recipes for such processes as soldering and splicing. This meteorograph was immensely strong and durable. Pressure was measured by five aneroid chambers, humidity by a bundle of hairs, and temperature by two Bourdon tubes in parallel, made of hardened steel and filled with "pure alcohol or ether." The lag coefficient at an unspecified wind speed ("ordinary" kite ventilation) seems to have been about thirty-seven seconds.[39] Provision was also made for the recording of wind speed by electrical means from a small separate cup anemometer.

[33] See G. Tissandier, *La Nature* (Paris), Vol. 24 (1896), p. 145.

[34] This can easily be seen from many illustrations in the work of K. Keil, *Uber Meteorographen für aerologische Zwecke* (International Meteorological Organization, Berlin, 1938).

[35] *Arbeiten aeron. Obs. Lindenberg 1900/1901* (Berlin, 1902), pp. 38–39.

[36] *Ann. Astron. Obs. Harvard College*, Vol. 42, part 1 (1897), p. 42.

[37] Donald R. Whitnah, *A History of the United States Weather Bureau* (Urbana, Ill., 1961), p. 101.

[38] C. F. Marvin, *Instructions for Aerial Observers*, U.S. Weather Bureau, Circular K, Instrument Division (Washington, D.C., 1898).

[39] Marvin, *U.S. Monthly Weather Rev.*, Vol. 27 (1899), pp. 458–61. The lag coefficient can be most simply defined as the time taken to indicate 63 per cent of a sudden change in temperature.

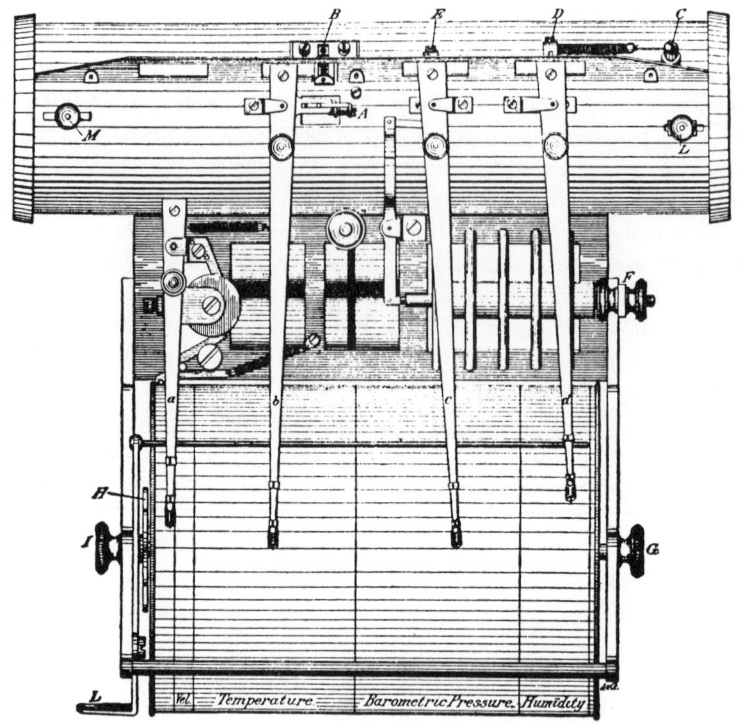

Fig. 9.6 The Marvin kite meteorograph.

The aerological literature of the time is full of praise for this meteorograph. The *Deutsche Seewarte* preferred it to the Richard instrument, after trying both,[40] and at Lindenberg it was paid the great compliment of imitation, Assmann's modification being shown in Figure 9.7. A small windmill anemometer was built into the duct. Assmann praised the workmanship of the New York firm, and confessed that when the instrument was made in Germany, it had proved impossible to duplicate the steel Bourdon tubes. He complained about how much it cost to buy them from the United States.[41]

Kites were being flown from small steam vessels with great success, for the steamers could "make their own wind." This was first done by H. Hergesell and Count Zeppelin on Lake Constance in 1900.[42] A program of this kind was also started at Crinan, on the west coast of Scotland, by W. H. Dines,[43] who incidentally developed a special kite reel for the purpose. At first a Richard meteorograph was

[40] Köppen, *Arch. dtsch. Seewarte*, Vol. 24 (1901), pp. 96–98.

[41] Assmann, *Arbeiten Obs. Lindenberg*, Vol. 7 (1912), p. XXIII.

[42] Hergesell, *Beitr. Phys. freien Atm.*, Vol. 1 (1904–5), pp. 1–34.

[43] Dines, *Quart. J. Roy. Meteorol. Soc.*, Vol. 29 (1903), pp. 65–85; *Collected Scientific Papers*, (London, 1931), pp. (126)–(143).

Fig. 9.7 The Marvin kite meteorograph as modified by Assmann.

used,[44] but in 1904 Dines invented an extremely simple and cheap kite meteorograph, which seems to have given good service in spite of its simplicity.[45] It is shown in Figure 9.8 with its cloth cover removed. The frame of the instrument is a light wooden board with curved side pieces. Above this a thin board 11 inches in diameter is arranged to be revolved on a brass pin; the motive power is supplied by a small, cheap clock driving a friction wheel. The circular board is covered by the chart paper, on which the pressure, temperature, and humidity are recorded. The barometer is an aneroid moving a pen by a single lever carried on a spring link. The thermometer consists of a helical coil of thin copper tube about 50 cm long and 6 mm in diameter, attached to a small metal capsule, the whole being filled with alcohol. The coil is at the back of the board. The hygrometer is a hair whose changes in length are magnified six times by a lever. The whole instrument is fastened in the kite by the four cords from its corners.

Later this instrument was modified slightly in detail, and the thermometer coil was made longer and thinner.[46] An anemometer was also added,[47] consisting of a light celluloid ball 7.6 cm in diameter hanging at the end of about 12 meters of fine sewing thread. The tension in the thread produced by the pressure of the wind on the ball

[44] W. N. Shaw and W. H. Dines, *Phil. Trans.*, *Ser. A*, Vol. 202 (1903), p. 124.
[45] W. H. Dines, *Quart. J. Roy. Meteorol. Soc.*, Vol. 31 (1905), pp. 218–22.
[46] W. H. Dines, Great Britain, Meteorological Office, *M.O.202* (London, 1909), pp. 15–38; *Collected Scientific Papers*, pp. (183)–(216).
[47] W. H. Dines, *Meteorol. Mag.*, Vol. 41 (1906), pp. 24–26.

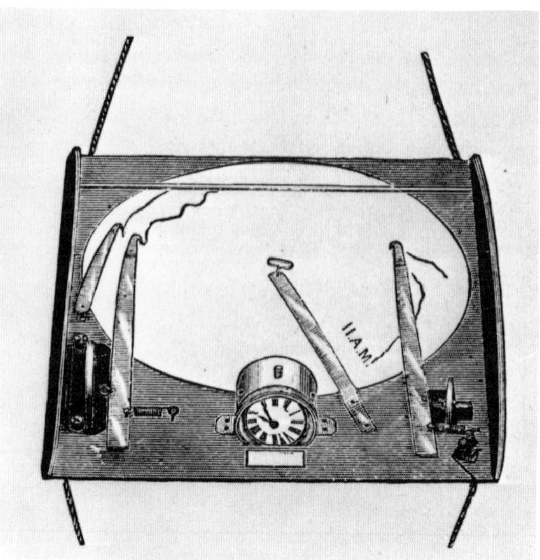

Fig. 9.8 The Dines kite meteorograph.

moved a pen at the end of a simple and ingenious system of levers that produced a scale that was almost linear in speed.

Plainly, the meteorographs used with kites could also be used with captive balloons, and indeed the two vehicles were employed interchangeably at some stations in accordance with the wind-speed. But the flying of meteorological kites and captive balloons came to rather a sudden end with the proliferation of electric power lines, although not before there had been some embarrassing accidents to various kinds of transport. The possibility of several miles of piano wire trailing over a city cannot be laughed off.

4. *Meteorographs for use with sounding balloons.* On November 14, 1892, a note on the possibility of scientific observations from unmanned balloons was sent to the Académie des Sciences in Paris by L. Capazza.[48] At the very next meeting Gustave Hermite told how he had sent up a free waxed-paper balloon 4 meters in diameter, inflated with illuminating gas, and carrying a minimum-registering mercury barometer.[49] This had been on September 17, 1892, the birthday of the sounding balloon. He got his instrument back intact, but it had weighed 1,200 gm, and he realized that such an approach had no future. He then devised a minimum-registering barometer carrying a piece of smoked glass, which moved past a fixed stylus. The whole apparatus weighed only 100 gm, and this weight was soon reduced

[48] *Compt. Rend.*, Vol. 115 (1892), p. 781 (no details).
[49] Hermite, *Compt. Rend.*, Vol. 115 (1892), pp. 862–64.

to 35 gm. One of these barometers, sealed, was taken on a long manned-balloon voyage by the aeronaut Besançon, and when it was examined later, it had agreed with the other instruments as to the minimum pressure. On October 4 this barometer and a Six's thermometer (Fig. 9.9) were sent up with a free balloon, but never found. Finally, on November 14, 1892, his ninth balloon, made of goldbeater's skin, carried similar instruments up to 7,600 meters, where the temperature was measured as −70°C. By December 10 he and Besançon had sent up and recovered a dozen balloons, one of which went as high as 9 km.[50]

Encouraged by this, they obtained from Richard the light barothermograph that we have already referred to (Fig. 9.4), and on

Fig. 9.9 Hermite's minimum barometer and thermometer, 1892.

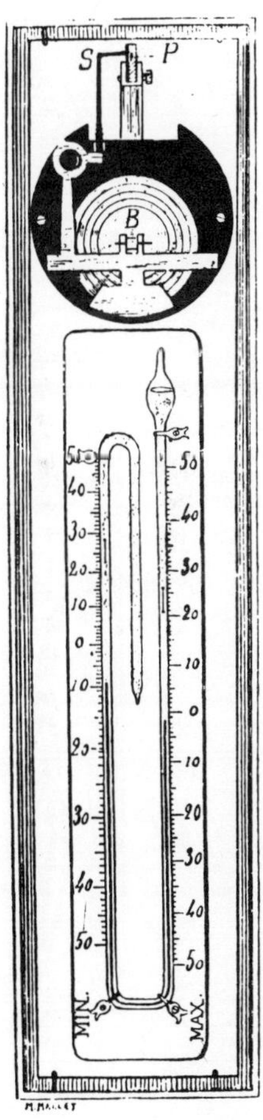

[50] W. de Fonvielle, *Les ballons-sondes de MM. Hermite et Besançon* (Paris, 1898), p. 11.

March 21, 1893, they sent it up with a goldbeater's-skin balloon with a volume of 113 cubic meters. Jules Richard, we are told, had devised a special nonfreezing ink. This apparatus went to 15 km and recorded a temperature of −51°C;[51] the altitude was greater than expected, and Hermite ascribed this to strong heating of the balloon by the sun.

In spite of Jules Richard's ink, pens had been replaced by tracing points, and the paper had been smoked, by October, 1895.[52] On the 20th of that month, the balloon also carried a minimum barometer, a photographic minimum thermometer, and an air sampler. The photographic thermometer was simply a thermometer filled with black spirit, with the tube in front of a slit, behind which was a piece of sensitive paper.

Shortly after this, Hermite and Besançon had a new model of barothermograph made (Fig. 9.10), with the smoked paper protected by a tube and the styli working through slits. This neat instrument is important because it has a helical bimetal for a thermometer, rather than the Bourdon tube used by Richard Frères. I know of no earlier thermograph of this type.

These French experiments naturally excited the interest of Richard

Fig. 9.10 Barothermograph of Hermite and Besançon, about 1895.

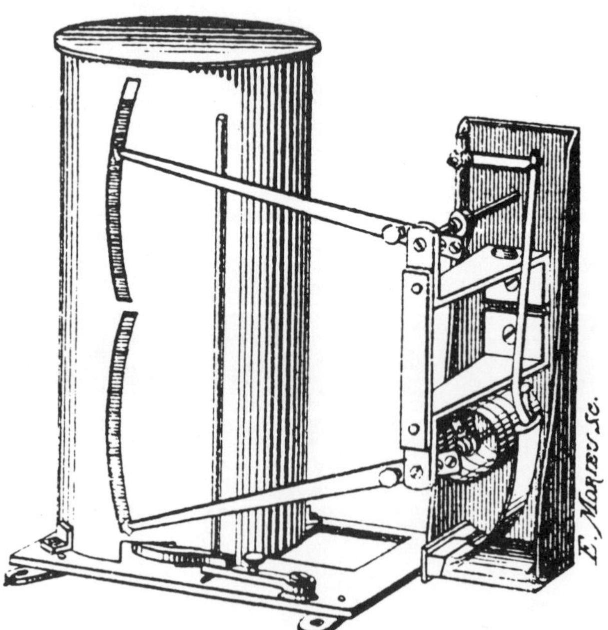

[51] *Ibid.*, pp. 18–22. Hermite (*Compt. Rend.*, Vol. 116 [1893], pp. 766–68) says to about 16 km.
[52] Fonvielle, *Les Ballons-sondes*, p. 29.

Assmann at the Royal Prussian Meteorological Institute, who had just published his important paper on the aspiration psychrometer,[53] and was acutely aware of the difficulty of making accurate measurements of temperature in the upper air. In 1894 he had the firm of R. Fuess build a photographically recording barothermograph in which the thermometer, filled with black spirit, was ventilated by a fan driven by a weight-driven clockwork. The stem of the thermometer was in front of a narrow slit, an image of which was projected on to a drum covered with sensitive paper. The pointer of the Bourdon barometer moved in front of the upper part of this slit.[54] This showed the barometric pressure as a white line, and the temperature as the boundary between black and white in the photographic image. It came into use in May, 1894. Assmann also used barothermographs of the usual type made by Richard Frères, which recorded on smoked paper. By September 1894, heights of over 18 km had been reached. Rather similar instruments were built by the firm of Bosch, especially for Hugo Hergesell of the Strasbourg Observatory (Fig. 9.11),[55] who, by 1896, had organized an international experiment in which, on the night of November 13–14 of that year, almost simultaneous unmanned balloon flights were made at Berlin, Paris, Strasbourg, Munich, Warsaw, and St. Petersburg.[56] This was the first of numerous such cooperative experiments.

The technique of sounding the upper air was influenced in more than one way by Léon Teisserenc de Bort, a dedicated amateur who devoted most of his private fortune to the subject at his farm at Trappes, near Paris, now a station of the Office National Météorologique. He had the advantage of a wide and deep knowledge of the meteorological theory of his time, as is clear from the notable book he wrote in collaboration with H. Hildebrandsson,[57] but he also had a grasp of instrumental problems.

The sounding-balloon flights at Trappes seem to have begun about 1896, and in July, 1898, he reported to the Académie des Sciences three ascents on June 8.[58] This was the first time, he said, that two balloons had been launched less than two hours apart and had given results that could be compared. They agreed very well, but a comparison of the ascending and descending traces of the first meteorograph showed that there were serious errors due to lag and to radiation. He noted that the registering apparatus weighed twenty-four times as much as the thermometer, so that the latter had to be protected from the radiation of the former. He then began to put all the

[53] *Abh. K. preuss. meteorol. Inst.*, Vol. 1 (1892), pp. 117–268.
[54] R. Assmann, *Wiss. Luftfahrten* (Brunswick), Vol. 2 (1900), pp. 671–73.
[55] R. Assmann, *Das aeronautische Observatorium Lindenberg*, p. 198.
[56] Hergesell, *Meteorol. Zeits.*, Vol. 14 (1897), pp. 121–22.
[57] *Les bases de la météorologie dynamique* (2 vols., Paris, 1898, 1900). Volume I of this work is of great value to the historian of meteorology.
[58] *Compt. Rend.*, Vol. 127 (1898), pp. 135–38.

Fig. 9.11 Sounding-balloon meteorograph of Hergesell.

meteorograph except the thermometer in a cork box. At this time he thought that the error due to lag was greater than that due to radiation, but he must have changed his mind almost at once, for we find him telling the International Aeronautical Commission[59] that it will not do to slow down the ascent of the balloon. The only cure is to reduce the mass of the thermometer. He seems to have been using an instrument rather like that shown in Figure 9.4, with a Bourdon barometer as well as a Bourdon thermometer, but they had been experimenting with bimetals at Trappes for some time. They had tried a thermometer consisting of a metallic strip a few hundredths of a millimeter thick, in a frame of *acier Guillaume.*[60] However, he came back to a curved bimetal, kept well away from the rest of the apparatus and isolated by a strip of ebonite, to prevent conduction (Fig. 9.12).

But he was worried, because, as a matter of fact, he had discovered the tropopause and could not believe it. He showed part of a night sounding on June 8, 1898, in which there appears an inversion of temperature gradient above about 100 mm of mercury (140 mb); but he ascribed this to the lag of the thermometer and the low ventilation. The balloon, which, we must remember, was of paper,

[59] *Int. Aeron. Comm., Strassburg, 1898, Protocol*, pp. 84–92.

[60] *Ibid.*, pp. 112–13. This is what we now call invar, invented by C. E. Guillaume at Sèvres.

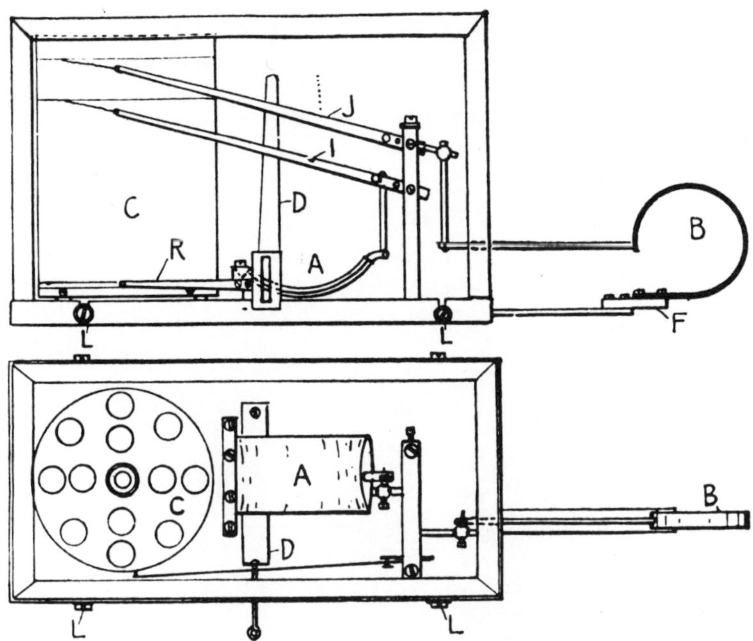

Fig. 9.12 Teisserenc de Bort's later meteorograph.

had floated at about 80 mm (105 mb) for half an hour.[61] By 1900 he had 140 records, 24 of them to at least 13 km, and still he would not believe what he found, but by April 1902, when he had 236 flights above 11 km, and 74 to at least 14 km, he made an announcement to the Académie.[62] It is interesting that in the first international flight of 1896, the Paris balloon had gone to 13.8 km, and the record had showed a sharp inversion at about 11.2 km. Hergesell[63] thought this quite incredible, and discussed the Richard meteorograph at length, giving his opinion that the protective casing was at fault because it enclosed the thermometer too much.

About 1900 Richard Assmann entirely changed the technique of the sounding balloon by introducing the use of closed rubber balloons to replace those of paper or goldbeater's skin.[64] They had the advantages of lower price, greater ease of handling, and above all the supreme advantage that they went up at a nearly constant rate until they burst. He calculated that if they could be stretched to 2½ times their original diameter (as they could) they might reach heights of more than 20 km. From his ascents with these balloons in 1901 he

[61] *Ibid.*, p. 91.
[62] *Compt. Rend.*, Vol. 134 (1902), pp. 987–89.
[63] *Meteorol. Zeits.*, Vol. 14 (1897), pp. 123–25.
[64] *Int. Aeronaut. Comm., Berlin, 1902, Verhandlungen*, pp. 81–83. In 1886 Paul Schreiber had suggested the use of rubber balloons for measuring upper winds (*Meteorol. Zeits.*, Vol. 3 [1886], p. 345), but this had been forgotten.

became convinced of the existence of the upper inversion. But there were many skeptics, and the efforts to put the matter beyond doubt led to a great deal of experimentation with meteorographs. Both Hergesell[65] and Assmann[66] built special instruments ventilated by small centrifugal fans, driven by electric motors. That of Assmann is shown in Figure 9.13. It is noteworthy that no attempt was made to ventilate the hair hygrometer that had been added by this time. A device had been made at St. Petersburg in 1898 to ventilate an entire meteorograph, not electrically, but by the weight of the instrument itself moving down a cord (Fig. 9.14).[67] It is hard to believe that a significant increase in ventilation could have been obtained.

Fig. 9.13 Assmann's electrically ventilated meteorograph.

[65] *Int. Aeronaut. Comm., St. Petersburg, 1904, Protocol*, p. 166.
[66] *Arbeiten Aeron. Obs. Lindenberg 1907* (Brunswick 1908), p. XIII.
[67] Rykatchew, *Int. Aeron. Comm., Strassburg, 1898, Report*, pp. 115–18.

Everyone suspected that if the lag of the thermometer could be made small enough, the ventilation caused by the ascent of a rubber balloon would be sufficient. Hergesell therefore made a meteorograph[68] (Fig. 9.15), in which the thermometer consists of a polished

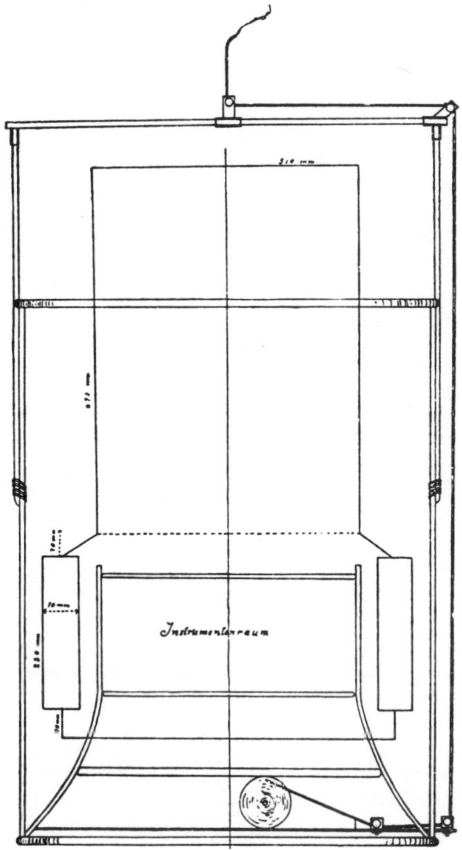

Fig. 9.14 Outline of Rykatchew's ventilation apparatus.

nickel-silver tube *R*, 15 mm in diameter and 172 mm long, weighing only 11.25 gm, held vertically in an invar frame *B*. It has a steel knife-edge at the bottom, acting on an agate plane at one end of a system of levers. An outer polished tube 30 mm in diameter surrounds the thermometer tube proper. A. de Quervain[69] made very careful measurements of the lag of various thermometers used in meteorographs for sounding balloons, and found Hergesell's tubular thermometer the most sensitive, but it must have been an expensive instrument.

[68] *Meteorol. Zeits.*, Vol. 21 (1904), p. 499; also *Zeits. für Instrum.*, Vol. 23 (1903), p. 312.
[69] De Quervain, *Beitr. Phys. freien Atm.*, Vol. 1 (1904–5), pp. 163–99.

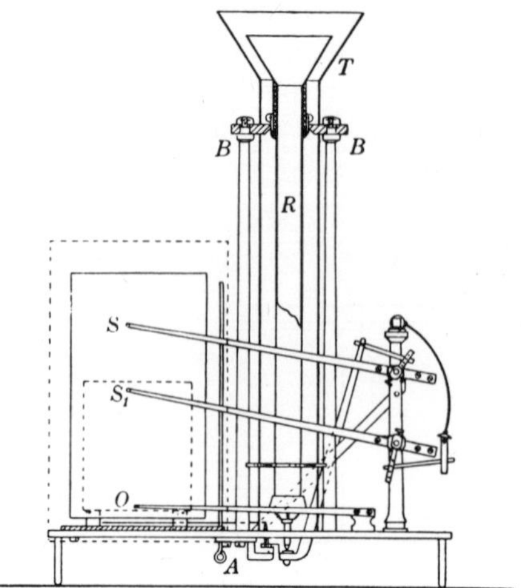

Fig. 9.15 Hergesell's meteorograph with tubular thermometer.

Clocks give trouble at low temperatures, no matter whether they are oiled with kerosene, as recommended by Rotch,[70] or run quite dry. In 1900 or 1901 Assmann constructed two instruments without clocks, one (Fig. 9.16) in which the drum was turned by the Bourdon thermometer, the other having it revolved by the aneroid barometer (Fig. 9.17), a better idea in view of the problem of ventilating the thermometer.[71] These designs did not prove satisfactory, and he later miniaturized the meteorograph with the clock, as shown in Figure 9.18.[72]

As far as meteorographs were concerned, Assmann seems to have been afflicted with *Bauwut.* The instrument with registration in rectangular coordinates shown in Figure 9.19 dates from about 1902.[73] Dr. Keil remarks with gentle truth that "in practice, because of its complicated construction, it was not in the long run successful."[74]

With his thoroughly practical mind, W. H. Dines realized that if sounding balloons were going to be useful in meteorology, they would have to be sent up in large numbers. What was needed was not complication but lightness and low cost, so Dines eliminated not only the clock but the drum as well. He also made use of the fact that a sharp point can scratch an extremely narrow but durable line on a

[70] *Ann. Astr. Obs. Harvard College*, Vol. 68, part 1 (1909), p. 13.
[71] Both are described in *Arbeiten aeron. Obs. Lindenberg 1900/1901* (Berlin, 1902), pp. 42–43.
[72] *Arbeiten aeron. Obs. Lindenberg 1911* (Brunswick, 1912), p. XXIII. The last conventional sounding-balloon meteorograph with a clock was probably that of S. P. Fergusson, *Monthly Weather Rev.*, Vol. 48 (1920), pp. 317–22.
[73] *Arbeiten aeron. Obs. Lindenberg 1901/02* (Berlin, 1904), p. III.
[74] K. Keil, *Über Meteorographen für aerologische Zwecke* (Int. Meteorol. Organization, Berlin, 1938), p. 15.

Fig. 9.16 Meteorograph with the drum turned by the thermometer (Assmann).

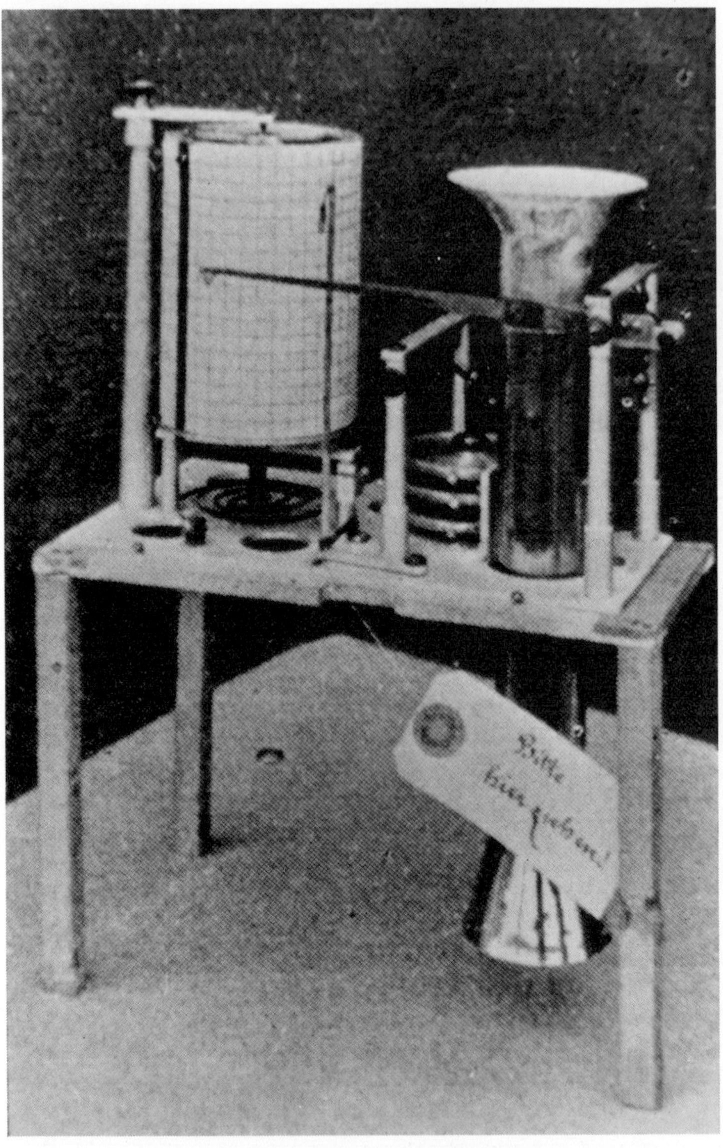

Fig. 9.17 Meteorograph with the drum turned by the aneroid barometer (Assmann).

Fig. 9.18 Assmann's miniature meteorograph with a clock.

Fig. 9.19 Assmann's meteorograph recording in rectangular coordinates.

polished surface of metal even though the force pressing the point on to the surface is very small. In this way a measuring microscope can be used to read the records, and very small deflections are sufficient. "This plan," Dines wrote, "was suggested by Dr. W. Watson some time since."[75]

The adoption of this principle made possible a remarkable degree of miniaturization. But before he reached this point Dines had designed the instrument shown in Figure 9.20 together with a diagram illustrating the linkage, which had four pivots at *A*, *B*, *C*, and *D*.[76] This was pivoted to the frame at *A*. An aneroid chamber pushed the frame at *G*, giving curves of pressure such as *LM* if the temperature was constant. "Some thermometric arrangement" pushes the point *B* sideways, so that constant-pressure curves of temperature are like *PQ*. What sort of thermometer was used is not stated, but the coil, presumably of tubing, suggests an arrangement similar to that of Dines's kite meteorograph described above. The record is made by the sharp point *E* on a piece of smoked metal. This instrument weighed 100 gm.

The second form (Fig. 9.21) weighed only 28 gm,[77] and had also the very great advantage that the four pivots, which would certainly have caused trouble, were eliminated by the use of one flexible joint and two knife-edges, at the cost of making the deflection due to change of temperature extremely small—about 1 mm for 60°C. What makes this usable is that there are two writing points *E* and *L* that draw lines of which the distance apart depends on the temperature. At first the record was made on copper-plated metal, but later the base was first plated with copper and then with silver, thus producing a corrosion-resisting surface without any scratches.

The temperature is measured by the difference in the expansion of a thin strip *M* of nickel-silver ("German silver"), 125 mm long, 10 mm wide, and about 0.15 mm thick, and a stiff invar wire about 1.5 mm in diameter. A steel knife-edge *D* is soldered to the strip, and another, *C*, to the invar wire; these work in nicks in the arm *DE*. Later, about 1908, the nicks and knife-edges were replaced by strips of thin flexible metal soldered to the arm, thus rendering the helical spring *N* superfluous and achieving the complete elimination of all pivots. The simple bending of the end of the frame at *B* provides an elastic joint to take care of the expansion of the aneroid chamber.

In 1906 the steel wire of the thermometer was made of ordinary steel, and Dines managed to persuade himself that this was better than invar for the purpose, by a strange argument[78] that he may have

[75] W. H. Dines, *Meteorol. Mag.*, Vol. 41 (1906), p. 101 (*Collected Scientific Papers* p. (180)). I am informed by Professor P. A. Sheppard that Watson was a professor of physics at the Imperial College.

[76] *Ibid.*

[77] *Ibid.* It is more fully described in: Great Britain, Meteorol. Office, *M.O.* 202 (1909), pp. 15–48 (*Collected Scientific Papers*, pp. (183)–(216)).

[78] *Meteorol. Mag.*, Vol. 41 (1906), p. 102.

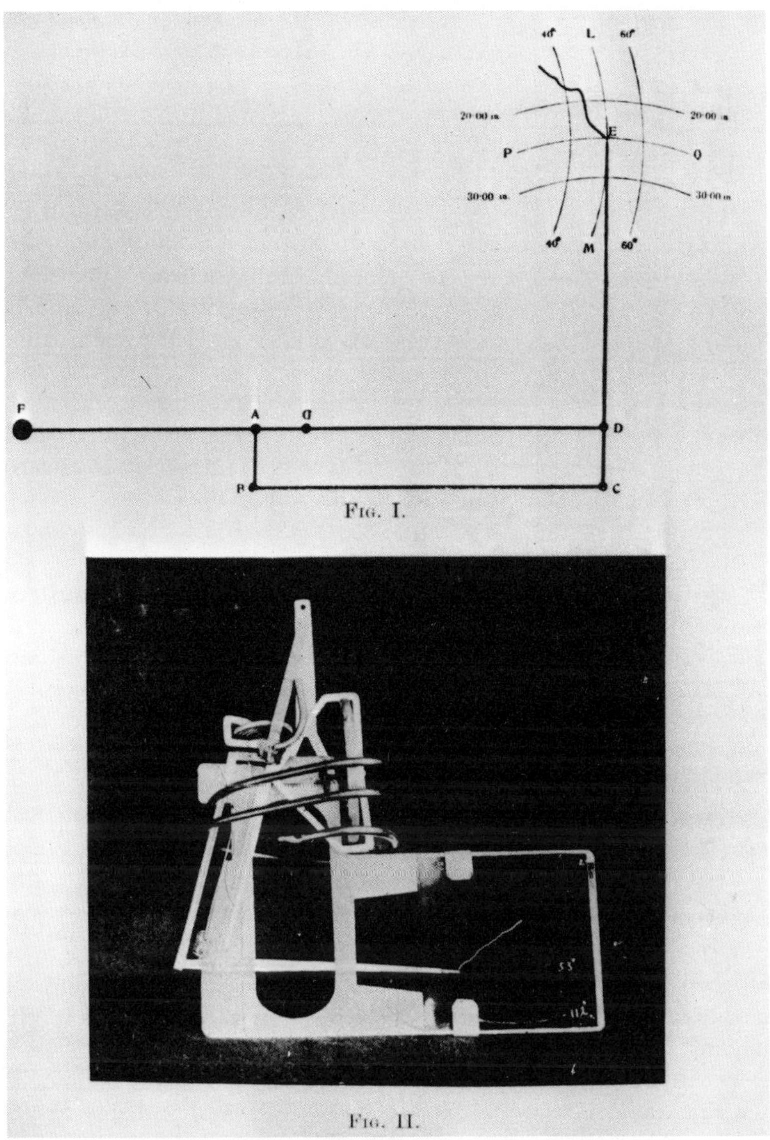

Fig. 9.20 Dines's earlier design of light meteorograph.

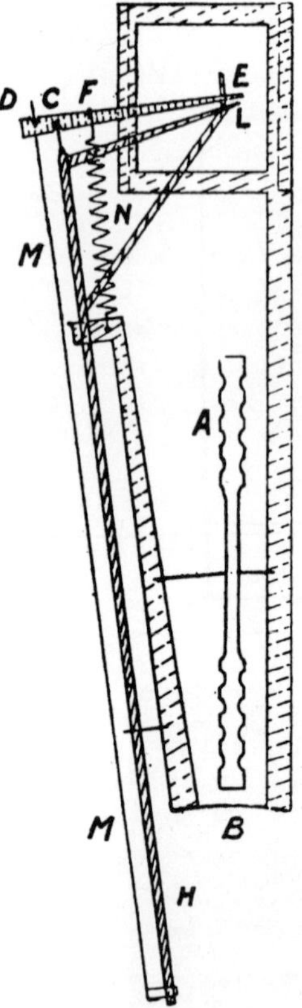

Fig. 9.21 The Dines balloon meteorograph, diagram.

regretted, because by 1909[79] only invar is mentioned. Another improvement introduced at this time was a device for lifting the styli off the record shortly before the instrument reached the ground on its descent, so that the trace would not be blurred by the shock of the landing.

W. H. Dines suggested that a hair hygrograph might be added, but this was done only by his son L. H. G. Dines, probably about 1927.[80] The resulting instrument[81] is shown in Figure 9.22, in which it will be seen that the hygrograph is carried on a separate frame, attached by screws. This was necessary because the hygrograph had to be calibrated separately. The meteorograph was suspended in a wire frame, attached only to the edge of the aneroid chamber at two points. This wire frame fitted inside a light aluminum radiation shield with flared ends. Before ascent, all this was mounted

[79] *M.O. 202*, p. 37 (*Coll. Sci. Papers*, p. (205)).
[80] I deduce this from item 50 in a collection of miscellaneous papers and letters left by L. H. G. Dines, and now in the possession of Dr. and Mrs. D. S. Poole of Esher, Surrey.
[81] Great Britain, Meteorol. Office, *M.O. 321* (1929).

in a large but very light shock-absorbing "spider" made of very thin bamboos and cotton threads. No parachute was necessary, and the whole thing, weighing only about 100 gm, was carried up by a 350-gm balloon inflated with hydrogen.

The Dines instrument led to several other designs of light meteorographs. The earliest was that of Kirchner, the instrumentmaker at

Fig. 9.22 The Dines balloon meteorograph, later form with hygrograph.

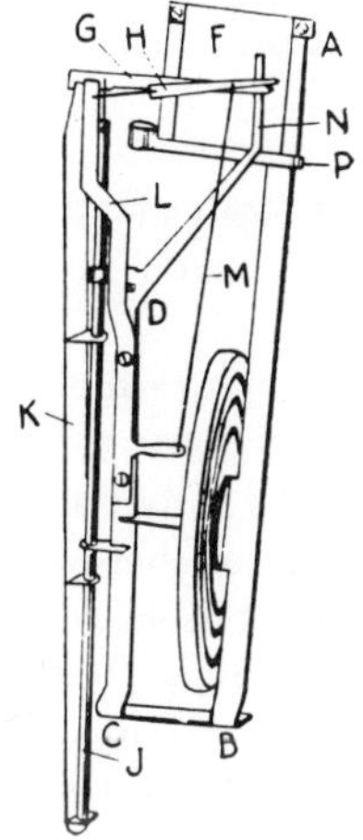

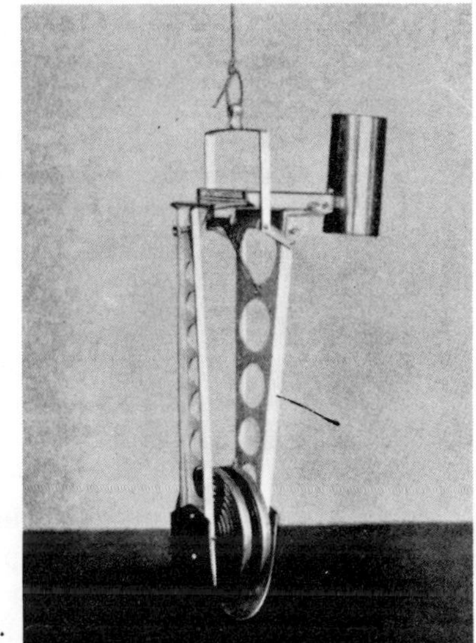

Fig. 9.23 Kirchner's meteorograph.

Lindenberg,[82] where they seem to have had bad luck with five Dines meteorographs, obtained from England.[83] Kirchner's meteorograph (Fig. 9.23) had a bimetallic thermometer of the usual form, giving a rather greater deflection than that of Dines. Much later, General Delcambre reported[84] on such a meteorograph, "inspired" by Dines's, and also on an extremely light one (110 gm) with a clock and a drum 4 cm in diameter. Both instruments were designed by the old instrumentmaker Geoffre, who had worked with Teisserenc de Bort. The Jaumotte meteorograph of 1930, recording temperature and humidity as a function of pressure, was another successful and very light instrument without a clock.[85] It was even smaller and lighter than the Dines meteorograph.

[82] *Arbeiten aeron. Obs. Lindenberg 1900* (Brunswick, 1902), pp. XXV–XXVI.
[83] *Ibid.*, p. XXIII.
[84] Delcambre, *Beitr. Phys. freien Atm.*, Vol. 14 (1928), pp. 63–64.
[85] J. Jaumotte, *Mém. Acad. roy. Belgique*, No. 1417 (1931), 44 pages.

5. Instruments for use on aircraft. At Lindenberg the indefatigable Richard Assmann seized every opportunity to explore the free air, and in 1912 we find him writing enthusiastically as follows:

The considerable heights recently attained by aviators—as much as 5720 m at the time of this report—make one think of providing aircraft with recording instruments, the more so because the high speed will ensure that these never have to suffer from lack of ventilation. If we could succeed in persuading even a fraction of the flyers operating almost daily in the various parts of Germany to take a Marvin kite meteorograph with them, an extremely rich and valuable collection of observations might be brought together, that could serve as some sort of equivalent for an increase in the number of aerological observatories, out of reach for a long time because of its cost. The author is at present busy implementing this plan.[86]

A year later, though he still believed in the future of aircraft soundings, he was obliged to confess that he had not aroused much interest among aviators.[87]

Assmann favored the Marvin type of meteorograph, which he modified a little for use on aircraft. After the 1914–18 war a large number of aircraft meteorographs were designed, and it is no part of the plan of this book to describe them all. Most of them were fundamentally heavier and more rigidly constructed versions of kite meteorographs, a good example being that described in 1926 by E. Calwagen,[88] shown in Figure 9.24 together with its aluminum case. In the 1930's numerous ingenious designs appeared in several countries; more attention was paid to good aerodynamic shape than before. The reader must be referred to the brochure by Keil already mentioned,[89] which has detailed descriptions and many excellent illustrations.

In England, aircraft soundings started in a somewhat different manner, as an offshoot of observations made in a rather amateur way from military aircraft during the war of 1914–18, for instance by C. K. M. Douglas, who accompanied the report of his observations with a plea for the regular use of the airplane in meteorology.[90] For readings of temperature and humidity the favored instrument was the "strut psychrometer," so named because it was normally attached to a strut of the single-engined biplanes common at the time, perhaps 120 to 150 cm from the center-line of the aircraft. At first very large spirit thermometers were used, but these had too great a lag, and so they were replaced by ordinary mercury thermometers, and a pair

[86] *Arbeiten aeron. Obs. Lindenberg 1911* (Brunswick, 1912), p. XIII.
[87] *Arbeiten, 1912* (Brunswick, 1913), pp. VII–VIII.
[88] *Geofysiske Publikasjoner*, (Oslo), Vol. 3 (1926), No. 10.
[89] *Über Meteorographen für aerologische Zwecke* (Int. Meteorol. Organization, Berlin, 1938).
[90] Douglas, *J. Scottish Meteorol. Soc.*, Vol. 17 (1916), pp. 65–73.

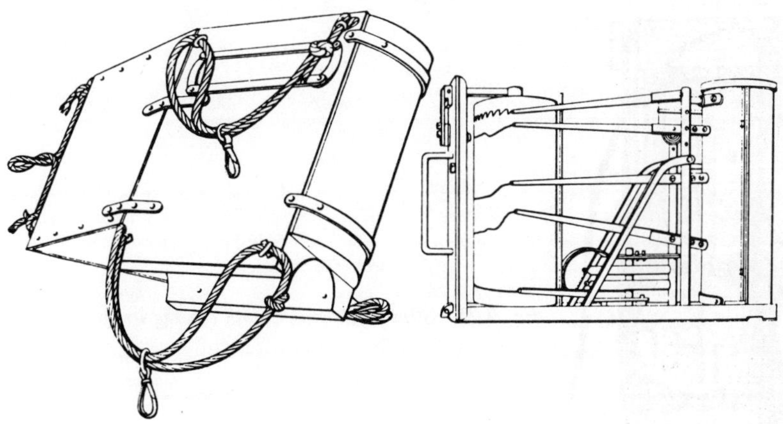

Fig. 9.24 Calwagen's aircraft meteorograph (courtesy of Julius Springer Verlag).

of lenses was provided, to be pulled up or down by cords extending to the cockpit (Fig. 9.25).[91] One of the difficulties was that of getting sufficient water to the wet bulb.

The early aircraft meteorographs used in England were of a design entirely different from those used on the Continent, and were much simpler, being derived from the meteorographs of Dines discussed above. That of G. M. B. Dobson is shown in Figure 9.26.[92] This instrument plots temperature against pressure as in the Dines balloon meteorograph, but on a much larger scale. The system of levers is designed so that the whole instrument is statically, and as far as possible dynamically, balanced, thus minimizing the effect of accelerations and vibration. The pivots are jeweled, and the records are made by a sharp point *P* on a smoked card.

In the 1930's the use of aircraft for sounding the upper air began to give way to the radiosonde, which will be dealt with in the following chapter.

[91] R. Glazebrook, ed., *A Dictionary of Applied Physics* (5 vols., London, 1923), Vol. III, p. 416. There is one of these instruments in the Science Museum, London, Inv. no. 1922–5.
[92] *Ibid.*, p. 517. Dr. Dobson informs me that he remembers no separate publication of this.

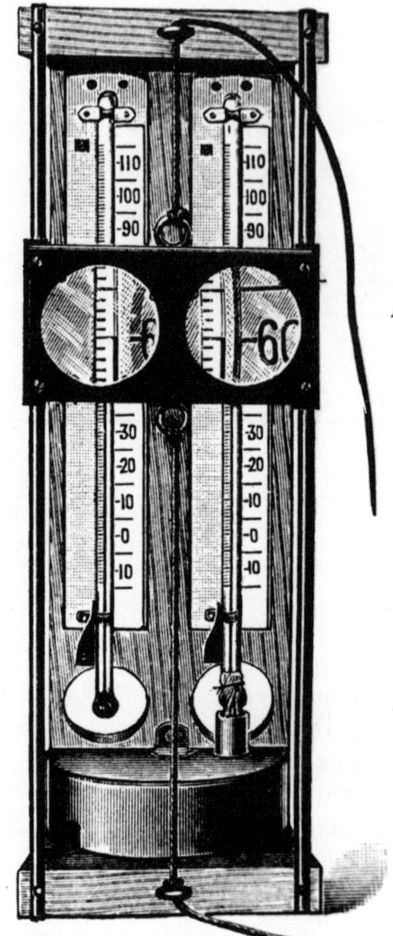

Fig. 9.25 Strut psychrometer, with reading lenses.

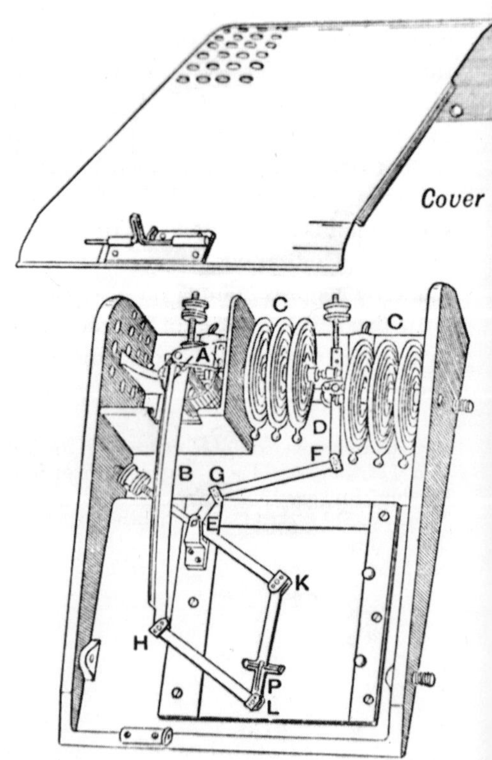

Fig. 9.26 Dobson's aircraft meteorograph (courtesy of Macmillan & Co.).

X

Telemeteorography and the radiosonde

1. The brief history of telemeteorographs for surface observations. As we saw in Chapter 7, the assembling of meteorological instruments into complex recording systems began very early, but it is noticeable that, except for special purposes like the exploration of the upper air, such systems have not proved popular, and have seldom justified the ingenuity and expense lavished upon them.

One of the best excuses for building a meteorograph is to put it on an island or on the top of a mountain, to record the meteorological elements between the periodic or occasional visits of an attendant. By the time that it became technically possible to build instruments that were sufficiently durable, and before nearly enough money could be found for such a purpose, it had become clear to meteorologists that what they really needed was not records of what had been happening, but simultaneous observations delivered at regular intervals to a central office: in short, synoptic observations, as it has become the fashion to call them.

The possibility of obtaining such observations came with the spread of the electric telegraph shortly after 1850. The familiarity with solenoids, resistors, switches and other electrical gear that was engendered by the telegraph led to an astonishing flood of ingenious devices in which the arts of the clockmaker and the electrician were often happily combined.[1]

The first telemeteorograph of which I am aware was described in 1843 by Charles Wheatstone,[2] who was one of the inventors of the electric telegraph itself. This instrument was designed to print the readings of a siphon barometer and the two thermometers of a psychrometer in figures on a paper tape at the receiving station. Platinum wires dipped into the open limb of the barometer and into the tubes of the mercury thermometers, which were open at the top. Every half hour these wires were raised in five minutes and lowered again in one minute by a clock. The breaking of the circuit caused a

[1] By 1862 the second edition of Théodose Du Moncel's *Exposé des applications de l'électricité* (5 vols., Paris, 1856–62) had run to over 2,000 pages, and it is by no means prolix.
[2] Wheatstone, B.A.A.S., Cork, 1843, *Report*, pp. xl–xlii.

hammer to fall and print the readings at the receiving station, where the type wheels were moved by another clock, presumably in synchronism with the one at the transmitter.

One of these instruments, transmitting temperatures only, was made for a Committee of the British Association which proposed to conduct experiments with captive balloons. T. R. Robinson[3] reported for the Committee that "Mr. Wheatstone's electric thermometer has been tried, and found to act in the most perfect manner at distances of some miles, and we have ordered the addition of another part for giving the hygrometric indications."[4] In an appendix, Wheatstone described a balloon thermometer, disclosing that with its case it weighed about four pounds.[5]

In 1867 Wheatstone invented another telemeter, applicable to any instrument with a rotating pointer.[6] This was not a recording instrument, but the observer, when he needed a reading at the distant station, turned a magneto, exactly as in the process of calling on an old-fashioned rural telephone, as some of my readers may remember. The pulses produced by the magneto revolved a contact at the distant instrument from a zero position, and also a pointer in front of the observer. When the contact touched the pointer of the instrument, both motions were stopped by a relay, and the distant contact was returned to zero, ready for the next inquiry. The observer brought the other pointer back by hand.[7]

An impetus to the construction of telemeteorographs seems to have been given by C. H. D. Buys-Ballot, the director of the Netherlands Meteorological Institute, after whom the rule for finding the wind from the barometric gradient is named. In 1868, in a pamphlet[8] on storm warnings, he pleaded for apparatus that would permit the readings of the barometers at four widely separated stations to be transmitted at frequent intervals to Utrecht, using special telegraph lines. This would allow a constant watch over the gradient. It seems that the existing commercial telegraph lines were too slow and too busy, but no such apparatus was available.

At about this time F. van Rysselberghe, a teacher of navigation at Ostend, was seized by the ambition to build himself a meteorograph.[9] He had read of the experiments of Secchi, Theorell, Wheatstone, and others. He was not a rich man and tried to make his instrument as simple as possible.

From his reading of Wheatstone and Theorell[10] he had taken the

[3] Regarding Robinson, see chapter 5, p. 213.
[4] B.A.A.S., Cork, 1843, *Report*, p. 128.
[5] *Ibid.*
[6] B.A.A.S., Dundee, 1867, *Sections*, pp. 11–13.
[7] The two parts of this instrument are at the Science Museum, London, Inv. no. 1950–197.
[8] Buys-Ballot, *De invoering en verklaring van den aeroklinoskoop* (Utrecht, 1868), p. 24.
[9] F. van Rysselberghe, *Bull. Acad. r. Belgique*, Vol. 36 (1873), pp. 346–74.
[10] See p. 261 above.

idea of having a platinum wire dip into a mercury surface and close a contact; but he seems to have had doubts in the case of the thermometer, and at first preferred a bimetallic spiral. Following Wheatstone's earlier paper, he chose to convert the readings of the instruments into intervals of time. Our Figure 10.1 helps to show how it was done. The drum *C* and the toothed sector *F* are revolved by a clock. In the position shown, the sector has just finished revolving the lead-screw *DE* that carries the stylus *B*, lowering the stylus by rather less than a millimeter. It will soon revolve the wheel *M*, which will let a platinum wire *R* down into the short limb of the siphon barometer, and later still it will turn the sector *V*, causing a contact to be made at a time corresponding to the temperature of the bimetal *Y*. Next comes the reading of the cup anemometer, and lastly that of the hygrometer, an ingenious device that plunges a bar *P′* into mercury in the tube *D*, making it touch a platinum contact at a time determined by the length of the hair *A′B′*. The windvane operates still another contact, not shown in the figure. A notable feature is that the circuit is never *broken* at the contacts operated by the instruments, but always by a contact on the ring *p*. This construction was derived from Theorell.

Whenever a contact is made an electromagnet presses the stylus against the drum, so that the record consists of a number of parallel lines of varying length, depending on the values of the meteorological elements. By means of a friction drive that we have not space to describe, the beginning of the anemometer trace is caused to be always on the same straight line parallel to the axis of the drum, and Van Rysselberghe notes that the straightness of this line gives a sure check on the proper functioning of the instrument.

He had built one at Ostend that gave records of the barometer, metallic thermometer, anemometer, and windvane, and he reproduced two of the weekly charts of this. He showed the instrument to the Brussels Academy on May 11, 1875, and by 1878 he had obtained a grant from the Belgian government for further work. The telemeteorograph that resulted from this was in operation on telegraph lines between Paris and Brussels during the Electrical Congress in Paris in 1881.[11] It differed from his original design by using two open-ended mercury thermometers, as Theorell had done, replacing the hygrometer as well as the bimetal. How the wet bulb was to be maintained at an unattended station we are not told. There was a mechanism to stop the descent of each platinum wire as soon as it touched the mercury.[12] In the opinion of S. Levy[13] this meteorograph, made by Schubart at the University of Ghent, was one of the most important pieces of scientific apparatus exhibited at the Congress.

[11] M. Snellen, *Meteorol. Zeits.*, Vol. 13 (1896), p. 367.

[12] The transmitter and receiver are described and illustrated in *Engineering* (London), Vol. 37 (1884), pp. 399–401.

[13] S. Levy, *Zeits. Instrum.*, Vol. 2 (1882), p. 233.

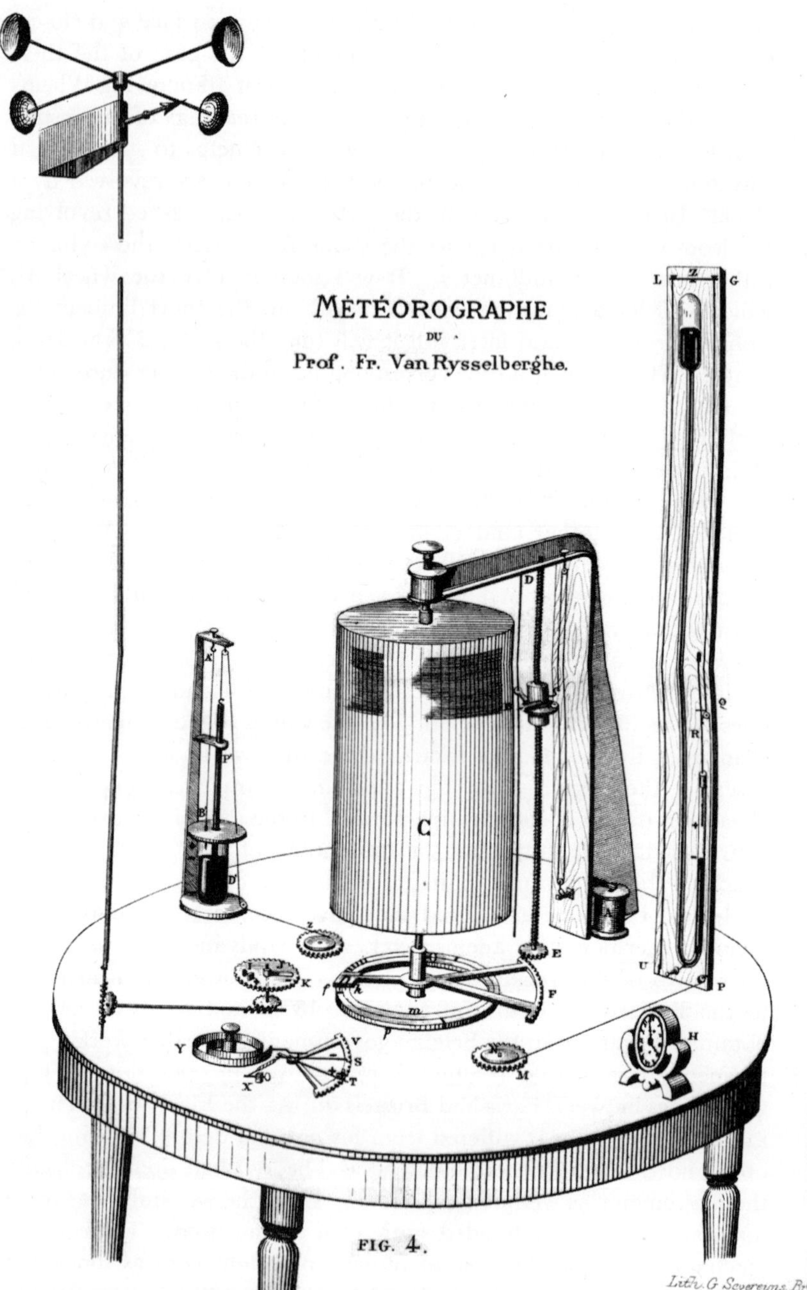

Fig. 10.1 Van Rysselberghe's meteorograph, 1873.

In 1872 the Société Hollandaise des Sciences à Haarlem[14] had offered a gold medal to anyone who should "find a satisfactory means of determining the temperature, the state of humidity, and the density of the air of the atmosphere at a considerable height above the surface of the ground. This method will have to permit the automatic recording of the observations, or at least their frequent repetition."[15] In other words, they wanted a telemeteorograph that could be used with a captive balloon or kite.

The offer produced no candidates, and expired on January 1, 1874; whereupon E. H. von Baumhauer, who had put himself *hors concours* because he was the secretary of the Society, came up with his own ideas,[16] not reduced to practice for lack of money. He does not seem to have seen Van Rysselberghe's paper until his own was about to be printed, and it is interesting that he nevertheless rejected the use of any instruments in which wires had to be plunged into mercury. In his arrangement, instruments that produce a rotary motion can be used, and thus the aneroid barometer, the metallic thermometer, and the hair hygrometer are indicated. The differences between his suggestions and those of Van Rysselberghe are mainly mechanical; the basis of both meteorographs is the translation of angular motion into intervals of time.

Besides pressure, temperature, and relative humidity, von Baumhauer proposed to measure wind direction by means of a windvane, wind speed with a cup anemometer, and rainfall by a tipping-bucket with an escapement that finally revolved a shaft. These three all demand continuous rotation, ingeniously transformed into linear motion by a device (Fig. 10.2) using a chain and three pointers.

Fig. 10.2 Von Baumhauer's conversion mechanism.

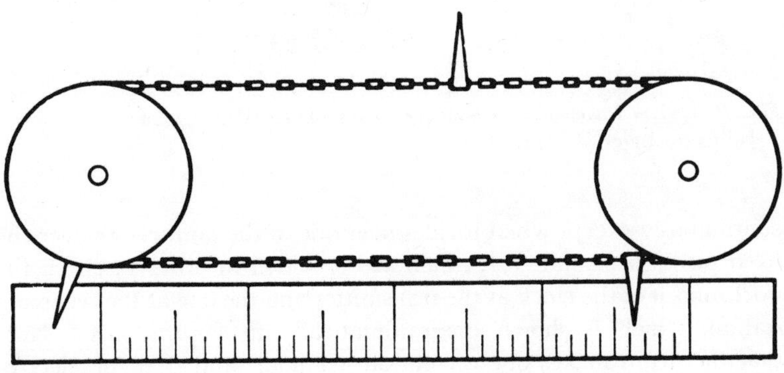

[14] I give their French title because their serial publication, *Archives Néerlandaises*, was in French.
[15] *Arch. Néerlandaises*, Vol. 9 (1874), p. 231.
[16] Baumhauer, *Arch. Néerlandaises*, Vol. 9 (1874), pp. 230–58.

The distance between the centers of the wheels is exactly equal to the circumference of either, and also to the spacing of the pointers.[17]

Three such motions *x*, *y*, and *z* (Fig. 10.3) are connected to the windvane, the anemometer, and the rain gauge, respectively. In addition there are three pointers *AA′*, *BB′*, and *CC′*, moved by the barometer, thermometer, and hygrometer. The ends of all the pointers, where they pass over the ring *M*, are of ebonite. The minute-hand *R* of a clock sweeps a tongue of gold around *M*, making a contact

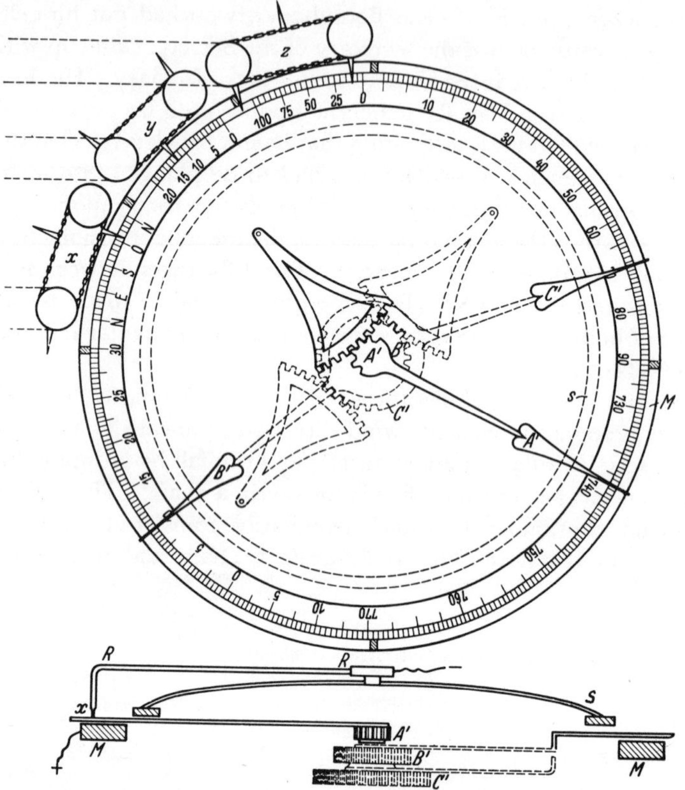

Fig. 10.3 Von Baumhauer's scheme for the transmitter (courtesy of Julius Springer Verlag).

continuously except when it rides over one of the pointers or over the fixed pieces of ebonite *e*, *e*, which are provided in order to check the isochronism of the clock at the transmitter and the one at the receiving station, which is shown diagrammatically in Figure 10.4. Note that the drum makes one revolution per hour, and that the electro-

[17] This idea was known to Jacob Leupold. See his *Theatrum machinarum, Pars III. Theatri statici universalis, sive theatrum aerostaticum* [etc.] (Leipzig, 1726), p. 305 and Plate XXII.

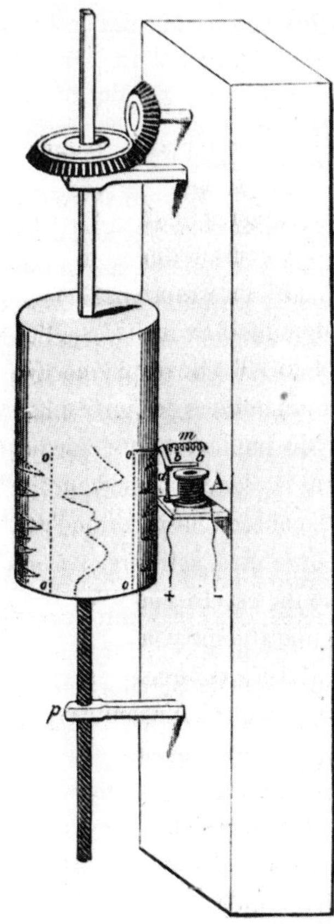

Fig. 10.4 Von Baumhauer's receiver.

magnet *A* keeps the stylus away from the drum except when the circuit is broken.

In order to keep the various observations almost simultaneous, von Baumhauer suggested that a signal could start clocks at the two stations; these would revolve the sweep hand and the drum in two or three minutes. Because the pointers of the meteorological instruments can be moved by small forces, he proposed a clamping ring *S* (Fig. 10.3) that would be depressed so that it held them down while the sweep hand was in motion.

I have dealt with von Baumhauer's instrument at length because it displays the essential features of the system of telemetry long known as the "Olland cycle." In a way this name is unjust to both von Baumhauer and Van Rysselberghe, for it seems certain that the Utrecht instrumentmaker H. Olland, who first made a successful telemeteorograph on this principle, got many of his ideas from von Baumhauer. Olland's work was described in 1879 by M. Snellen,[18]

[18] Snellen, *Arch. Néerlandaises*, Vol. 14 (1879), pp. 180–208.

who begins his paper by stating that "Reading [von Baumhauer's article] revived in Mr. Olland, maker of precision instruments at Utrecht, an old desire to execute an apparatus by whose aid the indication of an instrument might be transmitted automatically to any distance."[19]

In 1875 Olland started to make an instrument to send the readings of an anemometer and a barometer from a tower to the meteorological office in Utrecht, a distance of about 1 km, using only one telegraph wire. Von Baumhauer met Olland and persuaded him to make one such as von Baumhauer had previously suggested; this was sent to the Philadelphia Exhibition of 1876, where it received a medal.[20] But it required two connecting wires, and Olland, who had Buys-Ballot's requirements in mind, was determined to do it with one.

He had succeeded by July 1877, producing a most ingenious, if complicated, pair of instruments. He hit on the idea of synchronizing the two stations by means of conical pendulums, set in motion for a minute or two whenever an observation was required. His electromechanical programming system took Snellen[21] ten pages to describe, as well as a very large folding plate;[22] and Snellen's description could not usefully be shortened. I shall note only that there was just one clock, at the receiving station, which simply provided starting impulses at suitable intervals, say every fifteen minutes, setting both the conical pendulums in motion. The pendulum at the receiver then programmed the sequence of observations at both stations by means of relays and electromagnetic latches, finally stopping both pendulums in readiness for the next observation. What finally appeared on the receiving drum was a number of irregular lines of dots representing the readings of the various instruments, and several equally spaced lines whose straightness, or lack of it, would indicate whether isochronism had been obtained.

Olland used a balance barometer in which the cistern was weighed, a cup anemometer, a windvane with a very elaborate double ratchet-and-pawl arrangement, and a metallic thermometer of unusual design, the invention of F. W. C. Krecke, which von Baumhauer liked.

These instruments worked well for eight years after 1877, so well that in 1879 Buys-Ballot tried to get the government to install four of them at various points in the kingdom, fulfilling his cherished desire, but the project was deemed too costly.[23] Nevertheless, Olland made a telegraphic tide gauge that was installed at the Hook of Holland in 1877 and was still working to the complete satisfaction of the engineers

[19] *Ibid.*, p. 180. Olland (1825–1901) later founded a firm at De Bilt which still (1966) exists under the direction of his grandson, F. W. Olland. I am indebted to Mr. E. Engberts of the Leiden Museum for this information.
[20] These details are from Snellen, *Meteorol. Zeits.*, Vol. 13 (1896), p. 366.
[21] *Arch. Néerlandaises*, Vol. 14 (1879), pp. 185–95.
[22] One corner of this plate, showing the barometer, is reproduced in my *The History of the Barometer* (Baltimore, 1964), p. 333.
[23] Snellen, *Meteorol. Zeits.*, Vol. 13 (1896), p. 366.

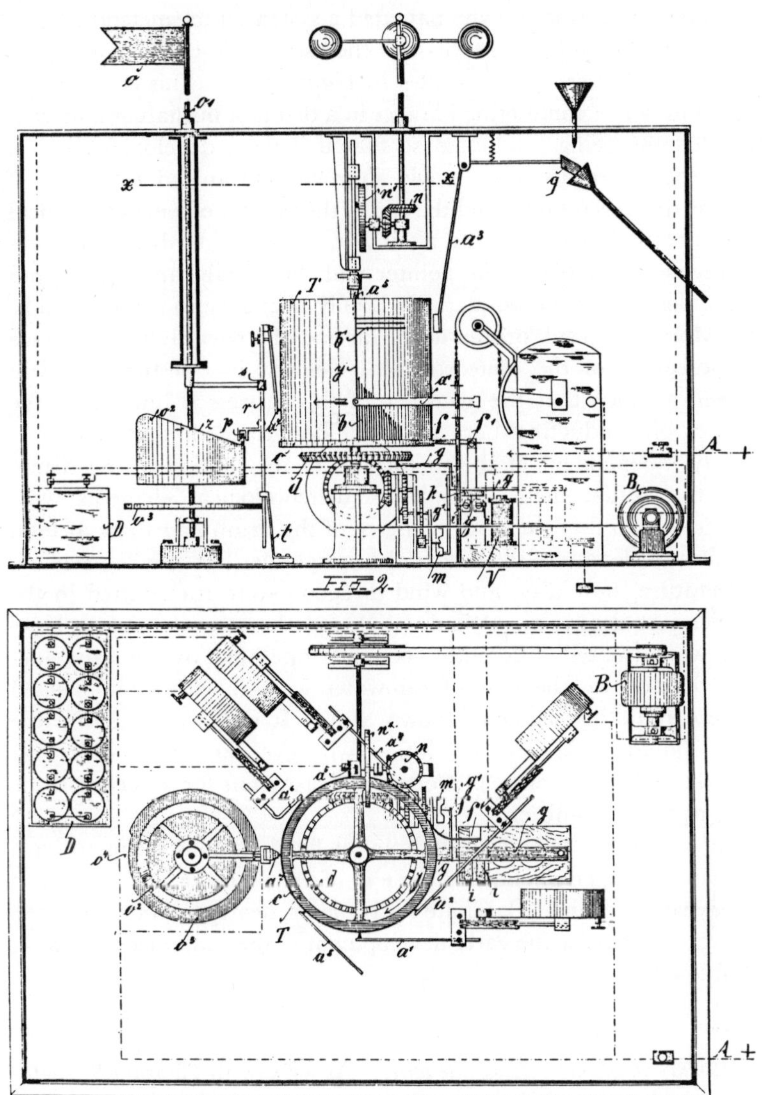

Fig. 10.5 The telemeteorograph of Cerebotani and Silbermann.

in 1894. Finally in 1895, after many difficulties, he got his telemeteorograph operating successfully over 160 km of wire between Flushing and Utrecht.

Both the Van Rysselberghe system and that of Olland and von Baumhauer demand clockwork motions that function at an exactly uniform speed, at least during the period of each observation. An attempt to avoid the necessity of this was made by Luigi Cerebotani and Albert Silbermann, who patented a system of telemeteorography in 1896.[24] The novelty in this was the use of a so-called "responder triangle" (*Meldedreieck*), seen at *b* in Figure 10.5. This consists of a large number of conducting bars set in a drum of insulating material, their lengths being graduated so that if it were developed on to a plane the entire assembly would form a right-angled triangle. If now the drum is rotated, and the end of the pointer of a meteorological instrument is supposed to rest against it, the total number of contacts per revolution between the pointer and the metal bars will depend on the position of the pointer. It is only necessary to count the dots on a Morse telegraph printer at the far end of a telegraph line to have an indication of the meteorological element in question. It will be seen at once that the accuracy of this process is limited by the number of conducting bars that can be provided, and that it is entirely independent of the uniformity of the motion of the drum.

In the actual meteorograph six meteorological elements were allowed for, so that the circumference of the drum had to be at least six times the length of the base of the "responder triangle." Pressure, temperature, humidity, and wind direction were transmitted in the manner just described. Wind speed and the rate of rainfall, however, were treated differently; contacts operated by the cup anemometer and by the flow of rainwater were in circuits connected to the conducting segments shown at *b′*. It was arranged that the drum should make one revolution whenever an observation was desired, and then stop. There was no provision for a record to be made at the transmitter.

The interest in telemeteorography at fixed stations seems to have died out almost completely at the end of the nineteenth century, but thirty years later these elaborate instruments, nearly forgotten, formed a source of ideas for the early development of the radiometeorograph, or radiosonde.

2. Attempts at telemetry down kite wires. As we saw in Chapter 9, meteorographs for use with kites and balloons were highly developed by 1914. It is not surprising that during the first world war attempts were made in both France and Germany to transmit information from the kite or balloon to the ground down the kite wire, using a method that would now be called carrier telegraphy.

In Germany Herath and Robitzsch found a way to do this in 1917.[25] A kite meteorograph seems to have been modified by replacing the

[24] *Apparat zum selbstthätigen Registriren des Standes meteorologischer Instrumente auf beliebige Entfernungen;* D.R. Patent 93032, Dec. 15, 1896. See also British Pat. 12098 of 1897, May 15, 1897.

[25] Manuscript, not identified, cited by P. Duckert, *Beitr. Phys. freien Atm.*, Vol. 18 (1932), pp. 73–74. Dr. Karl Keil informs me that this manuscript, which he once saw, has been lost.

recording drum by an insulating cylinder revolved once every six minutes; into this two metal strips were inserted, one parallel to the axis of the drum, the other forming a helix (Fig. 10.6). The styli of the meteorological instruments each made contact with a strip twice in each revolution, and each contact energized the primary of a small spark coil. The secondary of this was connected across a spark gap in the kite wire in parallel with a capacitor, and at the receiving station, a telephone was connected between the kite wire and the earth. We are not informed of the success or otherwise of the experiment.

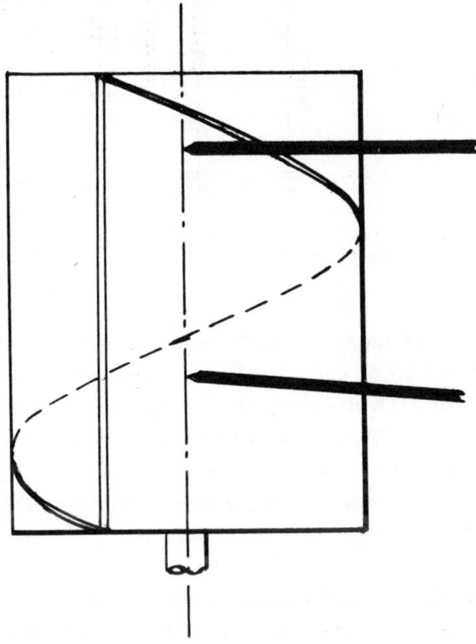

Fig. 10.6 The principle of the telemeteorograph of Herath and Robitzsch, 1917.

In September of the same year in France a remarkable junior officer named Pierre Idrac installed, on a captive balloon, an anemometer that was much more sophisticated in an electrical sense.[26] It made use of tuned circuits for both transmitting and receiving, as shown in Figure 10.7.[27] A buzzer *B*, operated by the contacts *A* of a windmill anemometer, drove an LC circuit that sent radio-frequency signals down the wire from the balloon to the series-tuned primary of a radio-frequency transformer. The secondary was parallel-tuned, and the *RF* signals were rectified by a galena detector *G* and heard in a telephone receiver *T*. It was noted that several instruments, each sending

[26] E. Rothé, *Compt. Rend.*, Vol. 170 (1920), pp. 1197–98. See also *La Météorologie*, Vol. 12 (1936), pp. 275–77.
[27] Adapted from *La Météorologie*, Vol. 12 (1936), p. 276.

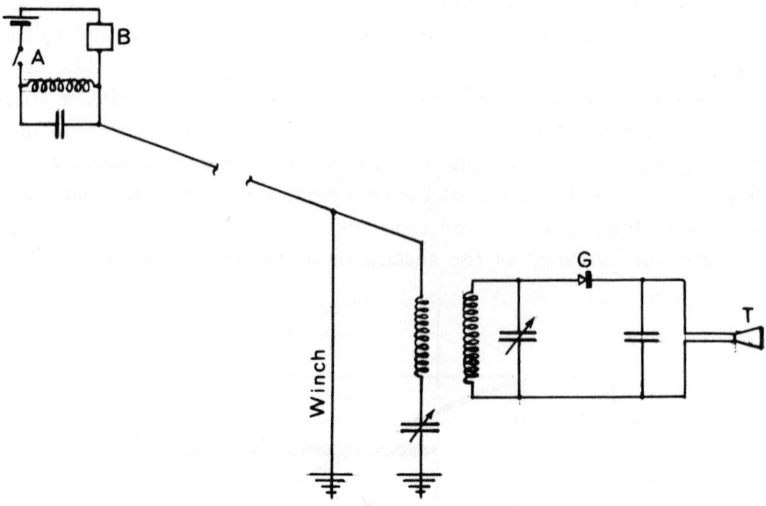

Fig. 10.7 Idrac's use of tuned circuits.

a different radio frequency, could be put at different heights along the wire.[28] Idrac went even further. Rothé wrote in 1920: "This way of transmitting is quite general. It applies, with variations, to the measurement of all the meteorological elements, and the possibility of having on the cable, at the same time, several transmitters, leads to the realization of a veritable observatory at a great height, giving indications at the ground over one wire only, and in all weathers, instantaneously and continuously."[29]

Reminiscing after Idrac's early death, Rothé recalled that the young officer had suggested, as one of the "variations," the use in the oscillating circuit of a capacitor designed to change its capacitance with temperature, in fact, a bimetal moving one plate of a capacitor. The frequency of the oscillation was to be measured by means of a wave meter on the ground.[30] This was during the 1914–18 war.

3. The radio-meteorograph or radiosonde. From what I have just said it will be seen that by 1920 only a suitable system of radio-communication was needed to make telemeteorography from free balloons a practical possibility. In 1921, Herath tried to track a pilot balloon by radio, sending up a small spark transmitter with it.[31] No accuracy was obtained, but the signal could be heard at the ground. At about this time there began a rapid development of short-wave technique,

[28] Rothé, *Compt. Rend.*, Vol. 170 (1920), p. 1198.
[29] *Ibid.*
[30] Rothé, *La Météorologie*, Vol. 12 (1936), p. 278.
[31] P. Duckert, *Beitr. Phys. freien Atm.*, Vol. 18 (1932), pp. 74–75.

engineered mainly by amateur operators who were being forbidden to use the longer waves then being preempted by broadcasting. Seldom has a greater virtue been made of necessity. It was found that surprisingly long distances could be reached with vacuum-tube transmitters radiating only a watt or two.

Experiments with a view to applying the new techniques to sounding balloons seem to have begun very early in Germany and the U.S.S.R., and also in France. It is difficult for the historian to assign priority, except in the matter of open publication, and here the honors go to the French. On March 7, 1927, a transmitter having a power of four watts was heard from well into the stratosphere by Idrac and Bureau, who at once reported their success to the Académie des Sciences.[32] They immediately began the development of radiosondes, and on June 10, 1929, Bureau reported[33] the ascent of the first model, a time-cycle device transmitting indications of temperature and pressure. An ordinary barothermograph was sent up with it and was recovered, and Bureau was pleased that the difference in reading between the two thermographs did not exceed 0.7°C. at any height.

Meanwhile, in 1928 P. Moltchanoff had written a somewhat cryptic paper[34] suggesting "the possibility of transmitting the readings of meteorological instruments to a distance by means of a wire or by radio."[35] He described the application of a time-cycle scheme to the problem, but gave no details of either the mechanical or electrical construction. Only in 1931 did Moltchanoff describe his first radiosonde,[36] which turned out to be of a completely different type, but in this paper he claimed to have started research on the subject in 1923.[37] In 1932 the radiosondes of the Lindenberg Observatory were described by Duckert and Thieme,[38] with a foreword by its Director H. Hergesell, stating that he had had researches on the telemetering of meteorological observations carried on under his direction since 1921. He had given the matter much attention himself, and "not unimportant" amounts of money had been spent. Such experiments were begun at Lindenberg, the Director claimed, "long before the researches of foreign, and especially Russian, scientists."[39]

One can only suppose that military policy may have had something to do with this tardiness of publication, and I shall not pursue the matter further, except to express my hope that some of my readers may be in a position to adjudicate. Meanwhile, the details of each

[32] P. Idrac and R. Bureau, *Compt. Rend.*, Vol. 184 (1927), pp. 691–92.
[33] *Compt. Rend.*, Vol. 188 (1929), pp. 1565–66.
[34] *Beitr. Phys. freien Atm.*, Vol. 14 (1928), pp. 39–48.
[35] *Ibid.*, p. 45.
[36] *Gerlands Beitr. Geophys.*, Vol. 34 (1931), pp. 36–56.
[37] From a paper about the radio transmitter by J. Freimann of Leningrad (*Petermanns Mitteilungen*, Erg.-Heft 201 [1929], pp. 56–58) it is clear that work was well under way by 1929.
[38] P. Duckert and B. Thieme, *Beitr. Phys. freien Atm.*, Vol. 18 (1932), pp. 50–67.
[39] *Ibid.*, p. 50.

development are of some interest, so let us start with the French radiosondes, especially because Robert Bureau of the Office National Météorologique was the first to explore the possibilities fully and logically, in a paper written in May, 1930, and published in 1931.[40]

Bureau saw clearly that there were several possible sorts of radiosonde.[41] In particular, the meteorological instruments could either modulate the signal mechanically, or themselves form part of the circuits of the transmitter. Information could be transmitted by interrupting the signal "in various rhythms," or by modulating the radio-frequency signal at a variable [audio] frequency, or by varying the radio frequency of the signal. Combinations of any or all of these methods are possible.

For his first radiosonde he went back to the idea of a time cycle, using a cylinder that was partly insulating and partly conducting (Fig. 10.8). If developed on to a plane, the metal part would form a triangle. The cylinder was rotated, at first by a windmill, later by a clockwork driven by the weight of the instrument itself. Contacts operated by a bimetal and a Bourdon barometer were used to modu-

Fig. 10.8 The principle of the first French radiosonde.

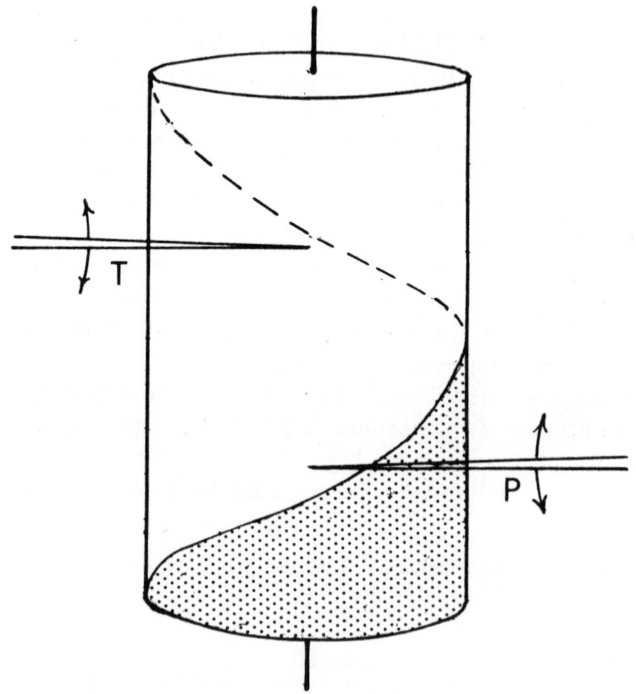

[40] Bureau, *La Météorologie*, Vol. 7 (1931), pp. 304–17.
[41] The term, in the form *radio-sonde*, was coined by Bureau (*ibid.*, p. 306).

late the radio signal, an auxiliary switch ensuring that three revolutions of the cylinder were devoted to the barometer and four to the thermometer, with a blank turn between.

It was difficult to get a uniform rotation of the cylinder, and Bureau adopted an ingenious solution that might be described as an electrical analogue of the apparatus of Cerebotani and Silbermann.[42] He provided a toothed wheel, geared to the cylinder and forming one plate of a small capacitor in the LC circuit of the transmitter, so that a frequency modulation was produced at a rate proportional to the instantaneous angular speed of the cylinder. This produced records with a number of "pips," followed by a blank space. The pressure or temperature was determined by counting the pips. Figure 10.9 shows the instrument at this stage, mounted in its protective frame, with the transmitter and batteries underneath. In describing it, Bureau gave great credit to two technicians, Gret and Geoffre, of whom the latter had once been the faithful collaborator of Teisserenc de Bort.[43] It is pleasant to record the impression that even today the enthusiasm aroused at Trappes by that great man has not vanished.

In May, 1930, Bureau replaced the cylinder by a disk driven by a spring clock. The toothed wheel was still used to frequency-modulate the signal, avoiding any appeal to constancy of rotation, and (an excellent idea) every tenth tooth was unsymmetrical, to facilitate counting.[44] It would have been easy, Bureau remarked, to introduce a measurement of humidity, "if any hygrometers were known that are good at low temperatures."[45] This stated a serious problem that was partially solved only several years later.

The next radiosonde to be publicly described was that of Moltchanoff. In this both temperature and pressure are transmitted in discrete steps by a contact mechanism, the appearance of which has given the instrument the name of "comb-apparatus" (*Kammgerät*).[46] Referring to Figure 10.10, we see that the pointer of the temperature-measuring element moves over four "combs" with teeth one-third as wide as the distance between them, and staggered as shown. Each comb is connected to a switch, p_1 to p_4, operated in turn by a wind-driven cam, so that the combs may be identified by one, two, three, or four impulses respectively. To avoid ambiguity if the signal is temporarily lost, a fifth comb is provided. This has only five teeth, q_1 to q_5, each opposite a missing tooth in one of the other combs, and as it is not connected to any of the switches p_1 to p_4, a continuous signal is emitted whenever the pointer of the thermometer is in contact with it. There is a separate comb for pressure, which operates through the switches p_0 and p_1, to lengthen a signal at known values of pressure.

[42] See p. 326 above.
[43] According to E. Delcambre, *Beitr. Phys. freien Atm.*, Vol. 14 (1928), pp. 63–64.
[44] Bureau, *La Météorologie*, Vol. 7 (1931), pp. 317–20.
[45] *Ibid.*, p. 319.
[46] Moltchanoff, *Gerlands Beitr. Geophys.*, Vol. 34 (1931), pp. 36–56.

Fig. 10.9 The s :ond French radiosonde (courtesy of M. L. Facy).

Details of the weight and size of this apparatus were not given in the 1931 paper, but it was fairly heavy. It was greatly improved later, and made lighter and more compact, although a much more elaborate circuit was developed.[47]

Moltchanoff also invented a time-cycle meteorograph that was put into production in Germany by the Askania-Werke.[48] This was an "Olland system" radiosonde with contacts driven by a clock, but provided with means of synchronization so that the signals could be received on a commercial facsimile receiver. It was rather elaborate and expensive, and of course depended for its accuracy on the uniform

[47] *Cf.* International Meteorological Organization (I.M.O.), *Über Radiosonde-Konstruktionen. Denkschrift*, ed. L. Weickmann (Berlin, 1937), pp. 45–54.
[48] Ludwig Heck and Günther Sudeck, *Gerlands Beitr. Geophys.*, Vol. 31 (1931), pp. 291–314.

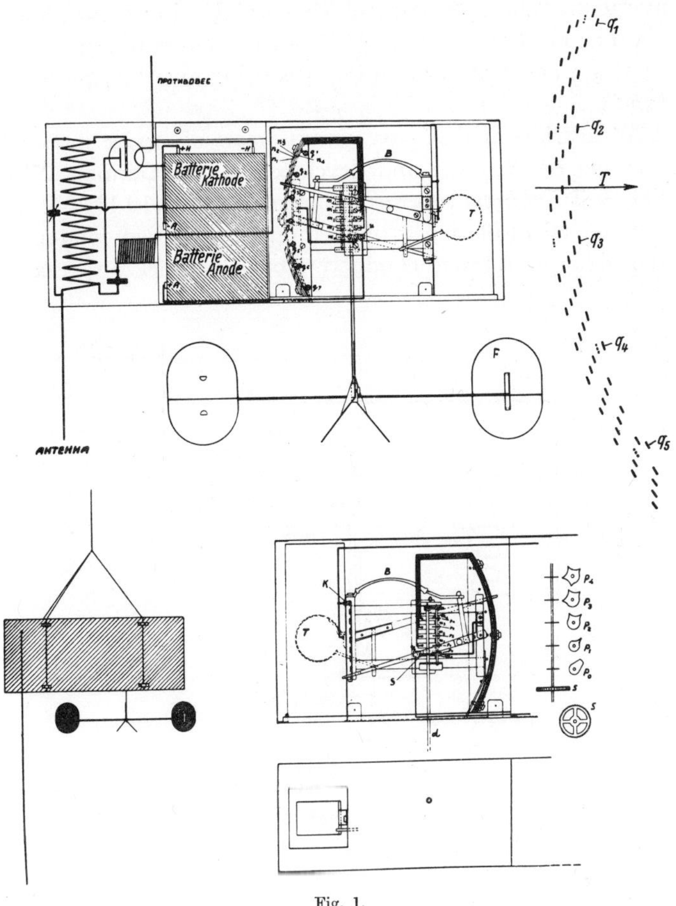

Fig. 1.

Fig. 10.10 Moltchanoff's radiosonde, diagram.

motion of the clock, as do all pure time-cycle instruments. It must also be mentioned that the motion of a clockwork is discontinuous, so that one limit to the precision of any reading is given by the number of "ticks" per revolution of the moving contact. Attempts to remove this limitation were made by substituting a tiny electric motor, for example by L. F. Curtiss of the National Bureau of Standards in Washington, D.C.,[49] and by R. C. Jacobsen of the Canadian meteorological service,[50] who used a tuned reed to stabilize the speed. Jacobsen's motor was in use for a number of years.

At least seven different constructions of time-cycle radiosonde, almost always referred to as "on the Olland system," had been described

[49] Curtiss, *Bull. Amer. Meteorol. Soc.*, Vol. 18 (1937), p. 391.
[50] R. C. Jacobsen, *Rev. Sci. Insts.*, Vol. 10 (1939), p. 315.

by the time the International Meteorological Organization published a summary of progress in 1937,[51] which demonstrated, if nothing else, what a great deal of attention was being given to the subject in the 1930's.

An instrument that gave several years of service in Germany was invented by A. Lang and developed by the German Weather Service.[52] Its general scheme is shown in Figure 10.11. A clock drives two coaxial insulating cylinders, one having two conducting helices and the other a broad conducting band parallel to the axis. Instead

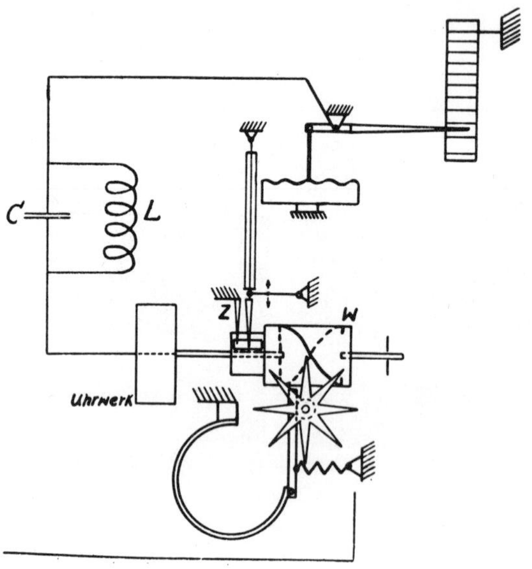

Fig. 10.11 The Lang–Reichsamt für Wetterdienst radiosonde, diagram.

of operating just one contact, the bimetal turns a star wheel, thus greatly expanding the temperature scale of the instrument. The hair moves a contact which, together with a fixed contact, shuts the transmitter off once during each revolution of the cylinders for a length of time depending on the relative humidity. The aneroid chamber operates a switch that interrupts the signal at known values of pressure.

Most of the other time-cycle radiosondes originated in the United States after about 1934, the first to be described being those of Blue Hill Observatory at Milton, Massachusetts.[53] At first an insulating disk was used, with a radial wire inset in its surface. When the disk was revolved once a minute by a clock, the wire made contact suc-

[51] I.M.O., *Über Radiosonde-Konstruktionen.* There is also a *Nachtrag* (Berlin, 1939). Both should be consulted for details.
[52] *Ibid* (1937), pp. 28–30.
[53] K. O. Lange, *Bull. Amer. Meteorol. Soc.*, Vol. 17 (1936), pp. 136–47.

cessively with two reference contacts and with styli attached to an aneroid chamber, a bimetal, and a hygrometric hair. Later the disk was replaced by a cylinder carrying a helical wire. This arrangement is a simpler equivalent of the scheme of Herath and Robitzsch of 1917 (Figure 10.6) with the axial wire replaced by fixed contacts. I have reproduced a picture of this model (Fig. 10.12) because it shows very clearly the essential parts of a time-cycle radio-meteorograph, though the design was soon modified at Blue Hill.[54]

The radiosonde of the Guggenheim Aeronautics Laboratory, California Institute of Technology (GALCIT) differed from the others

Fig. 10.12 One of the Blue Hill radiosondes of 1935, meteorograph.

in that it emitted a continuous signal, interrupted by the contacts.[55] This was done in the interests of radio direction-finding, and of course involved the use of larger batteries. Its designers also modified "an inexpensive pocket watch" to give 1,020 ticks per minute and run for one hour, giving plenty of torque to the output shaft.

The United States Weather Bureau[56] and the Canadian Meteorological Service made their moving contact in the form of a flat equiangular spiral across which the pointers of the meteorological instruments moved. The Canadian radiosonde, for the design of which R. C. Jacobsen was mainly responsible, gave excellent service for about two decades.

Let us go back to Germany, where Paul Duckert at Lindenberg had adopted, or rediscovered, Idrac's idea of using a thermometer to

[54] K. O. Lange, *Bull. Amer. Meteorol. Soc.*, Vol. 18 (1937), pp. 107–26.
[55] O. C. Maier and L. E. Wood, *J. Aeron. Sci.*, Vol. 4 (1937), pp. 417–22.
[56] D. M. Little, *Trans. Amer. Geophys. Union*, Vol. 18 (1937), pp. 138–41.

vary the capacitance of a condenser.[57] His first radiosonde (Fig. 10.13) sent out a continuous radio frequency that varied with the temperature, while the pressure was indicated at intervals by short interruptions of the signal, made by a Bourdon barometer *c* and a

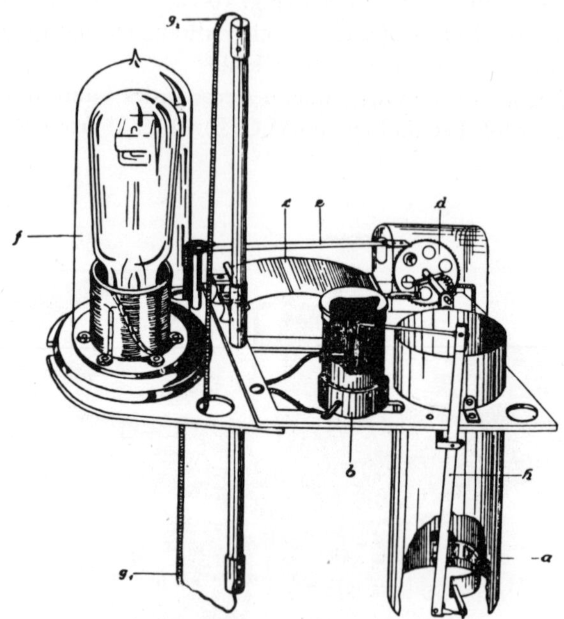

Fig. 10.13 Duckert's radiosonde, meteorograph and transmitter.

toothed wheel *d*. The oil-filled condenser *b* had one plate moved by a bimetal. Apart from this condenser, the entire radio transmitter was housed inside a glass tube *f*; and in a later model,[58] the Telefunken firm succeeded in sealing the entire circuit, except for *b*, in a vacuum, greatly improving the stability of the transmitter. But still no attempt was made to transmit indications of humidity. In this later model the pressure switch, instead of interrupting the signal, switched from the variable condenser to a fixed one in order to provide a reference signal as a check on the stability of the transmitter. Still later,[59] ceramic insulation was effectively used to make it more stable, and the thermometer was greatly improved by replacing the oil capacitor and the bimetal by a simple capacitor with a ceramic dielectric having a large temperature-coefficient of dielectric constant.

Vilho Väisälä was inspired to design a radiosonde when, in March, 1931, a Russian *Kammgerät* was found in Finland. It seemed to him

[57] P. Duckert and B. Thieme, *Beitr. Phys. freien Atm.*, Vol. 18 (1932), pp. 50–67.
[58] Duckert, *Beitr. Phys. freien Atm.*, Vol. 20 (1933), pp. 303–11.
[59] Duckert, *I.M.O. document* no. 21 (1935), pp. 118–19.

to be large, heavy, and complicated. At that time he knew only of Moltchanoff's work.[60] He decided to use the principle of variable radio frequency, making two bimetals vary the capacitance of an air condenser, and in 1931 he made an instrument with a thermometer only, using a "matchbox size" plate battery. The sonde, weighing 420 gm, was sent up on December 31 of that year, and followed to a height of 7 km. On November 20 he had read Duckert and Thieme's paper, referred to above.

Väisälä believed that it would be better to transmit all the meteorological elements in the same way, and built an instrument in which four condensers were connected successively into the tank circuit of the oscillator by a switch operated by a windmill. Three of the condensers were varied, by the thermometer—a silver wire in an invar frame—the hair hygrometer, and the barometer respectively. The last instrument consisted of two aneroid chambers supported a short distance apart by an insulating ring, so that it measured the pressure and simultaneously translated it into capacitance. The fourth condenser was a fixed reference.

In a later paper Väisälä gave a charming account of the development of his idea.[61] The silver wire was replaced by a bimetal and the windmill by a cup wheel, and there was only one aneroid chamber. Two fixed condensers were used, bracketing the range of capacitance assumed by the others (see Fig. 10.14), and thus permitting a valid correction for incidental changes in the frequency of the transmitter. The signal from such a radiosonde is compared with a calibrated local oscillator in the receiver, and for this purpose Väisälä designed a semi-automatic recorder in which the shaft of the local oscillator condenser was geared to a recording drum.

In his 1932 paper, Väisälä had made a proposal for an automatic weather station, to report by radio from an isolated spot without any attention for a long time. It was to have a generator driven by an internal-combustion engine, started automatically when a storage battery fell to a certain voltage, and contained in a hut that would be heated when necessary. By a technique similar to that of his radiosonde, the station would transmit pressure, temperature, humidity, wind direction and speed, and precipitation. A quarter of a century later, automatic weather stations were being developed by various services, but these efforts are beyond the scope of this book.

By about 1936 it was becoming obvious in the United States, and to some extent in the United Kingdom, that a serious and properly financed attack on the radiosonde problem was in order. In the United States the Navy Department asked the National Bureau of Standards (NBS) to develop a suitable system, first publicly described

[60] Väisälä, *Mitt. meteorol. Insts. Univ. Helsinki*, no. 20 (1932); *Soc. Sci. Fenn., Comm. phys.-math.*, Vol. 6 (1932), no. 2.
[61] Väisälä, *Mitt. meteorol. Insts. Univ. Helsinki*, no. 35 (1937); *Soc. Sci. Fenn., Comm. phys.-math.*, Vol. 9 (1937), no. 9.

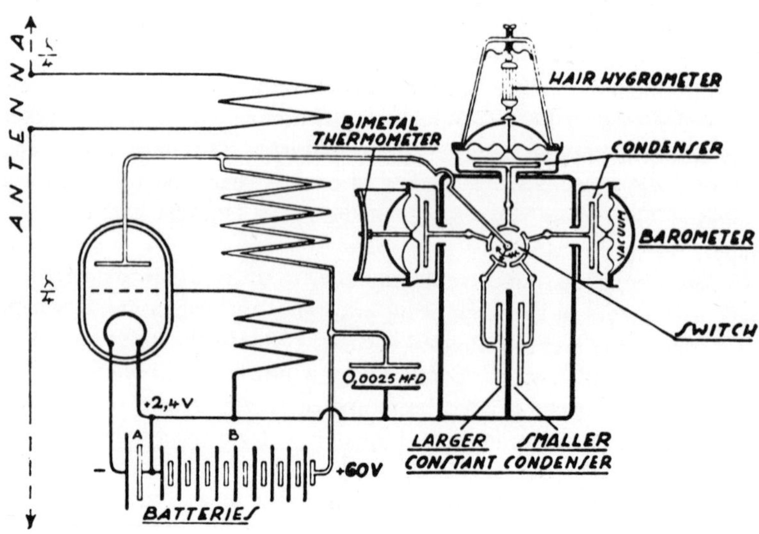

Fig. 10.14 Väisälä's radiosonde, diagram.

in 1937 by Diamond and his associates.[62] The design requirements of the Navy are interesting: The unit cost "in reasonable quantities" was to be about $25; the entire instrument was to weigh about 1 lb.; and it was to measure pressure to ±1°C between +30° and −70°C, and relative humidity to ±3 per cent at all values. Furthermore, it was to emit signals suitable for radio direction finding.

Diamond and his colleagues, after canvassing all existing and possible systems, decided that no clock or motor drive should be used, thereby putting the time-cycle principle entirely out of court. Meteorologists expressed a preference for a record in which temperature and humidity appeared as functions of pressure, and this preference was of course supported by the expectation that as the balloon ascended, the pressure would always decrease in a smooth and featureless curve. The designers therefore provided what they called a "pressure drive," in which an arm operated by an aneroid barometer makes successive contacts at calibrated values of pressure.

For the transmission of the other elements, they modulated the radio frequency signal at a rate of a few hundred cycles per second, and designed the thermometer and hygrometer to produce a calibrated variation in this audio frequency. This was done by using a "relaxation," or RC, oscillator to modulate the signal, the resistance R of the circuit being varied by the changes in temperature or humidity. We may note that the use of a variable audio frequency had been mentioned as a possibility by Bureau in 1931.[63] I suspect that the

[62] H. Diamond, W. S. Hinman, Jr., and F. W. Dunmore, *Bull. Amer. Meteorol. Soc.*, Vol. 18 (1937), pp. 73–99; also in *J. Aeron. Sci.*, Vol. 4 (1937), pp. 241–48.
[63] Bureau, *La Météorol.*, Vol. 7 (1931), p. 307.

decision of the group at NBS to adopt this method may have been influenced by the commercial availability of an excellent new audio-frequency meter, but at any rate, the method had three considerable advantages over that of varying the radio frequency: it used up less of the crowded radio-frequency spectrum; it was not much affected by interference; and a signal of constant radio frequency is almost a necessity for direction-finding.

Thus the thermometer and hygrometer had only to produce changes of electrical resistance. Assuming for the moment that we have ways of doing this, the reader is asked to refer to Figure 10.15, in which *P* is the arm operated by the barometer, sweeping over a

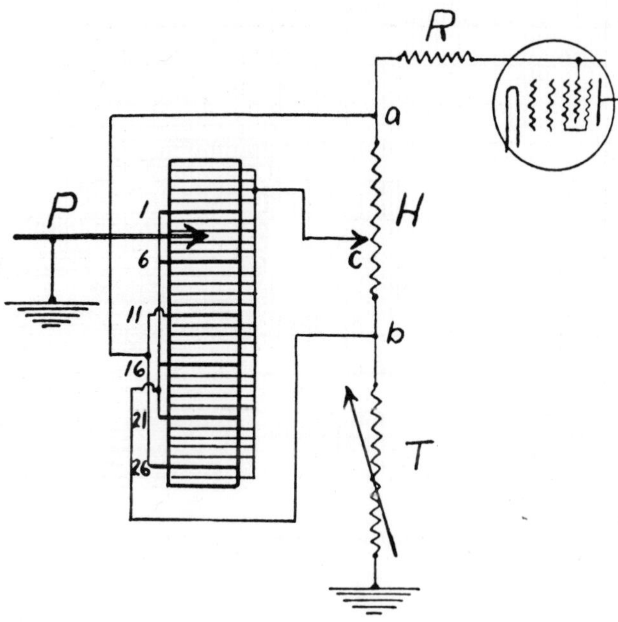

Fig. 10.15 The measuring circuit of Diamond et al., *1937.*

number of conducting strips in an insulating block. Certain of the strips, numbered *1*, *6*, *11*, and so on in the figure, are wider than the rest. Connected in series between one of the grids of a multi-electrode tube and the frame of the instrument are a fixed resistor *R*, a resistor *H* with a contact at *c* operated by a hygrometer, and a resistance *T* that varies with temperature.

When the pointer *P* is not touching any of the contacts, the only path from the grid to the frame is through all the resistors *R*, *H*, and *T* in series. In this condition the total resistance will be a function of the temperature, and so will the audio frequency that is emitted. When the pointer touches any of the narrower, unnumbered contacts, the total resistance will be R plus the portion of H between a and c. This is a function of the humidity. When it touches strips *11* and *26*, the point a is connected directly to the frame, resulting in an upper fixed frequency. When it touches numbers *1*, *6*, *16*, and *21*, the point b is connected to the frame, and a lower fixed frequency is produced. If the instantaneous values of the resulting frequency are automatically recorded, the result is like that shown in Figure 10.16. The record consists of discrete indications of audio frequency, the horizontal lines between them being made by the pen of the recorder

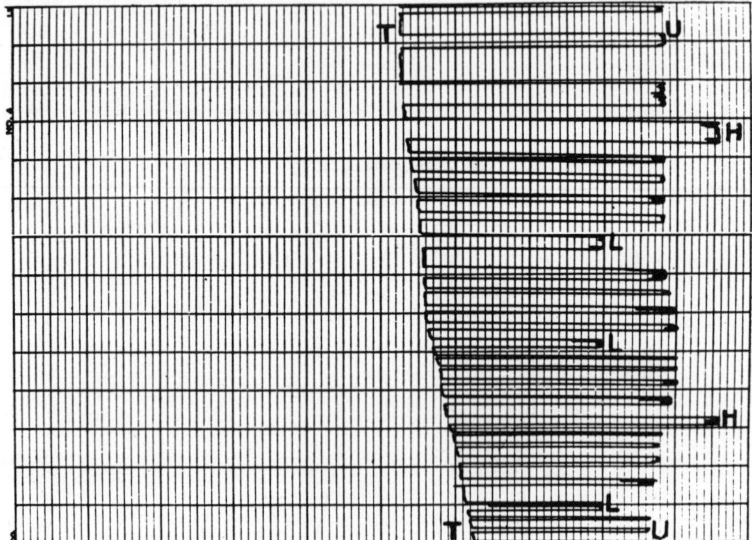

Fig. 10.16 The NBS–Navy radiosonde, appearance of the record.

in its rapid motion from one position to another, each time the barometric switch on the radiosonde passes on to a contact or off it. The small fluctuations are due to interference. The low reference contacts, adjusted to 160 cycles per second, are shown at L, and the high reference contacts at H; the humidity contacts are at U-U, and the temperature is recorded at T-T in the intervals. Each discontinuity in the record represents a certain pressure, determined by calibration, and recognizable by its position in the series of short and long contacts and of high and low reference frequencies.

Let us return to the thermometer and hygrometer. The thermometer adopted by Diamond and his colleagues was a small V-shaped glass tube filled with a nonfreezing electrolyte, and with an electrode at each end. This device was replaced some years later by a solid temperature-sensitive resistor known as a "thermistor." The hygrometer was more troublesome. In the first radiosondes of this type the hygrometer was a bundle of hairs operating a contact on a resistance coil (Fig. 10.17*a*). As I have already noted, the hair hygrometer is far from satisfactory at low temperatures, and in the 1937 paper it was announced that attempts were being made to find a suitable humidity-sensitive resistor. As a matter of fact this problem had been attacked long before by H. E. F. Goold-Adams and others,[64]

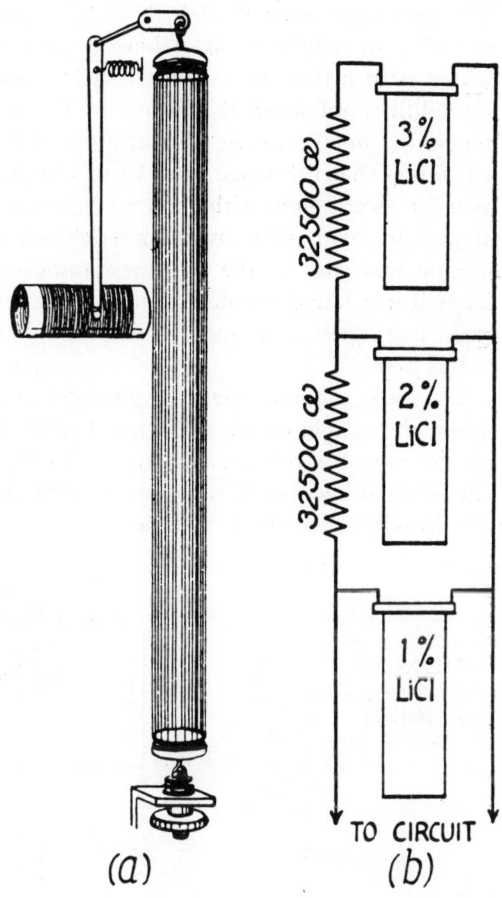

Fig. 10.17 Hygrometers for the NBS–Navy radiosonde.

one of whose "detectors" consisted of two fine platinum wires wound around an insulating rod on which a thin film of some kind of salt had been deposited. This seems to have been more of a water-vapor alarm than a hygrometer. Apparently independently, F. Albrecht[65] tried silk threads moistened with a solution of salt and glycerine, wound between two rectangular electrodes, and seems to have ob-

[64] H. E. F. Goold-Adams, W. R. Bousfield, and G. W. Todd, British Patent 137,547, March 10, 1920.
[65] Albrecht, *Beitr. Phys. freien Atm.*, Vol. 11 (1924), pp. 164–78.

tained good results in an aircraft instrument. But Francis W. Dunmore took the matter much further, arriving by 1938[66] at the use of a roughened glass tube with a dual winding of tinned copper wire, the whole coated with a standardized solution of lithium chloride in water. This gave encouraging results, but showed effects of ageing at an unacceptable rate. It was quickly improved[67] by the use of a thin aluminum tube (0.01-inch wall) coated with polystyrene and wound with a bifilar coil of bare palladium wire. This was then coated with a thin film of partly hydrolyzed polyvinyl acetate with the addition of a small amount of lithium chloride, the amount depending on the range of relative humidity to be covered, but it was found that the range from 10 to 100 per cent could be obtained by using several units with different concentrations of lithium chloride in parallel, with series resistors as shown in Figure 10.17*b*. This scheme was used in the improved radiosonde described in 1940.[68] Later it was found possible to produce simpler and even better electrical hygrometers for radiosondes, but these are beyond the scope of this book.[69]

The use of a humidity-sensitive resistor of course made it impossible to use the simple circuit of Figure 10.15. The problem was solved by the use of a relay, as in Figure 10.18, which shows the entire circuit of the improved radiosonde, using only one vacuum tube, a twin triode. The triode *C* generates oscillations at 65 Mc/sec; the

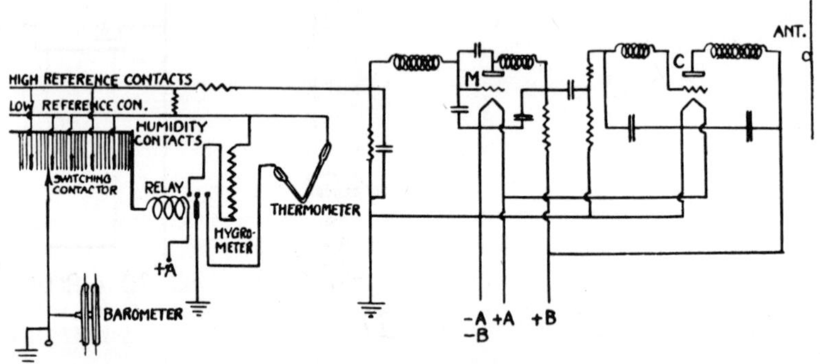

Fig. 10.18 NBS–Navy radiosonde, measuring and transmitter circuit, 1940.

[66] Dunmore, *J. Res. Nat. Bur. Standards*, Vol. 20 (1938), pp. 723–44 (RP 1102); see also *Bull. Amer. Meteorol. Soc.*, Vol. 19 (1938), pp. 225–43.
[67] Dunmore, *J. Res. Nat. Bur. Standards*, Vol. 23 (1939), pp. 701–14 (RP 1265). See also *Bull. Amer. Meteorol. Soc.*, Vol. 21 (1940), pp. 249–56.
[68] H. Diamond, W. S. Hinman, F. W. Dunmore, and E. G. Lapham, *J. Res. Nat. Bur. Standards*, Vol. 25 (1940), pp. 327–67 (RP 1329).
[69] A. Wexler, *Circ. Nat. Bur. Standards*, no. 586 (Washington, D.C., 1957), gives 86 references to electrical hygrometry, including many patents, most of which date from after 1939. See also Donald A. Matthews in *Humidity and Moisture, Measurement and Control in Science and Industry* (4 vols., New York & London, 1965), I, 219–46.

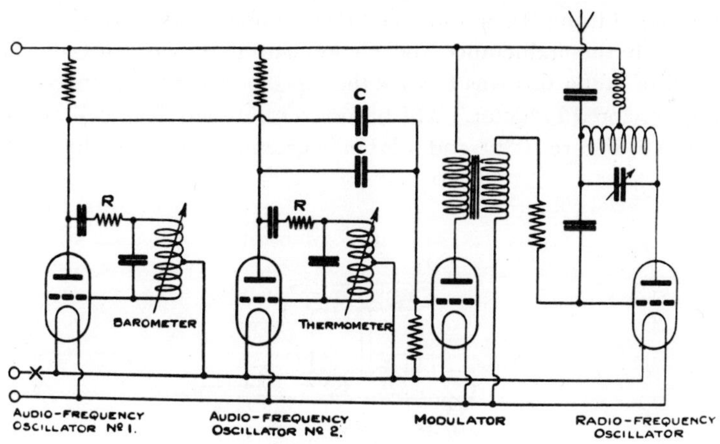

Fig. 10.19 NPL radiosonde, 1938, transmitter circuit (courtesy of The Royal Society and H. A. Thomas).

triode *M* oscillates at 1 Mc/sec and has a special grid circuit that causes it to "block" at an audio frequency that depends on the total resistance between grid and cathode. This radiosonde was designed for quantity production and calibration and was immensely successful. It was flown in an outer case made of corrugated cardboard covered with foil, so arranged that necessary adjustments could easily be made before the ascent. The fact that temperature and humidity were translated into electrical resistance suggested its application to other problems, as for example the measurement of ultraviolet radiation in the stratosphere.[70] For this, a phototube was put in a Wheatstone bridge that was part of a direct-current amplifier.

While these developments were going on in the United States, another sort of audio-frequency-modulation radiosonde was taking shape at the National Physical Laboratory in England.[71] H. A. Thomas, like the Americans, had reviewed all the possibilities and independently decided on the modulated signal. He rejected the use of contacts operated by the barometer, however, apparently on the basis of tests he had made on a Duckert instrument. His solution to the problem was to modulate the signal with two frequencies, 700 to 1,000 cycles per second for the barometer and 1,400 to 1,700 cycle per second for the thermometer, simultaneously. It was easy to provide separate local oscillators at the receiver, connecting them alternately so that the operator could adjust either one to obtain the zero-beat condition. This solution made for a rather complex electrical circuit in the radiosonde (Fig. 10.19).

[70] R. Stair, *J. Res. Nat. Bur. Standards*, Vol. 22 (1939), pp. 295–300 (RP 1181).
[71] H. A. Thomas, *Proc. Royal Soc., Sec. A*, Vol. 167 (1938), pp. 227–50.

Instead of using RC audio oscillators, he used conventional LC circuits, with the inductance the component that was varied. The method of doing this was to vary the air-gap in the magnetic circuit of an iron-cored inductor. The barometer unit with its metal bellows is shown in Figure 10.20, and it is only necessary to explain that a thin

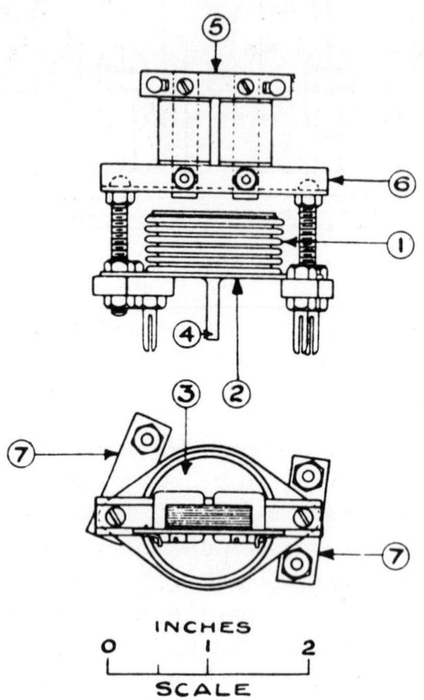

Fig. 10.20 NPL radiosonde, barometer (courtesy of The Royal Society and H. A. Thomas).

iron plate is soldered onto the flat top of the bellows. The thermometer (Fig. 10.21) makes use of a brass wire *1* in an invar frame *2*. The invar spring *3* carries an iron plate *4*, which moves in front of the iron core *5* of the inductor. The thermometer unit is shielded from radiation by the shield at *3* in Figure 10.22, which also shows the very symmetrical and clean arrangement of the components of the radiosonde. Unfortunately it weighed 2,390 gm when ready for launching, of which the battery accounted for 935 gm and the parachute 350.

Later a humidity unit and a third audio oscillator were added, and mu-metal was substituted for silicon iron in the cores of the inductors; but this model was even heavier, and it seems that the temperature and humidity units were not really satisfactory. At any rate, an entirely new design was developed by E. G. Dymond and others at

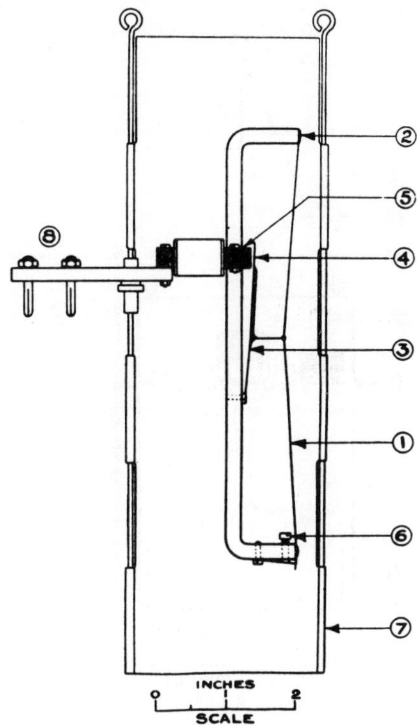

Fig. 10.21 NPL radiosonde, thermometer (courtesy of The Royal Society and H. A. Thomas).

Kew in 1940, though there was no published description until after the war.[72]

In this "Kew radio-sonde, Mark 1," the continuous transmission of all the meteorological elements was sacrificed in the interests of lightness. In fact the pressure, temperature, and humidity elements were switched into the circuit by a switch operated by a cup wheel, very much as in the radiosonde of Väisälä referred to earlier.[73] This meant that only one audio-frequency oscillator was needed. The measuring units were also redesigned. For the barometer, a single aneroid capsule was substituted for the metal bellows. A bimetal replaced the brass wire in its invar frame, as this had been found to be sensitive to the swinging of the radiosonde. A piece of goldbeater's skin was used in the humidity element.

This general design proved highly satisfactory. Although improvements have been made in the transmitter circuit to increase the

[72] E. G. Dymond, *Proc. Phys. Soc. London*, Vol. 59 (1947), pp. 645–66.
[73] On page 337.

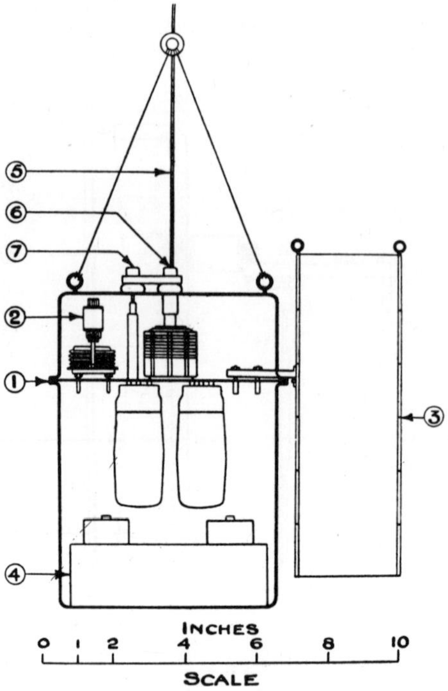

Fig. 10.22 NPL radiosonde, general arrangement (courtesy of The Royal Society and H. A. Thomas).

stability of the radio and audio frequencies, it is still in use.[74]

I shall not follow the later history of meteorological telemetry, which has become extremely sophisticated, especially in devices that combine radiosonde techniques with the measurement of the upper winds. As a result of the 1939–45 war, enormous resources became available for the improvement of meteorological services, particularly in the United States, and the steadily increasing requirements of civil aviation have continued the process. It is fitting to recall the pioneers of the 1920's and 1930's, who laid the foundations of the subsequent edifice on budgets that would not support the canteen of a present-day establishment.

[74] Further information is readily available in: Great Britain, Meteorological Office, *Handbook of Meteorological Instruments, Part II* (M.O. 577) (London, 1961), chap. 8. This has an excellent bibliography.

XI

Epilogue

As far as we know, the instrumental observation of weather phenomena and climatic elements began in India about the fourth century B.C. with attempts to measure rain. About 100 B.C. there was a windvane in the form of a Triton on the "Tower of the Winds" in Athens. Fifteen centuries went by until a hygroscopic hygrometer was described by Leone Battista Alberti about 1450, and, probably independently, by Nicolas of Cues in the same decade. Alberti also invented a simple anemometer.

Nevertheless the history of meteorological instruments can fairly be said to begin in the seventeenth century and in Italy, where the air thermoscope was invented in the first decade, the normal-plate anemometer a few years later, the barometer in 1644, and the sealed liquid-in-glass thermometer—in the form that we know it today—in 1653 or a little earlier. After about 1660 the initiative in these matters passed to the more northern countries of Europe, and before the end of the century the meteorologist's armament had been increased by several hygrometers (none of great merit), the windmill anemometer, and at least two remarkable instruments that recorded several of the meteorological quantities on moving paper charts. More important, people had begun to speculate on the complicated relations between the indications of their instruments—especially the barometer—and the weather. The seventeenth century also saw the beginning of the process of refinement that transformed the barometer from the simple Torricellian tube into an instrument of high precision, as well as the first attempts at standardizing the thermometer. In the last decade of the century the first atmometers were installed.

No new kinds of meteorological instrument, unless the balloon can be called one, were invented in the eighteenth century, but great improvements were made in them all. The jungle of thermometer scales was hacked away, leaving only three in use by 1800. A great deal of experiment in hygrometry produced the hair hygrometer and laid the foundations of the dew-point hygrometer and the psychrometer. The pressure-tube anemometer was invented, but scarcely used, and nothing very momentous happened in the measurement of rain. As to recording instruments, there was one remarkable but costly anemograph, and several elaborate barographs were designed by clockmakers, but otherwise these devices in the eighteenth century were ingenious rather than practical, and in any case not very numerous. One event of general meteorological importance was the establishment of a very extensive, although short-lived, network of reporting stations by the Meteorological Society of the Palatinate, the observers being supplied at the expense of the Elector Palatine with instruments of standard, but unfortunately not very admirable, design.

In the nineteenth century all the existing instruments were further refined. Sunshine recorders were invented, and toward the end of the century a beginning was made on the systematic investigation of the upper air. Between about 1850 and 1880 there was a spate of large, complex, and costly recording instruments and meteorographs, for the most part designed and installed at important astronomical and meteorological observatories as a result of the enthusiasm of the directors of these institutions. In the 1880's the situation was entirely changed by the introduction of small, convenient, and cheap recording instruments that could be furnished to many stations without too great a strain on the budgets of governments that then had little interest in science.

Almost at the beginning of the twentieth century the introduction of rubber sounding balloons and light meteorographs led to the discovery of the tropopause and a sudden increase in our knowledge of the upper air. Improved pilot-balloon techniques permitted the mapping of the upper winds. The "ceiling projector" made possible the rapid measurement of the height of clouds at night, and later in the daytime. On the ground, the improved pressure-tube anemograph enlarged our understanding of wind structure. In the 1930's the rapid technical advances in the use of short radio waves led to the radiosonde and radio direction-finding, making synoptic maps of the upper air possible for the first time. These developments were very greatly accelerated by the advent of commercial aviation and still more by the war of 1939–45, which initiated the application of such advanced techniques as radar, more recently still, lasers. But at this point I leave the story to more knowledgeable pens.

Index

Where possible, the dates of birth and death of the people concerned with the development of meteorological instruments have been given. For help in finding some of these I wish to thank the following: Mr. David Bryden of the Royal Scottish Museum; Miss Marjorie A. Clark of the ESSA Atmospheric Sciences Library; Mr. W. Garriock of the Meteorological Office, Bracknell, England; Dr. Johannes Grunow, Peiting, Germany; Mr. L. P. Harrison of the U.S. Weather Bureau; Herr Max Schlegel of the Deutscher Wetterdienst; and Dr. Birger Svensson of the Meteorological Institute at Uppsala, Sweden.

Designed by Gerard A. Valerio

Composed in Baskerville
by Monotype Composition Company

Printed offset by Universal Lithographers, Inc.
on 60 lb. Warren's 1854

Bound by L. H. Jenkins, Inc.
in Interlaken AV 3-530

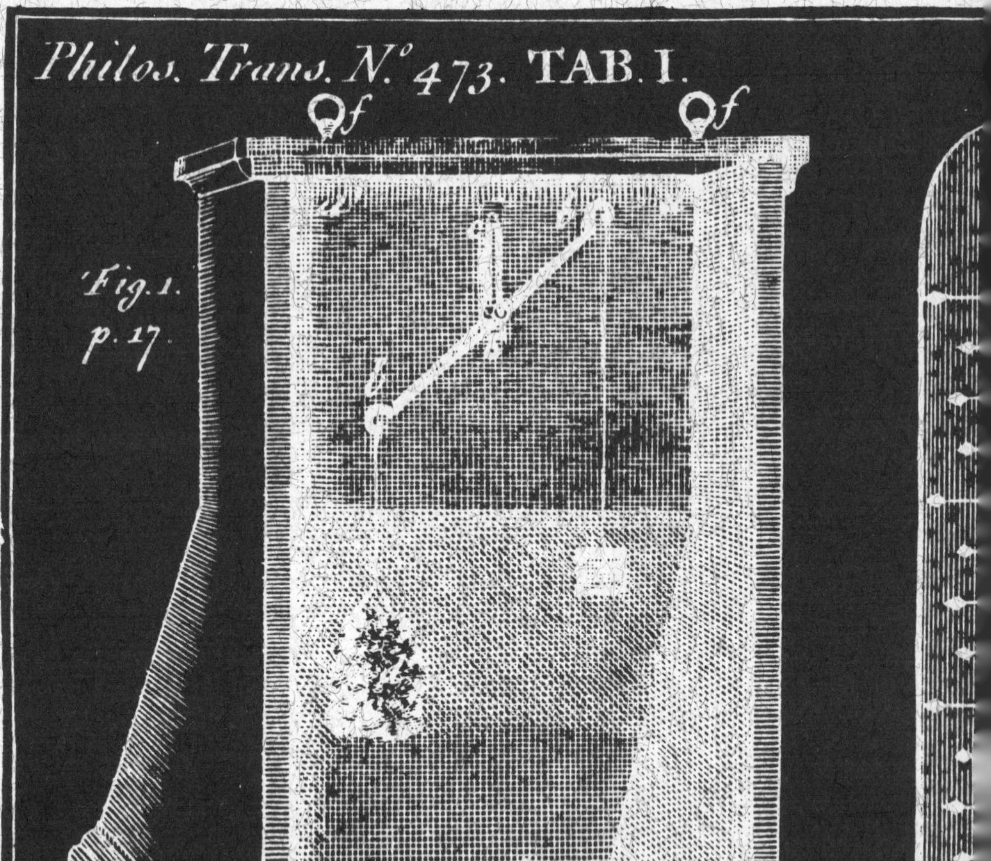

Philos. Trans. N° 473. TAB. I.

Fig. 1. p. 17.

Fig. 2. p. 17.

Fig. 6. p. 18.